SELECTED TOPICS ON OPTICAL FIBER TECHNOLOGIES AND APPLICATIONS

Edited by **Fei Xu** and **Chengbo Mou**

Selected Topics on Optical Fiber Technologies and Applications
http://dx.doi.org/10.5772/intechopen.68456
Edited by Fei Xu and Chengbo Mou

Contributors

Erick Estefen Reyes Vera, Gloria Margarita Varon Duran, Juan Usuga Restrepo, Pedro Torres, Andres Macho, Roberto Llorente, Dora Juan Juan Hu, Ping Shum, Rebecca Yen-Ni Wong, L.S. Supian, Dr. Mohd Syuhaimi Ab-Rahman, Norhana Arsad, Chew Sue Ping, Nani Fadzlina Naim, Nurdiani Zamhari, Syed Mohd Fairuz Syed Mohd Dardin, Yong Sheng Ong, Ian Andrew Grout, Elfed Lewis, Waleed Mohammed, Chaoyang Ti, Minh-Tri Ho Thanh, Yao Shen, Qi Wen, Yuxiang Liu, Zhengyong Liu, Hwa-Yaw Tam, Vasily Spirin, Cesar López-Mercado, Patrice Mégret, Andrei Fotiadi, Ramon Jose Perez, Keisuke Kojima, Toshiaki Koike-Akino, Tsuyoshi Yoshida, David. S. Millar, Kieran Parsons

Published by InTech

Janeza Trdine 9, 51000 Rijeka, Croatia

Publishing Process Manager Edi Lipovic
Technical Editor SPi Global
Cover InTech Design team

First published February, 2018
Printed in Croatia
Legal deposit, Croatia: National and University Library in Zagreb

Additional hard copies can be obtained from orders@intechopen.com

Selected Topics on Optical Fiber Technologies and Applications, Edited by Fei Xu and Chengbo Mou
p. cm.
Print ISBN 978-953-51-3813-6
Online ISBN 978-953-51-3814-3

Meet the editor

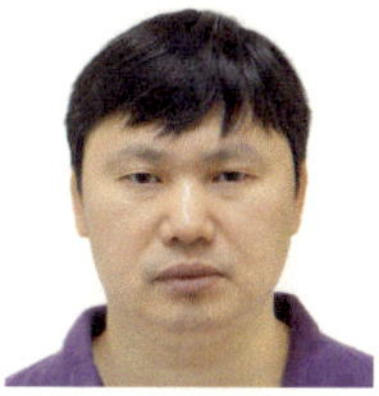

Dr. Fei Xu received his BE degree from the Nanjing University, China, in 2000; MSc degree from the Nanjing University, China, in 2005; and PhD degree in Optoelectronics from the Optoelectronics Research Centre, University of Southampton, UK, in 2008. He is currently a full professor at the College of Engineering and Applied Sciences, Nanjing University, China. His current research interests include the development of novel fiber devices for ultrasmall sensor and laser systems and the optomechanical effect in nanoscale waveguide systems. To date, he has authored or coauthored 8 book chapters, granted more than 30 patents (China and US), and published more than 100 peer-reviewed articles in academic journals in the previously mentioned areas.

Dr. Chengbo Mou received his BE degree from the Tianjin University, China, in 2004; MSc degree from the University of St Andrews, UK, in 2005; and PhD degree in Photonics from the Aston University, UK, in 2012. He is currently working with the Key Laboratory of Specialty Fiber Optics and Optical Access Networks, Shanghai Institute for Advanced Communication and Data Science, Joint International Research Laboratory of Specialty Fiber Optics and Advanced Communication at the Shanghai University, China. He investigates the design and fabrication of novel fiber grating devices and their applications in fiber lasers and sensors, nanomaterial-based mode-locked fiber lasers, and polarization dynamics of ultrafast fiber lasers. He has authored or coauthored over 100 papers in international journals and conferences. He was a recipient of the Shanghai Young Eastern Scholar Fellowship from Shanghai in 2015 and the National Young Thousand Talent Plan Award from China in 2016.

Contents

Preface

Glass fibers are fabricated since the 1920s, but their use remains restricted to medical applications because of their high losses. In 1965, Kao et al. suggested that thin-glass waveguides could be a practical medium for optical communication if their losses are reduced, and then an intense technological progress on the purification of fused silica and the reduction in fiber loss was followed. In 1970, fabrications of low-loss fibers led to a revolution in the long-haul and high-bandwidth communication systems. Optical fibers are now being widely used in communications, sensing, biology, medicine, and so on.

Today, the most common optical fiber is the standard telecom single-mode fiber, which has an outer diameter of 125 μm with a core radius of ~5 μm. Besides the SMFs, multimode fibers, birefringence-modified fibers, and rare-earth-ion-doped fibers are widely developed and utilized. The excellent light-guiding capabilities of optical fibers resulting from a simple silica glass-based step-index solid structure derive many types of in-fiber photonic devices and configurations, such as couplers, interferometers, amplifiers, lasers, and broadband sources. Nevertheless, the performance and functionality of standard optical fibers found limitations in rapidly grown technological demands from information technology and various industrial sectors.

Recently, specialty fibers with modified geometrical structures and nonsilica materials, for example, microstructured optical fibers, multicore fibers, soft-glass fibers, and polymer fibers, have attracted much more attention because of their great potential in diverse functionality and applications in communication, sensing, and laser.

This book is a collection of contributions by selected active researchers in the optical fiber fields highlighting the design, fabrication, and application of optical fibers and fiber systems and covering various topics such as microstructured optical fibers, polymer fibers, nonlinear effects, optical tweezers, and gyroscopic systems. The goal of the book is to provide an updated overview of the current research trends in the optical fiber fields, serving as a general reference for the recent development in optical fiber technologies, though inevitably many topics are not covered.

Finally, we would like to thank all the authors for their contributions and the InTech editorial staff for the help to edit this book.

Fei Xu
Nanjing University, China

Chengbo Mou
Shanghai University, China

Chapter 1

Fabrication and Sensing Applications of Special Microstructured Optical Fibers

Zhengyong Liu and Hwa-Yaw Tam

Additional information is available at the end of the chapter

http://dx.doi.org/10.5772/intechopen.70755

Abstract

This chapter presents the fabrication of the special microstructured optical fibers (MOFs) and the development of sensing applications based on the fabricated fibers. Particularly, several types of MOFs including birefringent and photosensitive fibers will be introduced. To fabricate the special MOFs, the stack-and-draw technique is employed to introduce asymmetrical stress distribution in the fibers. The microstructure of MOFs includes conventional hexagonal assembles, large-air hole structures, as well as suspended microfibers. The birefringence of MOFs can reach up to 10^{-2} by designing the air hole structure properly. Fiber Bragg gratings as well as Sagnac interferometers are developed based on the fabricated special MOFs to conduct sensing measurement. Various sensing applications based on MOFs are introduced.

Keywords: microstructured optical fibers, MOFs, fiber sensor, interferometry, FBG

1. Introduction

Microstructured optical fibers (MOFs), which have air-hole structure along the fiber length, have attracted tremendous attention in the development of novel fiber sensors for a multitude of industry applications. Majority of the effort has been focused on silica photonic crystal fibers (PCFs), which are composed of many air channels arranged in hexagonal (honeycomb) shape [1–3]. Since Phillip St. Russel proposed the use of silica/air structure to design the fiber and realized all-silica single mode PCF in 1990s [4, 5], significant advances were made in supercontinuum generation [6–8], fiber lasers [9–11], as well as fiber-optic sensor [12–14]. More flexible fiber design and different kinds of materials are used in MOFs, for example, using tellurite or chalcogenide glass rather than pure silica glass [7] and suspended-core structure instead of honeycomb arrangement [13]. Light is guided in PCFs by two principles, namely, the modified total internal reflection (M-TIR) and the photonic bandgap (PBG). Typically, fibers employing the M-TIR principle have a solid core made of pure silica or silica doped with metal ions (e.g., Ge, Er, Yb, Tm, Co), whereas PCFs using the PBG principle have hollow core. Even though the fiber core is free of dopants or hollow, light can be confined in the core by cladding with periodic air-hole structure due to the lower effective refractive index of cladding or the bandgap effect. The air-hole structure or dopants of MOFs modified the mechanical and optical properties of the fibers, and thus, the MOFs can be tailored to suit specific sensor requirements.

Physical parameters, such as strain, temperature, pressure, vibration, torsion, etc., have been measured accurately using PCF-based sensors [15, 16]. Basically, the air-hole structure in MOFs for sensing applications is designed to give the desired stress distribution in the fiber so that the stress is enhanced by any external physical perturbation to the fiber, leading to change of the effective index of the guided mode in the core. The stress change can be induced either thermally or mechanically such as pulling, compression, and twist. Various kinds of sensors can be implemented using such approach. For instance, polarization-maintaining PCFs with an asymmetrical stress distribution provide an excellent option to construct Sagnac interferometer for the measurement of oil pressure with very high sensitivity of 3.4 nm/MPa [17]. On the other hand, the core in hollow-core PCFs can be filled with gas or analytes to enhance the interaction between the materials in the hollow core and light to realize highly sensitive gas and chemical sensors [18–20]. Alternatively, the air channels in the cladding of index-guiding PCFs confining light via the M-TIR principle can be filled with the materials to be sensed. The materials in the air holes change the index or stress distribution in the cladding and modify the guided light in the core [21–23]. By selectively filling liquid into some of the air holes, refractive index sensor with extremely high sensitivity = 12,750 nm/RIU was reported [23]. The ease and flexibility of fabricating MOFs with different structures bring tremendous opportunities in the development of fiber-optic sensors suited for a wide range of applications.

In this chapter, we present the fabrication techniques of some special MOFs and demonstrate the sensing applications based on the fabricated fibers. The stack-and-draw approach is used to make the fibers; however, specific modifications were introduced to obtain different air-hole structures, especially to induce the asymmetrical stress distribution in the fiber to realize highly sensitive pressure sensors. Various kinds of MOFs including twin-core PCF, high birefringence PCF, suspended core fiber, suspended microfiber, as well as two semicircle holes fiber were fabricated for sensing applications. MOF-based pressure sensors were fabricated using fiber Bragg grating (FBG) and/or interferometry technique. Section 2 will present the fabrication techniques used to fabricate our special MOFs. Section 3 describes the use of MOF sensors for the oil and gas industry. The summary is given at the end of this chapter.

2. Fabrication and characterization of special microstructured optical fibers

2.1. Fabrication of MOFs

Several approaches are being used to fabricate MOFs, including extrusion [24], casting/molding [25], mechanical drilling [26], and stack-and-draw [27]. The stack-and-draw technique is the most versatile and flexible. This is because by stacking small capillaries, not only various structures can be implemented but also different materials besides silica glass can be utilized. Extrusion, casting, and drilling are widely used to make polymer or soft-glass MOFs due to the much lower softening temperature of these materials. In contrast, it is easier to stack different structures regardless of materials. In addition to fabricate conventional hexagonal air-hole structures, more complicated superlattice structures can be fabricated using circular capillaries that are arranged in patterns to approximate triangular holes, square holes, and elliptic holes [27, 28]. We demonstrated the fabrication of superlattice PCFs using the stack-and-draw technique [27] to realize elliptical holes in the cladding and its optical characteristics are comparable to the design with ideal elliptical holes. Various designs of MOFs with either side hole, two core, or suspended core were used to fabricate optical fiber sensors for different sensing applications. Some of the circular capillaries arranged in hexagonal pattern can be replaced by capillaries/rods of different diameters. The stack-and-draw method offers the flexibility to fabricate a large variety of MOF sensors suited for different industries.

Figure 1 illustrates the process of fabricating a twin-core PCF (TC-PCF) using the stack-and-draw technique. Hexagonal structure is the natural pattern when stacking circular capillaries of equal diameter together. Typically, pure silica capillaries and rods are first drawn before the stacking process. To fabricate PCFs with n-layer of air hole in hexagonal pattern, $3\,n\,(n+1)$ of capillaries plus one rod for the central core are required. The stacked assemble can be secured using tungsten wire and then fixed in position by melting some of the capillaries. Alternatively, the entire assemble can be directly stacked inside a jacket tube, and the gaps are filled with silica rods of various outer diameters.

Basically, two drawing stages are employed to fabricate MOFs, particularly for complicated structures that have many layers of air holes. The first drawing stage is to draw the stacked assembly into cane with the desired outer diameter of 1–2 mm, as illustrated in **Figure 1(c)**. The

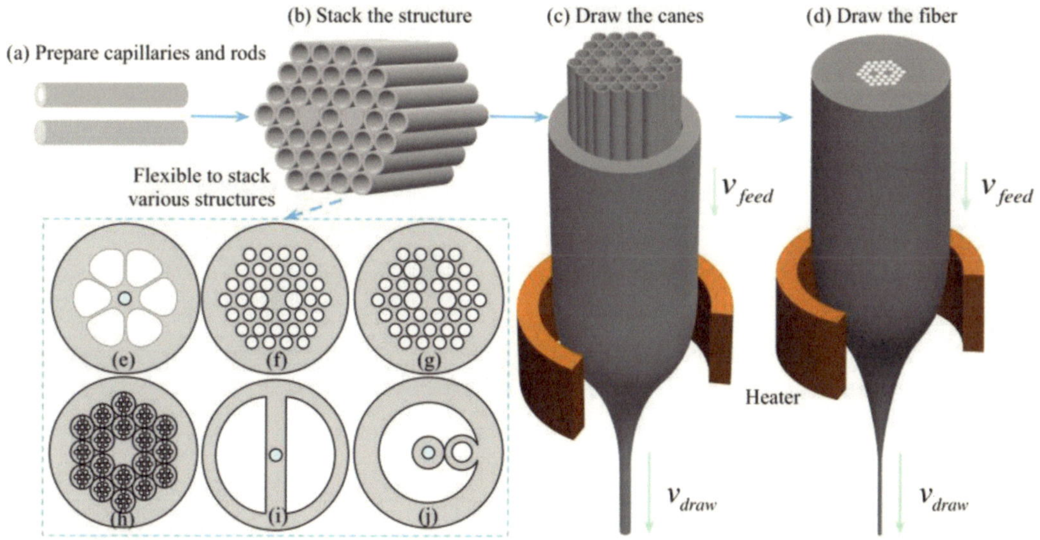

Figure 1. Illustration of the stack-and-draw method used in microstructured optical fibers (MOFs) fabrication.

cane is then inserted into a jacket tube, forming a second preform. The final fiber is drawn from the second preform, as shown in **Figure 1(d)**. By using two drawing stages, the total reduction ratio of diameter can be divided into two parts. Using smaller reduction ratio during the first drawing stage, the stacked assemble can be drawn with minimal distortion from the desired structure. For simple structures or structures with only a few large air holes, one drawing stage is normally adequate. For the superlattice structure reported in Ref. [27], three drawing stages were used, where the first drawn cane was used to make a second stack. To draw the MOFs properly, the drawing temperature ≈ 1900°C is lower than that of pulling all silica single mode fiber (SMF). The air hole would collapse if the drawing temperature is too high [29]. The collapse ratio is inversely proportional to the viscosity and the ratio of drawing velocity over feeding velocity (i.e., v_{draw}/v_{feed}). High temperature leads to low viscosity and ultimately large collapse ratio. Collapse ratio equal to 1 means air holes completely collapse. Relatively fast drawing velocity is preferable to maintain a certain tension to sustain the air-hole structure. However, large tension also results in poor fiber strength, which makes the fiber fragile. Thus, there is a trade-off to adjust the drawing temperature and tension. The control of tension and temperature is particularly important for drawing MOFs with large air holes.

Figure 2 shows the scanning electronic microscopic (SEM) photos of the cross-section of some fabricated MOFs in our lab using the stack-and-draw technique. By introducing air holes with different diameters, the mechanical properties of the fiber can be modified. In particular, the noncircularly symmetrical air-hole structure provides some degree of tailoring the stress

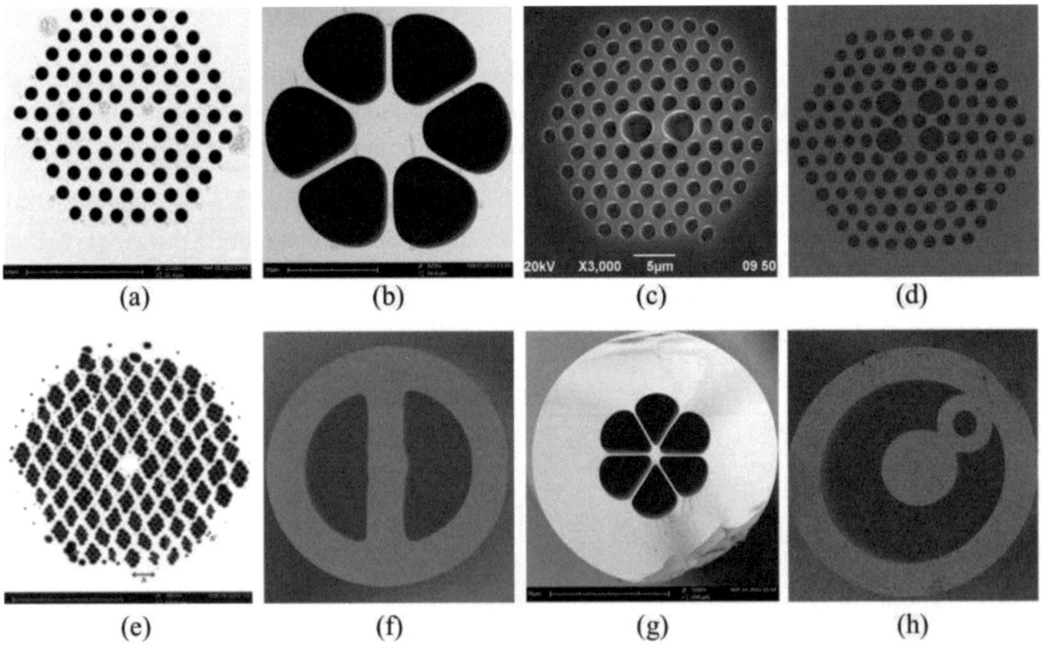

Figure 2. Scanning electronic microscopic photos of the cross section of the fabricated MOFs by using various stacking strategies. The examples shown are (a) twin-core photonic crystal fiber (TC-PCF) [12], (b) six-hole suspended-core fiber (SH-SCF), (c) high birefringence PCF (HB-PCF), (d) two-core HB-PCF, (e) superlattice PCF [27], (f) two semicircle hole fiber, (g) high birefringence suspended-core fiber (HB-SCF) [13], and (h) single-ring suspended fiber [30].

distribution in the fiber. Basically, the stress distribution of circularly symmetrical structures, including the standard single-mode optical fiber, is identical along the two polarization axes. MOFs with noncircularly symmetrical cross-section are birefringent. **Figure 2(a)** shows a two-core MOF, where the optical coupling between the cores is affected by pressure. **Figure 2(b)** shows a 6-hole MOF for the measurement of refractive-index changes of fluid. The cores of the MOFs shown in **Figure 2(c)–(g)** are elliptical, and the fibers are highly birefringent. **Figure 2(h)** shows a small-diameter fiber suspended inside a fiber and is designed for vibration measurement.

2.2. Twin-core photonic crystal fiber

A twin-core PCF was developed as an alternative sensor to standard single-mode optical fiber for pressure measurement. The air-hole structure of the TC-PCF is shown in **Figure 2(a)**, where two capillaries are replaced by two pure silica rods in the stacked assembly. The fabrication process is described schematically in **Figure 1**, and details can be found in Ref. [12]. The fiber was drawn at a temperature close to 1900 C. The outer diameter of the fiber is 125 μm, and diameter of two cores is ~2.5 μm. The hole diameter is ~1.1 μm, and pitch is ~1.85 μm. As the distance between the two cores is so close (~4 μm), the coupling effect between each other is strong. The modes guided in each core are combined and known as supermodes, i.e., even and odd modes [12, 31, 32]. Particularly in the fabricated TC-PCF, there are x-polarized even and odd modes and y-polarized even and odd modes. **Figure 3** shows the simulated mode profile of these modes.

According to the coupling theory, the coupling length of even and odd modes at each polarization (e.g. x polarization) can be written as

$$L_c(\lambda) = \frac{\lambda}{2\left|n_{p,x-even} - n_{p,x-odd}\right|} = \frac{\lambda}{2\left|\Delta n_{x,p}\right|}, \tag{1}$$

where $n_{p,x\text{-}even}$ and $n_{p,x\text{-}odd}$ represent the phase effective index of the x-polarized even and odd modes respectively, and $\Delta n_{p,x}$ stands for their difference. The effective index of each mode changes with the stress distribution induced by the external environment (such as pressure and strain), which eventually causes the corresponding change of coupling coefficient. As for

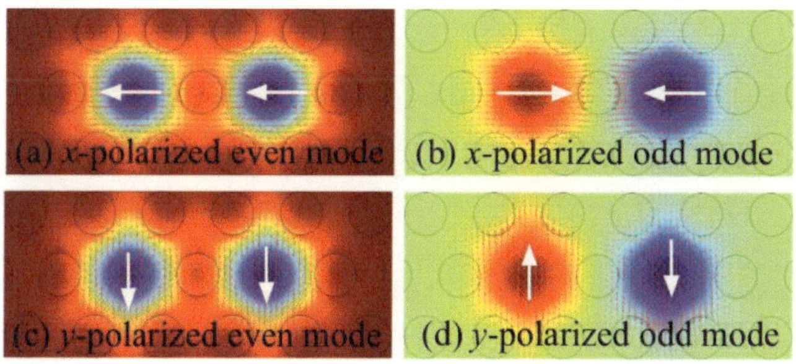

Figure 3. Simulated profiles of the electrical field for the x-polarized even and odd modes and y-polarized even and odd modes supported in TC-PCF, where the arrows show the direction of the field (adapted from ref. [12], OSA).

one polarization direction, the power coupled from one core to the other after propagating a distance of L can be expressed as

$$I_x(\lambda) = \sin^2\left[\frac{\pi}{2L_c(\lambda)}L\right] = \sin^2\left(\frac{\pi}{\lambda}\Delta n_{x,p}L\right). \tag{2}$$

Since the x and y polarization are orthogonal, the total output power of the transmission is the sum of both polarizations, which is then expressed as

$$I(\lambda) = I_x(\lambda) + I_y(\lambda) = 1 - \cos\left[\frac{\pi}{\lambda}\left(\Delta n_{x,p} + \Delta n_{y,p}\right)L\right] \cdot \cos\left[\frac{\pi}{\lambda}\left(\Delta n_{x,p} - \Delta n_{y,p}\right)L\right]. \tag{3}$$

It can be seen that the transmission spectrum of the TC-PCF is modulated due to the mutual influence of the two polarizations. If considering a short wavelength range, the fringe spacing can be approximated as

$$\Delta\lambda = \frac{2\lambda^2}{\left(\Delta n_{x,g} + \Delta n_{y,g}\right)L}. \tag{4}$$

Here, $\Delta n_{x,g}$ and $\Delta n_{y,g}$ are the group effective index difference of the even and odd modes of x- and y-polarizations, respectively. The group index (n_g) and phase index (n_p) have a relationship of $n_g = n_p - \lambda dn_p/d\lambda$. The calculated $\Delta n_{x,g}$ and $\Delta n_{y,g}$ are 3.745×10^{-3} and 3.386×10^{-3}, respectively. The calculated fringe spacing for a 110-cm long TC-PCF is ~0.613 nm. In experiment, the TC-PCF was spliced to two SMFs using manual splicing mode [33, 34]. A broadband source (BBS) and optical spectrum analyzer (OSA) were utilized to monitor the transmission spectrum of the output. To obtain high fringe contrast, an offset in the alignment and repeated arc discharges are needed. **Figure 4** shows the calculated and experimental transmission spectrum. Both results show modulation on the interference. Besides, the fringe spacing obtained experimentally is about 0.676 nm, which is also in good agreement with the calculated value. The obtained transmission spectrum of TC-PCF is sensitive to external physical perturbation and can be employed as pressure sensors.

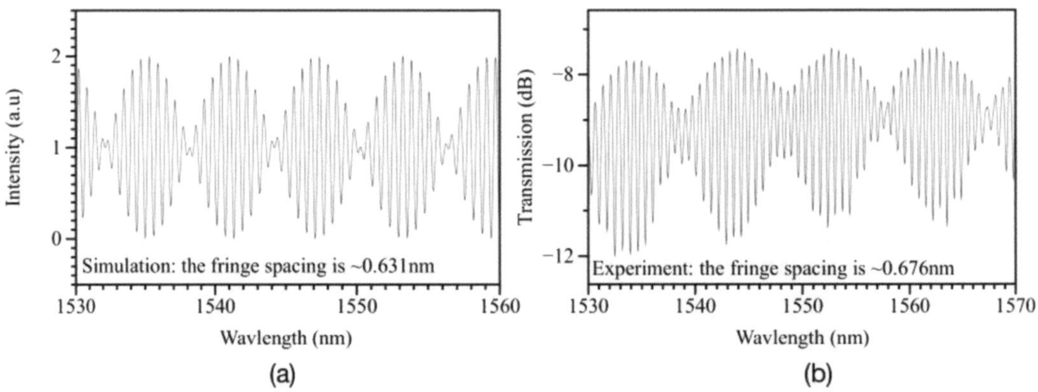

Figure 4. (a) Simulated and (b) experimental transmission spectra of a 110-cm long TC-PCF, adapted from ref. [12], OSA.

2.3. High birefringence microstructured optical fiber

In addition to the multiple cores achieved by replacing certain capillaries during stacking, capillaries with different inner diameters can be used to fabricate noncircularly symmetric air-hole structures to introduce birefringence in MOFs. The effective index of the x-polarized mode and y-polarized mode is different in birefringent MOFs. The birefringence of the fiber is defined as the index difference of the two polarization modes (i.e., $B = |n_x - n_y|$). In optical fiber communications, birefringent fiber is utilized to realize polarization maintaining. Commonly used polarization-maintaining fibers (PMFs) have a bowtie, elliptical, or PANDA structures [35]. The main feature of these fibers is that there are two heavily doped parts in the cladding, e.g., *PANDA* fiber [36] or an elliptical core designed to induce noncircularly symmetric stress to the core. The stress applying part in the cladding or the asymmetrical geometry of the fiber is regarded as exterior stress to the fiber core, whereas the thermal expansion of the noncircularly symmetric core yields interior stress to the core. The total birefringence of the fiber is composed of the contributions from exterior stress, interior stress, or both, depending on the fiber type.

Similar concept is employed in MOFs. Unlike conventional PMFs, high birefringence MOFs (HB-MOFs) exhibit the flexibility of modifying the fiber geometry and inducing exterior stress to the core. MOF fabrication permits the ease of introducing various air-hole structures in the cladding and breaks the circular symmetry of the fiber. Birefringence is an important property in fiber sensors for the measurement of many physical parameters. PM-PCF has been demonstrated to be a good candidate in the measurements of pressure [17], strain [37], temperature [38], torsion [39], etc. Most of these applications are based on the commercially available PM-PCF from NKT Photonics (PM-1550-01), which has a birefringence of ~4×10^{-4} at the wavelength of ~1550 nm.

The birefringence of MOFs allows for the construction of Sagnac interferometer (SI) by simply splicing PM-PCF between two output ports of a 3-dB coupler. The 3-dB coupler splits the light from one input port into two counter-propagating beams that interfere with each other at the 3-dB coupler after propagating through the PM-PCF, as shown in **Figure 5(a)**. The phase difference (ϕ) of the two polarization is $2\pi BL/\lambda$, where B and L are the birefringence and length of the fiber, respectively. At the valleys of the spectrum, the phase difference is always equal to $2k\pi$ (k is an integer). As for two adjacent valleys, the change of phase difference is 2π and equals $(d\phi/d\lambda)*\Delta\lambda$, where $\Delta\lambda$ is the wavelength difference of these two adjacent valleys. Thus, the birefringence of one HB-MOF can be expressed by

$$G = \frac{\lambda_1 \cdot \lambda_2}{\Delta\lambda \cdot L}, \quad (5)$$

where λ_1 and λ_2 represents the two adjacent valley wavelengths and G is the group birefringence, which can be measured via the experimental SI spectrum. **Figure 5(b)** plots the spectrum of a typical SI constructed using in-house HB-PCF (shown in **Figure 2(c)**) having a length of 5.5 cm. The wavelengths of two valleys close to 1550 nm are 1547.16 nm and 1550.66 nm. The group birefringence of this fiber is calculated to be ~1.25×10^{-2}.

The birefringence of the MOFs offers an alternative to conventional PMFs in the construction of interferometric sensors using phase difference of two polarization modes propagating in the

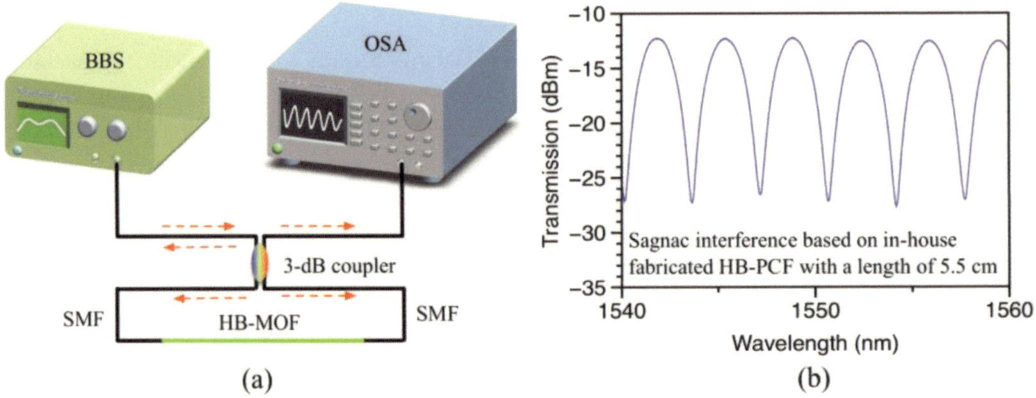

Figure 5. (a) Schematic figure of the configuration of a typical Sagnac interferometer realized by HB-MOF and (b) interference spectrum of a SI based on a 5.5-cm long in-house fabricated HB-PCF.

MOFs. The sensitivity of interferometric sensors is highly dependent on the birefringence change induced by external perturbations. Therefore, the arrangement of the air-hole structure that transfers the external perturbation to the optical mode in the core is important. We designed and fabricated several types of MOFs to measure oil pressure and refractive index of fluid for biomedical sensing. The sensitivity of the sensors varies greatly with the geometry of the core and cladding, as well as the sensing principle employed in the sensors. In general, interferometry-based sensors exhibit better sensitivity than grating-based sensors. However, in interferometry-based sensors, the sensing information is encoded in the spectral dips/peaks, which is much broader than the reflection peaks of grating-based sensors. Detecting the shift of the dips/peaks of a broad spectrum accurately is more difficult, therefore could affect the measurement accuracy. The sensing performance of different MOFs will be presented and compared in the following two sections.

Figure 2(c)–(g) show the SEM photos of some of the MOFs with high birefringence fabricated in our lab. As the air-hole structure is different, the birefringence of the fibers is not the same. The phase and group birefringence of HB-PCF we fabricated (show in **Figure 2(c)**) are measured to be 1.1×10^{-2} and 1.25×10^{-2}. Both values are in good agreement with the calculated values of 1.4×10^{-2} using finite element method. The two large holes in the cross section of the fiber shown in **Figure 2(c)** are slightly elongated, resulting in the desired noncircularly symmetric air-hole structure for high birefringent fiber. For index-guiding MOFs made of silica, this fiber possesses the highest birefringence in reported literatures. **Table 1** lists the comparison of our various HB-MOFs and others reported in the literatures.

Basically, elliptical core leads to higher birefringence because the two polarization modes are along with the major and minor axis of the elliptical core, individually. Due to the feature that optical mode is confined in the air for hollow-core PCF, even slight imperfection can influence the light propagation significantly, such as the loss and modal profile. Thus, hollow-core PCF with elliptical core exhibits very high birefringence even for small ellipticity. For example, the elliptical core with an aspect ratio of 1.16 exhibits a group birefringence of 2.5×10^{-2} [52]. MOFs

PM fiber type	Description of the structure	Reported date	Measured birefringence
HB PM-PCF, **Figure 2(c)**	Two large holes close to an elliptical core	May, 2013 [40]	1.25×10^{-2}
Superlattice PCF, **Figure 2(e)**	Superlattice with rhombic cell of 9 holes	June, 2014 [27]	8.5×10^{-4}
6-hole suspended-core fiber, **Figure 2(g)**	Elliptical core suspended by 6 large air holes	June, 2014 [13]	5×10^{-4}
Semicircle hole fiber, **Figure 2(f)**	Two large semicircle side holes	August, 2017	1×10^{-4}
Low-loss PM-PCF	Two large air holes close to the core	December, 2011 [41]	1.4×10^{-3}
PANDA PMF	Stress-applying parts with PANDA shape	July, 1981 [42]	8.5×10^{-5}
Bow-tie PMF	Stress-applying parts with bow-tie shape	November, 1982 [43]	4.87×10^{-4}
First fabricated HB PCF	Twofold rotational symmetry	September, 2000 [44]	3.85×10^{-3}
Side-hole fiber	Two side holes with elliptical Ge-core	September, 2008 [45]	1.39×10^{-4}
HB MOF	Irregular air holes with elliptical core	July, 2007 [46]	1.1×10^{-2}
Butterfly-type MOF	Twofold air-hole structure with butterfly shape	June, 2010 [47]	1.8×10^{-3}
Squeezed lattice PCF	Squeezed air holes with two big holes	February, 2010 [48]	5.5×10^{-3}
HB index-guiding PCF	Elliptical core by achieved by replacing two holes	June, 2001 [49]	9.3×10^{-4}
Fiberized glass ridge waveguide	Borosilicate glass-based fiberized ridge waveguide	April, 2015 [50]	9.5×10^{-3}
HB nonlinear MOF	Small elliptical core	July, 2004 [51]	7×10^{-3}
Hollow-core PBGF	Elliptical hollow core	August, 2004 [52]	2.5×10^{-2}
SF57 glass MOF at 1.06 μm wavelength	Asymmetric structure with elliptical air holes	May, 2017 [53]	9×10^{-2}
Chalcogenide glass PCF at 7.5 μm wavelength	Two large air holes close to the core	April, 2016 [54]	1.5×10^{-3}

Table 1. Comparison of birefringent MOFs reported in the literatures.

made of other type of glass (e.g. SF57, Chalcogenide glass) also have large birefringence in the order of 10^{-2}, if the structure is optimized. Sensors with high birefringence allow the implementation of compact sensing configuration because shorter fiber can be used. The sensing information of Sagnac interferometer is encoded in the wavelength shift of the interference spectrum. The phase difference at one valley is equal to $2k\pi$, and the sensitivity, which is defined as the wavelength shift per unit change in the measurand, can be expressed by

$$\frac{d\lambda}{dX} = \frac{\lambda}{G}\frac{\partial B}{\partial X}, \tag{6}$$

where X is the measurand, such as pressure, temperature, refractive index, etc., B is the modal (phase) birefringence, and $\frac{\partial B}{\partial X}$ is the polarimetric sensitivity. Note that the length effect is neglected in Eq. (6), which is not applied to the strain measurement. The strain is mainly caused by elongation and thus needs to be taken into account to estimate the sensitivity, which can be formulated by considering the length, and the equation becomes

$$\frac{d\lambda}{d\varepsilon} = \frac{\lambda \cdot [\partial B_1(\lambda,\varepsilon)/\partial\varepsilon + B_1(\lambda,\varepsilon)]}{G_1(\lambda,\varepsilon) + (1/\alpha - 1) \cdot G_2(\lambda)}, \quad (7)$$

where $B_1(\lambda,\varepsilon)$ and $G_1(\lambda,\varepsilon)$ are the phase and group modal birefringence for the MOF section under stressed, $G_2(\lambda)$ is the group modal birefringence for the MOF section without strain applied, and α is the ratio of fiber section under strain over total length of the MOF. Different air-hole structures give different values of birefringence derivative with respect to the measurands; therefore, it is important to design the MOFs with the desirable birefringence derivative to optimize the sensing performance. In terms of the form of wavelength shift, the sensitivity is inversely proportional to the birefringence. The key consideration to achieve high sensitivity is to have large-phase birefringence change with measurands but relatively low-group birefringence.

2.4. Sensors based on fiber Bragg grating inscribed in MOFs

The ease and flexibility of fabrication of birefringent MOFs result in the increasing use of these fibers in Sagnac interferometric sensors. Fiber Bragg gratings can be inscribed in MOFs to increase their sensing capabilities. However, it is more difficult to inscribe FBGs in the MOFs compare to SMFs due to the existence of the air holes that diffract the UV light needed to write gratings. UV light at 193, 248, 213, and 266 nm, as well as femtosecond laser [55–57], is being used to inscribe FBGs in various types of MOFs. 193 nm lasers [56] and femtosecond lasers were used to write FBGs in nonphotosensitive MOFs. These lasers induced physical deformation in the fiber cores. FBG written on MOFs using femtosecond laser can be used for high temperature measurement up to 800°C [56]. There are two reflective peaks instead of one peak in the reflection spectrum of an FBG inscribed in birefringent fibers [40, 58, 59]. The separation of these two peaks is directly related to the fiber birefringence.

It is easy to introduce photosensitivity in MOFs by using a germanium-doped rod during the stacking stage. Gratings can be inscribed in MOFs using UV light, and the ease of grating inscription depends on the air-hole patterns. **Figure 6** shows the reflection spectrum of an FBG inscribed in a six-hole suspended-core fiber (SCF) and HB PCF fabricated in our lab. The SEMs of their cross-section are shown in the insets. By optimizing the inscription system, very strong FBGs can be achieved. Owing to the ultrahigh-phase birefringence (~1.1×10^{-2}), the two reflective peaks of FBG inscribed in HB PCF show very large wavelength separation of more than 10 nm which is much larger than that in the commercial PM-PCF (0.5 nm) [59] and the HB MOFs (2.1 nm) reported in Ref. [58]. Two peaks with large separation allow for simultaneous measurement of temperature and pressure because the stress transferred to the core along the two polarization axes is different when subjected to pressure.

The FBG inscribed in the six-hole suspended-core fiber can be employed to measure strain and temperature. The measured strain sensitivity is about 0.96 pm/μm, which is close to that of SMF. Since the germanium is doped in the core as in SMF, similar temperature sensitivity of

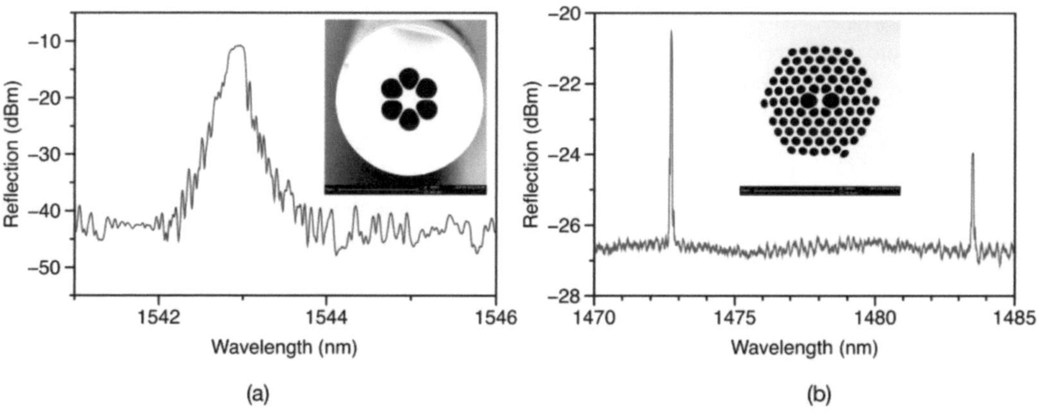

Figure 6. Reflection spectrum of an FBG inscribed in the (a) photosensitive six-hole SCF and (b) HB PCF, adapted from ref. [40], OSA. The insets show the SEM photos of the cross section of the fabricated fibers.

10.7°C was measured. However, the pressure sensitivity of the FBG inscribed in the fiber was measured to be 8.2 pm/MPa, which is higher than that of SMFs, which is ~3.1 pm/MPa [60]. This improvement is due to the six air holes, and further increase in the sensitivity is possible by using larger air holes, as demonstrated in [61]. We designed and fabricated a single-ring suspended fiber, as shown in **Figure 2(h)**, to increase the pressure sensitivity. The measured pressure sensitivity using FBG on this fiber is about 18 pm/MPa, more than 5 times higher than that of standard SMFs.

In addition, the large air holes of MOFs permit materials to be filled in the cladding to functionalize MOFs into a large variety of sensors. For example, low-temperature melting point metal was filled into the air holes of the six-hole SCF to function as anemometer to measure wind speed [22]. The metal absorbs energy from light propagating in the six-hole SCF, and the FBG inscribed in the core measures the temperature change. The cooling rate of the metal/FBG is directly related to the wind speed. Due to the large optical absorption of metal, the heating efficiency is very high, i.e., ~7.3°C/mW. Laser power as low as 14 mW is sufficient to heat the metal/FBG up to 100°C. This is more efficient than the heating process using Co^{2+}-doped fiber as we have demonstrated in [62].

3. Sensing applications of MOFs in the oil and gas industry

The demand for fiber optic sensors in the oil/gas industry comes from the harsh downhole conditions and the depth of oil wells which can be as deep as 12 km. The great distance coupled with the high pressure (up to 100 MPa) and temperature (more than 200°C) of oil in downholes restricts the use of conventional sensors. However, the intrinsic features of optical fiber sensors such as long distance transmission, immune to EMI, and high operating temperature make it a promising candidate for the oil industry. The use of multiple FBGs distributed along a single strand of SMF has been employed in oil monitoring. Key parameters like pressure, temperature, and flow speed are widely measured in oil wells [63, 64]. In such circumstances, high pressure can cause large irretrievable disasters during oil exploitation and oil transportation. Therefore, it is of great importance to measure pressure, and MOFs are potential candidates to enhance the capabilities of optical fiber sensors used in the oil industry.

The stress distribution around the core of MOFs can be tailored via the air-hole pattern to enhance the MOFs' sensitivity to pressure. The application of pressure to MOFs leads to large compression stress in the fiber core where optical light propagates. Good performance in terms of sensitivity, resolution, as well as fast response can be achieved by properly designing the structure of the MOFs.

The basic principle that permits fiber sensors to measure pressure is the photoelastic effect of silica glass. When subjected to pressure, fiber sensors regardless of the use of different operating principles, the silica fiber shows the corresponding dependence on the change in pressure. Particularly, the refractive index of the core/cladding of MOFs varies with applied pressure, which can be expressed as

$$\begin{bmatrix} n_x \\ n_y \\ n_z \end{bmatrix} = n_0 - \begin{bmatrix} C_1 & C_2 & C_2 \\ C_2 & C_1 & C_2 \\ C_2 & C_2 & C_1 \end{bmatrix} \begin{bmatrix} \sigma_x \\ \sigma_y \\ \sigma_z \end{bmatrix}, \tag{8}$$

where n_i is the index component in i_{th} direction ($i = x, y, z$), σ_i represents the corresponding stress component, and n_0 is the silica index without pressure applied. The constants C_1 and C_2 are the stress-optic coefficient for silica glass and have values of 6.5×10^{-13} and 4.2×10^{-12} m^2/N, respectively. The change in refractive index induced by pressure is determined by the above equation. The pressure-induced change in index is taken into account to calculate the guided mode in MOFs. Either FBGs or interferometry is employed to make pressure sensors, and the stress transfer mechanism is similar. However, in terms of the polarimetric approach based on high birefringence MOFs as introduced in Section 2.3, the resultant pressure sensors that employ the index difference of two polarized modes exhibit better sensitivity.

Figure 7 plots the results of oil pressure measurement using FBGs written on conventional SMF and single-ring suspended fiber (shown in **Figure 2(h)**) with various outer diameters. The single-ring suspended fiber differs from SMF because of the large air region in the fiber, resulting in high air-filling ratio (AFR). AFR of SMF can be regarded as 0, as no air holes exist in the fiber. Typically, higher AFR gives better sensitivity [61]. The measured results using FBG on single-ring suspended fiber show a large improvement (five times) of pressure sensitivity compared to that obtained on SMF. Furthermore, by etching the cladding of the fiber, smaller outer diameter of this MOF means higher AFR and consequently further increases the pressure sensitivity from −18 to −21 pm/MPa.

In addition to improvement in sensitivity, MOFs enable simultaneous measurement of pressure and temperature by using FBG inscribed in high birefringence fiber. As shown in **Figure 6(b)**, the two distinct peaks occurred in the reflection spectrum have a wide wavelength separation due to the large fiber birefringence. The separation is more than 10 nm for this MOF, which has a measured birefringence of 1.1×10^{-2}. Two-parameter measurement tends to be easier when utilizing two peaks with large separation. **Figure 8(a)** shows the pressure responses of the two modes polarized in fast and slow axes, corresponding to the two FBG peaks located at shorter and longer wavelength, which gives sensitivities of −1.9 pm/MPa for the fast axis peak and −5.1 pm/MPa for the slow axis peak [40]. The different pressure responses obtained from the fast and slow axis peaks are due to the asymmetrical air hole structure, which breaks the uniformity of the pressure-induced stress. Therefore, the stress change along the fast axis is

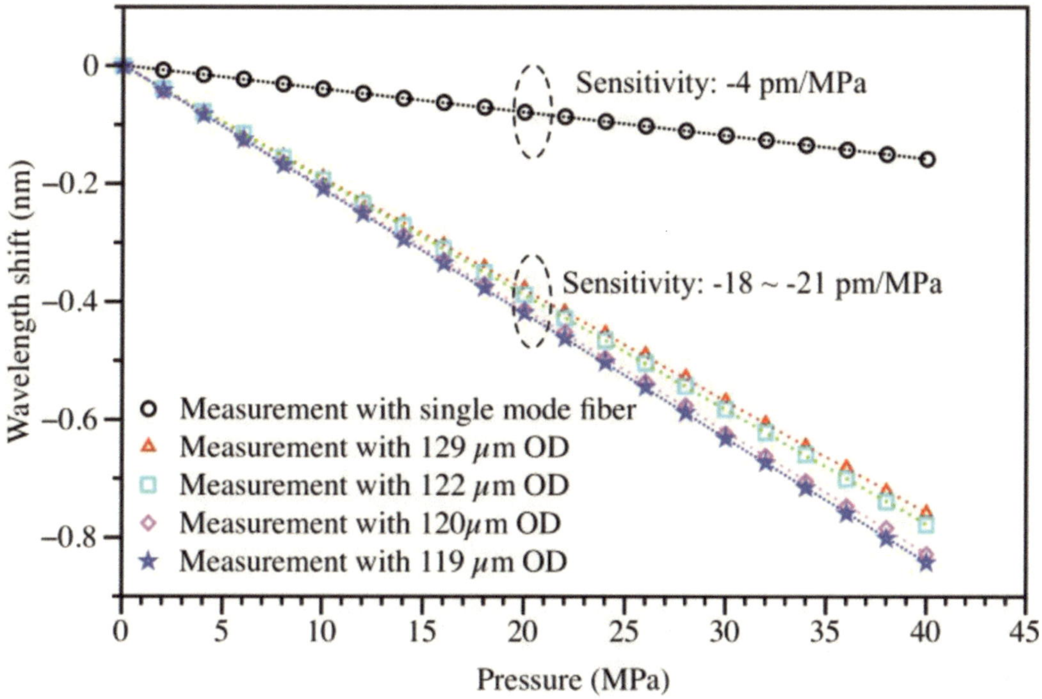

Figure 7. Pressure response of FBGs inscribed in SMF and MOFs.

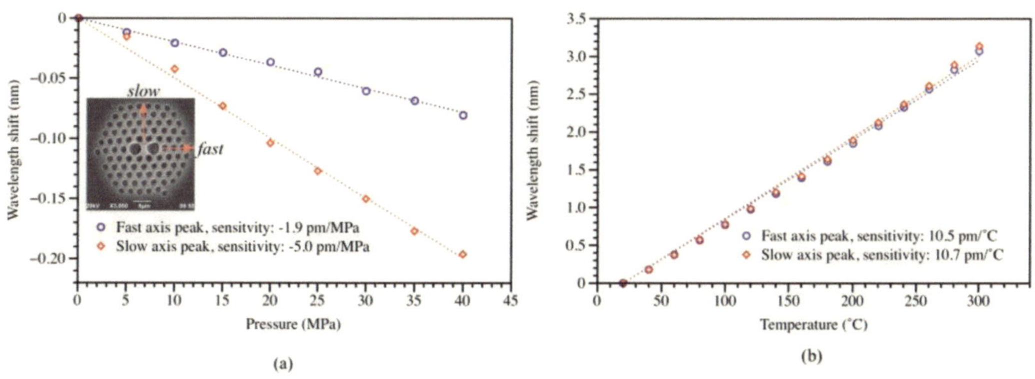

Figure 8. (a) Pressure response of FBG written in the HB-PCF fabricated in our laboratory, where the fast and slow axis peaks show different responses, (b) temperature dependence of the two axes (adapted from Ref. [40], OSA).

smaller than that in the slow axis due to the existence of two large air holes. However, the temperature dependence of the two polarized modes are the same, about 10 pm/°C, as shown in **Figure 8(b)**. Such discrimination in pressure and temperature allows the simultaneous measurement of these two parameters. The change in pressure and temperature can be calculated according to the total wavelength shift of the fast and slow axis grating peak via the following equation

$$\begin{bmatrix} \Delta T \\ \Delta P \end{bmatrix} = \frac{1}{32.17} \begin{bmatrix} -1.9 & 5.0 \\ -10.5 & 10.7 \end{bmatrix} \begin{bmatrix} \Delta\lambda_s \\ \Delta\lambda_f \end{bmatrix}, \tag{9}$$

where $\Delta\lambda_s$ and $\Delta\lambda_f$ are the total wavelength shift for the slow and fast axis grating peaks due to changes in the applied pressure and temperature, respectively.

The pressure sensitivity of FBG sensors is very small, varying from a few to tens of pm/MPa even if special MOFs are employed. The low sensitivity is due to the slight modal index change with respect to pressure. On the other hand, the sensitivity can be improved significantly by applying the polarimetric approaches based on high birefringence MOFs. For example, SI or rocking filter pressure sensors exhibit much higher sensitivity in pressure measurement [65, 66]. The construction of a SI sensor is shown in **Figure 5(a)**, where the MOF is subjected to oil pressure. The oil pressure sensitivity demonstrated by a PM-PCF

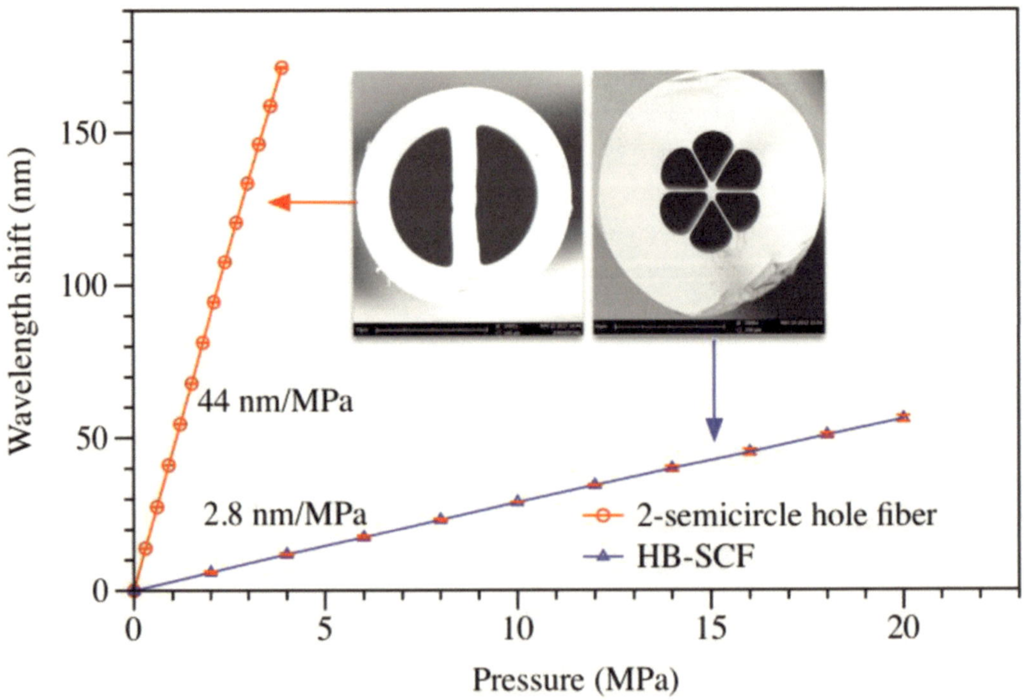

Figure 9. Hydrostatic oil pressure response of Sagnac interferometer realized using HB-SCF [13], (2014 IEEE) and two semicircle hole fiber.

commercially available from NKT Photonics, Inc. is 3.24 nm/MPa at ~1550 nm [67]. The pressure sensitivity of a SI pressure sensor can be written as

$$\frac{d\lambda}{dP} = \frac{\lambda}{G}\frac{dB}{dP}, \tag{10}$$

where the derivative dB/dP represents the phase birefringence change with respect to applied pressure, which means the index difference of two polarized modes is considered rather than only one mode as in the case of FBG sensor. This value can be improved easily by using noncircularly symmetric structure. **Figure 9** shows the wavelength shift as a function of the applied pressure on the highly birefringent MOFs, where (a) plots the results for HB-SCF and (b) is the pressure response for the two semicircle hole fiber. Both measurements are based on SI configuration. The measured pressure sensitivity is ~2.8 nm/MPa using HB-SCF, which is comparable to the commercial PM-PCF. However, the fabrication of HB-SCF is much easier than PM-PCF that has a lot of air holes in honeycomb arrangement [13]. On the other hand, the sensitivity improvement is significant by using the design of two semicircle hole structure, which was measured to be 44 nm/MPa, about 13 times larger than that achieved with commercial PM-PCF. Such great increase owes to the relatively low birefringence (i.e., small G), as well as the cutoff of fast axis by two large semicircle holes (i.e., large dB/dP).

4. Conclusion

To conclude the chapter, several novel structures of optical fibers are proposed and demonstrated for sensing applications. The sensing performance is comparable to and better than most sensors developed based on traditional single-mode fibers (SMFs). The basic fabrication method is briefly reviewed, especially to introduce the asymmetrical stress distribution to MOFs. Due to the novel structure of MOFs, it also exhibits unique feature that SMF does not possess, for instance, the high birefringence. Different air hole structures of MOFs exhibit various mechanical and optical properties that are employed to develop the sensors. The ultrahigh birefringence of HB-PCF can be ~1.2×10^{-2}, which is the highest one for the fabricated index-guiding PCFs. The pressure sensor based on the fabricated two semicircle hole MOF shows very high sensitivity >40 nm/MPa by constructing a Sagnac interferometer. Those sensors can find good applications in oil and gas industry, as well as the biomedical detection. The demonstrated MOFs can give full understanding of developing sensors to measure physical and biomedical parameters, in terms of the design, fabrication of the MOFs, and the approaches to configure sensors.

Acknowledgements

The authors like to acknowledge the financial support by the Hong Kong Polytechnic University under the project of 1-ZVGB.

Author details

Zhengyong Liu* and Hwa-Yaw Tam

*Address all correspondence to: zhengyong.liu@connect.polyu.hk

Photonics Research Centre, Department of Electrical Engineering, The Hong Kong Polytechnic University, Hung Hum, KLN, Hong Kong

References

[1] Knight JC. Photonic crystal fibres. Nature. 2003;**424**:847-851

[2] Russell P. Photonic crystal fibers. Science. 2003;**299**:358-362

[3] Russell PSJ. Photonic-crystal Fibers. Journal of Lightwave Technology. 2006;**24**:4729-4749

[4] Knight J, Birks T, Russell P. All-silica single-mode optical fiber with photonic crystal cladding. Optics Letters. 1996;**21**:1547-1549

[5] Birks TA, Knight JC, Russell PS. Endlessly single-mode photonic crystal fiber. Optics Letters. 1997;**22**:961-963

[6] Dudley JM, Coen S. Supercontinuum generation in photonic crystal fiber. Reviews of Modern Physics. 2006;**78**:1135-1184

[7] Toupin P, Brilland L, Renversez G, Troles J. All-solid all-chalcogenide microstructured optical fiber. Optics Express. 2013;**21**:14643

[8] http://www.nktphotonics.com/lasers-fibers/en/

[9] Kong F, et al. Polarizing ytterbium-doped all-solid photonic bandgap fiber with ~1150μm^2 effective mode area. Optics Express. 2015;**23**:4307

[10] Groothoff N, Canning J, Ryan T, Lyytikainen K, Inglis H. Distributed feedback photonic crystal fibre (DFB-PCF) laser. Optics Express. 2005;**13**:2924-2930

[11] Canning J, et al. All-fibre photonic crystal distributed Bragg reflector (PC-DBR) fibre laser. Optics Express. 2003;**11**:1995-2000

[12] Liu Z, et al. Intermodal coupling of supermodes in a twin-core photonic crystal fiber and its application as a pressure sensor. Optics Express. 2012;**20**:21749-21757

[13] Liu Z, Wu C, Tse M-LV, Tam H-Y. Fabrication, characterization, and sensing applications of a high-birefringence suspended-core fiber. Journal of Lightwave Technology. 2014;**32**: 2113-2122

[14] Monro TM, et al. Sensing with microstructured optical fibres. Measurement Science and Technology. 2001;**12**:854-858

[15] Liu Z, Tam H-Y, Htein L, Tse M-LV, Lu C. Microstructured optical Fiber sensors. Journal of Lightwave Technology. 2017;**35**:3425-3439

[16] Pinto AMR, Lopez-Amo M. Photonic crystal Fibers for sensing applications. Journal of Sensors. 2012;**2012**:1-21

[17] Fu HY, et al. Pressure sensor realized with polarization-maintaining photonic crystal fiber-based Sagnac interferometer. Applied Optics. 2008;**47**:2835-2839

[18] Jin W, Cao Y, Yang F, Ho HL. Ultra-sensitive all-fibre photothermal spectroscopy with large dynamic range. Nature Communications. 2015;**6**:6767

[19] Bykov DS, Schmidt OA, Euser TG, Russell PSJ. Flying particle sensors in hollow-core photonic crystal fibre. Nature Photonics. 2015;**9**:1-14

[20] Cubillas AM, et al. Photonic crystal fibres for chemical sensing and photochemistry. Chemical Society Reviews. 2013;**42**:8629-8648

[21] Wu C, et al. In-line microfluidic refractometer based on C-shaped fiber assisted photonic crystal fiber Sagnac interferometer. Optics Letters. 2013;**38**:3283-3286

[22] Wang J, et al. Fiber-optic anemometer based on Bragg grating inscribed in metal-filled microstructured optical Fiber. IEEE Xplore: Journal of Lightwave Technology. 2016;**34**: 4884-4889

[23] Sun B, et al. Microstructured-core photonic-crystal fiber for ultra-sensitive refractive index sensing. Optics Express. 2011;**19**:4091-4100

[24] Ebendorff-Heidepriem H, Warren-Smith S. Suspended nanowires: Fabrication, design and characterization of fibers with nanoscale cores. Optics Express. 2009;**17**:2646-2657

[25] Large MCJ, et al. Microstructured polymer optical fibres: New opportunities and challenges. Molecular Crystals and Liquid Crystals. 2006;**446**:219-231

[26] El-Amraoui M, et al. Microstructured chalcogenide optical fibers from As2S3 glass: Towards new IR broadband sources. Optics Express. 2010;**18**:26655-26665

[27] Tse M-LV, et al. Superlattice microstructured optical Fiber. Materials (Basel). 2014;7:4567-4573

[28] Chen D, Vincent Tse ML, Tam HY. Super-lattice structure photonic crystal fiber. Progress in Electromagnetics Research. 2010;**11**:53-64

[29] Fitt A, Furusawa K, Monro T. The mathematical modelling of capillary drawing for holey fibre manufacture. Journal of Engineering Mathematics. 2002;**43**:201-227

[30] Htein L, Liu Z, Tam HY. Hydrostatic pressure sensor based on fiber Bragg grating written in single-ring suspended fiber. Proc. SPIE 9916, Sixth Eur. Work. Opt. Fibre Sensors 99161R; 2016. DOI:10.1117/12.2235842

[31] Chen D, Hu G, Chen L. Dual-core photonic crystal Fiber for hydrostatic pressure sensing. IEEE Photonics Technology Letters. 2011;**23**:1851-1853

[32] Zhang L. Polarization-dependent coupling in twin-core photonic crystal fibers. Journal of Lightwave Technology. 2004;**22**:1367-1373

[33] Tse MLV, et al. Fusion splicing holey Fibers and single-mode Fibers: A simple method to reduce loss and increase strength. IEEE Photonics Technology Letters. 2009;**21**:164-166

[34] Xiao L, Jin W, Demokan MS. Fusion splicing small-core photonic crystal fibers and single-mode fibers by repeated arc discharges. Optics Letters. 2007;**32**:115-117

[35] Noda J, Okamoto K, Sasaki Y. Polarization-maintaining fibers and their applications. Journal of Lightwave Technology. 1986;**4**:1071-1089

[36] Zhu M, Murayama H, Wada D, Kageyama K. Dependence of measurement accuracy on the birefringence of PANDA fiber Bragg gratings in distributed simultaneous strain and temperature sensing. Optics Express. 2017;**25**:4000

[37] Dong X, Tam HY, Shum P. Temperature-insensitive strain sensor with polarization-maintaining photonic crystal fiber based Sagnac interferometer. Applied Physics Letters. 2007;**90**:151113

[38] Chen T, et al. Distributed high-temperature pressure sensing using air-hole microstructural fibers. Optics Letters. 2012;**37**(6):1064

[39] Weiguo C, et al. Highly sensitive torsion sensor based on Sagnac interferometer using side-leakage photonic crystal Fiber. IEEE Photonics Technology Letters. 2011;**23**:1639-1641

[40] Liu Z, Wu C, Tse M-LV, Lu C, Tam H-Y. Ultrahigh birefringence index-guiding photonic crystal fiber and its application for pressure and temperature discrimination. Optics Letters. 2013;**38**:1385-1387

[41] Suzuki K, Kubota H, Kawanishi S, Tanaka M, Fujita M. Optical properties of a low-loss polarization-maintaining photonic crystal fiber. Optics Express. 2001;**9**:676-680

[42] Edahiro T, Sasaki Y, Hosaka T, Miya T, Okamoto K. Low-loss single polarisation fibres with asymmetrical strain birefringence. Electronics Letters. 1981;**17**:530-531

[43] Birch RD, Payne DN, Varnham MP. Fabrication of polarisation- maintaining fibres using gas-phase etching. Electronics Letters. 1982;**18**:1036-1038

[44] Ortigosa-Blanch A, Knight J, Wadsworth W. Highly birefringent photonic crystal fibers. Optics Letters. 2000;**25**:1325-1327

[45] Frazão O, et al. Simultaneous measurement of multiparameters using a Sagnac interferometer with polarization maintaining side-hole fiber. Applied Optics. 2008;**47**:4841-4848

[46] Kim S, et al. Ultrahigh birefringence of elliptic core fibers with irregular air holes. Journal of Applied Physics. 2007;**102**:16101

[47] Martynkien T, et al. Highly birefringent microstructured fibers with enhanced sensitivity to hydrostatic pressure. Optics Express. 2010;**18**:15113-15121

[48] Beltrán-Mejía F, et al. Ultrahigh-birefringent squeezed lattice photonic crystal fiber with rotated elliptical air holes. Optics Letters. 2010;**35**:544-546

[49] Hansen TP, et al. Highly birefringent index-guiding photonic crystal fibers. IEEE Photonics Technology Letters. 2001;**13**:588-590

[50] Shi J, Feng X, Horak P, Poletti F. A fiberized highly birefringent glass micrometer-size ridge waveguide. Optical Fiber Technology. 2015;**23**:137-144

[51] Ortigosa-Blanch A. Ultrahigh birefringent nonlinear microstructured fiber. IEEE Photonics Technology Letters. 2004;**16**:1667-1669

[52] Chen X, et al. Highly birefringent hollow-core photonic bandgap fiber. Optics Express. 2004;**12**:3888-3893

[53] Pei T-H, Zhang Z, Zhang Y. The high-birefringence asymmetric SF57 glass microstructured optical fiber at 1060.0 um. Optical Fiber Technology. 2017;**36**:265-270

[54] Caillaud C, et al. Highly birefringent chalcogenide optical fiber for polarization-maintaining in the 3-8.5 μm mid-IR window. Optics Express. 2016;**24**:7977

[55] Hill KO, Meltz G. Fiber Bragg grating technology fundamentals and overview. Journal of Lightwave Technology. 1997;**15**:1263-1276

[56] Jewart CM, et al. Ultrafast femtosecond-laser-induced fiber Bragg gratings in air-hole microstructured fibers for high-temperature pressure sensing. Optics Letters. 2010;**35**: 1443-1445

[57] Groothoff N, Canning J, Buckley E, Lyttikainen K, Zagari J. Bragg gratings in air-silica structured fibers. Optics Letters. 2003;**28**:233-235

[58] Geernaert T, et al. Fiber Bragg gratings in germanium-doped highly birefringent microstructured optical fibers. IEEE Photonics Technology Letters. 2008;**20**:554-556

[59] Guan B, Chen D, Zhang Y. Bragg gratings in pure-silica polarization-maintaining photonic crystal fiber. IEEE Photonics Technology Letters. 2008;**20**:1980-1982

[60] Xu M, Reekie L, Chow Y, Dakin JP. Optical in-fibre grating high pressure sensor. Electronics Letters. 1993;**29**:398-399

[61] Wu C, Guan B-O, Wang Z, Feng X. Characterization of pressure response of Bragg gratings in grapefruit microstructured Fibers. Journal of Lightwave Technology. 2010;**28**: 1392-1397

[62] Liu Z, Tse MV, Zhang AP, Tam H. Integrated microfluidic flowmeter based on a micro-FBG inscribed in Co2+−doped optical fiber. Optics Letters. 2014;**39**:5877-5880

[63] Kersey AD. Optical fiber sensors for permanent downwell monitoring applications in the oil and gas industry. IEICE Transactions on Electronics. 2000;**83**:400-404

[64] Nakstad H, Kringlebotn JT. Oil and gas applications: Probing oil fields. Nature Photonics. 2008;**2**:147-149

[65] Anuszkiewicz A, et al. Sensing characteristics of the rocking filters in microstructured fibers optimized for hydrostatic pressure measurements. Optics Express. 2012;**20**:23320

[66] Kakarantzas G, et al. Structural rocking filters in highly birefringent photonic crystal fiber. Optics Letters. 2003;**28**:158-160

[67] Fu HY, et al. High pressure sensor based on photonic crystal fiber for downhole application. Applied Optics. 2010;**49**:2639-2643

Chapter 2

Photonic Crystal Fiber–Based Interferometric Sensors

Dora Juan Juan Hu, Rebecca Yen-Ni Wong and
Perry Ping Shum

Additional information is available at the end of the chapter

http://dx.doi.org/10.5772/intechopen.70713

Abstract

Photonic crystal fibers (PCFs), also known as microstructured optical fibers, are a highlighted invention of optical fiber technology which have unveiled a new domain of manipulating light in engineered fiber waveguides with unparalleled flexibility and controllability. Since the report of the first fabricated PCF in 1996, research in PCFs has resulted in numerous explorations, development and commercialization of PCF-based technologies and applications. PCFs contain axially aligned air channels which provide a large degree of freedom in design to achieve a variety of peculiar properties; numerous PCF-based sensors have been proposed, developed and demonstrated for a broad range of sensing applications. In this chapter, we will review the field of research on design, development and experimental achievement of PCF-based interferometric sensors for physical and biomedical sensing applications.

Keywords: photonic crystal fibers, interferometry, fiber optic sensors, Fabry-Perot interferometer, Mach-Zehnder interferometer, Michelson interferometer, Sagnac interferometer

1. Background

Optical fiber interferometric sensors have been widely used for various sensing applications and characterization of physical magnitudes. The advantages provided by optical fibers have been well recognized and utilized in interferometric sensor applications, which include compactness, alignment freedom from free space optics, high sensitivity, high reliability etc. [1]. Photonics crystal fibers (PCFs), a highlighted fiber technology that was first invented and demonstrated in 1996, have brought breakthroughs in communications, sensing, defense and medicine [2]. These fibers have demonstrated superior features in many applications and created substantial scientific and industrial impact in recent years. In the past decade, PCFs have received intensive and continuous attention, and undergone rapid development from design and fabrication to device realization and commercialization. Compared to conventional optical fibers, PCFs represent a more versatile platform to construct interferometry sensors because of enhanced flexibility in manipulating optical properties and light-medium interactions. Various PCF structures, such as polarization-maintaining (PM) PCFs, photonic bandgap (PBG) PCFs including hollow core (HC) PCFs and all-solid PBG PCFs, Bragg fibers, large mode area (LMA) PCFs and highly nonlinear PCFs have been demonstrated with good potential in developing interferometric fiber sensors. PCFs can provide a platform for integration of materials such as gas, fluid or metals for additional functionality. For example, PCFs have been exploited for optofluidic sensing and gas sensing applications utilizing the selective or unselective infiltration of fluid or gas in the holey structures [3]. In addition, they are a desirable platform for the incorporation of plasmonic structures that can enhance application opportunities in terms of performance as well as versatility. Integration of plasmonic structures such as metal nanoparticles, metal nanowires, and metal thin films in PCF structures have proven to substantially improve sensor performance, e.g. sensitivity [4]. The continuing development and maturation of PCF technologies and PCF-based interferometric sensors are expected to make more contributions to optical fiber technology and real world applications [5].

2. Overview of PCF-based interferometric sensors

In this chapter, various configurations of interferometry sensors based on PCFs and their sensing applications are demonstrated, namely Fabry-Perot interferometer (FPI), Mach Zehnder interferometer (MZI), Michelson interferometer and Sagnac interferometer. Compared to standard optical fibers, PCF structures possess many interesting characteristics and tunable properties which are highly desirable when constructing interferometric sensors with enhanced performance.

2.1. Fabry-Perot interferometer (FPI)

A Fabry-Perot interferometer is comprised of a cavity (or etalon) made of two highly reflective surfaces/mirrors which enable light propagating down the fiber to be partially reflected. The transmitted, and subsequently, reflected beams will form an interference pattern due to the difference in phase delay.

The reflection coefficients, R_i, at the mirrors can be defined by [6]:

$$R_i = \left(\frac{n_i - n_{i+1}}{n_i + n_{i+1}} \right)^2, i = 1, 2, 3, \ldots \quad (1)$$

where n_i is the refractive index of the cavity and surrounding medium.

The phase difference, δ, of the interferometer can be represented by [1]:

$$\delta = \frac{2\pi}{\lambda} n2L \quad (2)$$

where λ is the incident light wavelength and L is the physical length of the cavity.

FPIs can typically be classed as either extrinsic or intrinsic, depending on their make-up. Extrinsic FPIs (EFPI), as shown in **Figure 1(a)**, use the air gap between two fibers and reflects light between the cleaved ends. The cavity of intrinsic FPIs (IFPI) is formed within the fiber itself, where the two reflectors lie along the length of the fiber [1], as shown in **Figure 1(b), (c)**. IFPIs can have advantages over EFPIs such as higher coupling efficiency.

For IFPIs, the etalon/cavity can be formed by fusion splicing a section of HC-PCF, which acts as the cavity, between two lengths of single mode fiber with cleaved end surfaces [7]. This configuration allows for a customizable cavity length, which can be a few micrometers or a few centimeters long [7]. Villatoro . presented a spherical FP cavity by means of a microscopic air bubbles (20–58 μm diameters) fabricated via arc discharge between a standard SMF and a PCF. This technique can reduce the number of steps required for fabricating FPIs [8]. Favero et al. [9] pressurized the holes in the PCF to produce reproducible elliptical bubbles with controllable cavity dimensions.

Hu et al. [6] were able to realize a refractive index tip sensor used in reflection mode. This sensor was based on a hollow silica sphere with a thin silica wall being formed at one end of a simplified HCF via means of arc discharge, as shown in **Figure 2**. The reflected spectrum was modulated by the interaction between the sensor head and the environment (refractive index (RI) and temperature). A RI resolution of 6.2×10^{-5}, using fringe visibility (**Figure 2 (c)**), was determined and the temperature sensitivity for the high and low frequency fringes were 1.3 and 17 pm/°C, respectively (**Figure 2 (d)**), at temperatures up to ~1000°C.

Micro FP cavities can offer low cross sensitivity with temperature, yet high RI sensitivity. A RI sensor was realized by drilling micro-holes into a simplified hollow core (SHC) PCF micro

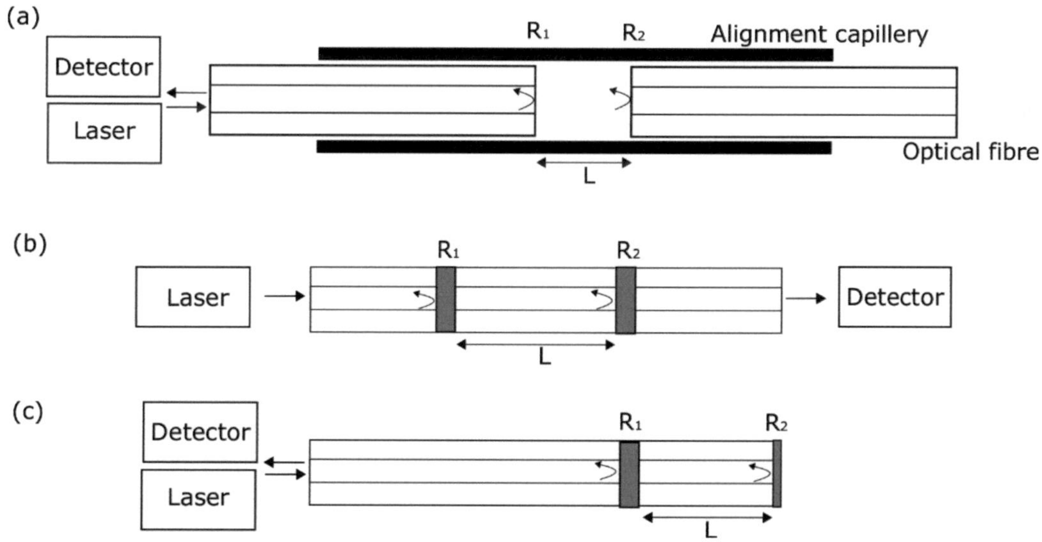

Figure 1. (a) The schematic of an extrinsic Fabry-Perot interferometer; (b) the schematic of an intrinsic Fabry-Perot interferometer; (c) the schematic of an intrinsic Fabry-Perot interferometer working in reflection mode. R_i represents the reflective surfaces and L is the length of the cavity.

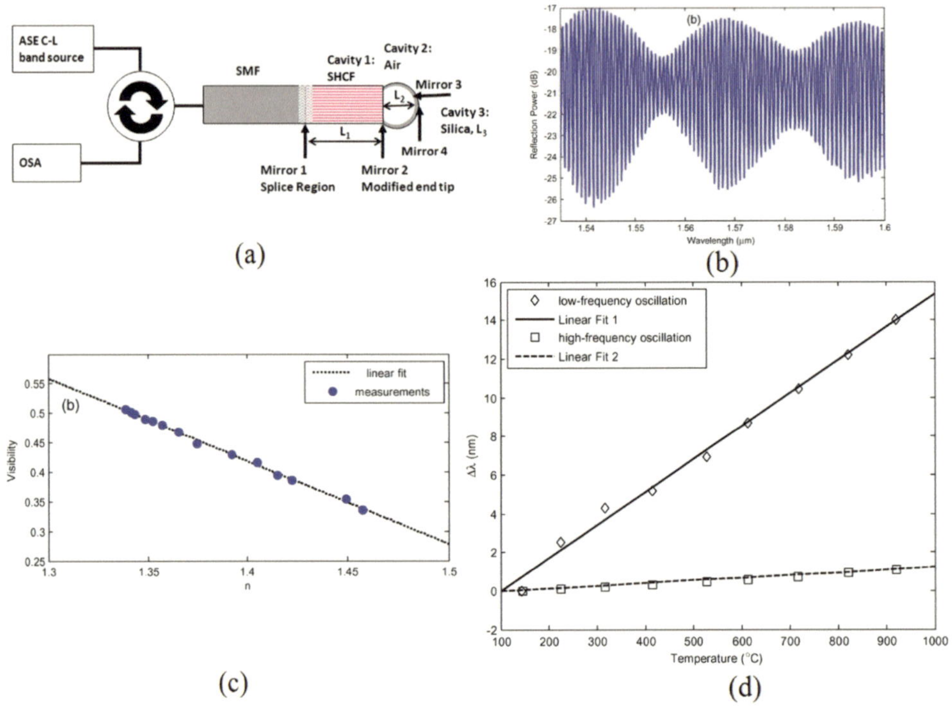

Figure 2. (a) Configuration of the simplified HCF based IFPI working in reflection mode; (b) interference spectrum of the sensor head; (c) refractive index against fringe visibility. (d) Temperature sensitivity against wavelength shift for low and high frequency oscillations. © 2012 IEEE. Reprinted, with permission, from Ref. [6].

cavity using a femtosecond laser to allow the analyte to enter the cavity [10]. A RI sensitivity of ~851 nm/RIU and a low cross sensitivity with temperature of ~3.2×10^{-7} RIU/°C was obtained. A short section of hollow fiber had been sandwiched between single mode fiber (SMF) and solid core (SC) PCF [11]. By taking advantage of the air holes in PCF, air was allowed to infiltrate the PCF cavity and the RI changes under different pressures were measured. When creating hole collapse regions between a section of SMF and PCF, the length of the region can affect the sensitivity of the sensor to RI and also temperature. A longer collapsed area will lead to more cladding modes being excited and in turn larger changes in the interference pattern. Dash and Jha [12] found that as they increased the length of the collapsed region from 180 to 270 μm, the RI sensitivity also increased from 30 to 53 nm/RIU with a RI resolution of 1.18×10^{-4} RIU.

A Microbubble FPI has been shown as a strain and vibration sensor with a spheroidal cavity achieving a strain sensitivity of ~10.3 pm/με and high fringe visibility (~38 dB) [13]. A 157 nm laser was used for micromachining an in-line etalon, with two smooth and parallel reflecting sides, in an endlessly single mode PCF. This was demonstrated for strain measurements in a high temperature environment with a fringe contrast of ~26 dB [14]. Shi et al. [15] were able to produce a multiplexed strain sensor system using different lengths of HCF spliced between SMF. Due to their wide free spectral range, the signals could be easily demodulated using fast Fourier transform (FFT). A strain insensitive IFPI has been developed by splicing one end of a solid PCF to a SMF and the other end to a HC-PCF to form a micro cavity. A large portion of the strain sensitivity comes from changing the size of the micro cavity, but in this case, this was at the end of the sensor and remained fixed in size [12].

Due to the all silica structure of PCFs, they are able to withstand high temperatures [6, 16], often for long periods of time [17]. The sensitivity to temperature is often based on the thermal-optic effect of silica and is therefore proportional to the length of the PCF cavity [18]. Frazão et al. [18] characterized the strain and temperature sensitivities of suspended core fibers with three and four holes. The normalized temperature sensitivities were found to be similar at 67.8 and 67.6 rad/m°C for three and four holes, respectively. The strain response was found to be greater for the three hole fiber. This was because the strain was applied to the cladding region (supporting walls of the fiber), and this cross-section was smaller for the three hole fiber.

Wu et al. were able to successfully demonstrate a high pressure (up to 40 MPa) and high temperature FPI (up to 700°C) sensor. A SC-PCF was spliced to one end to a SMF and hole collapse was carried out at the other end, to improve the reflectivity of the second mirror. The pressure and temperature sensitivities were −5.8 and ~13 pm/°C, respectively [19]. PCF based IFPIs may also have potential use in photonic integrated circuits [20].

2.2. Mach-Zehnder interferometer (MZI)

The Mach-Zehnder interferometer (MZI) works where an incident beam from a single light source is split into two arms and later recombined, forming an interference pattern [1]. When there is a perturbation in one arm, the difference in optical path length changes and is conveyed by the variation in the interference signal.

A MZI can be fabricated using two fibers with the light being split and recombined using fiber couplers, as shown in **Figure 3(a)**. One beam is referred to as the reference arm, and the other the sensing arm. To fabricate an in-line MZI using a singular fiber, as in **Figure 3(b)**, modal dispersion is used and the propagating modes are coupled into the cladding as well as the core. Though the physical length of the two arms is the same the phase velocity is different as the effective indices of the core and cladding are not the same.

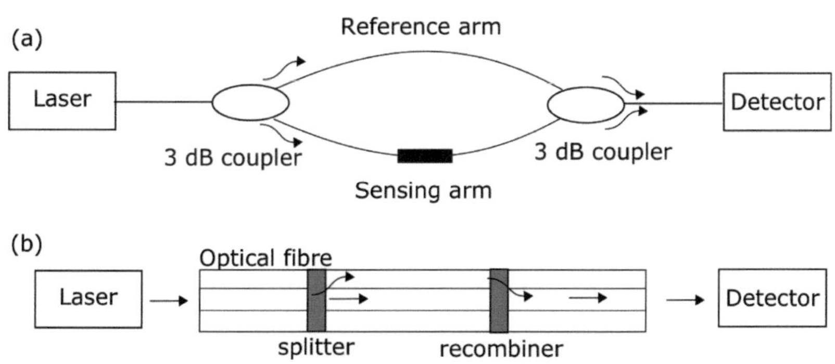

Figure 3. (a) The schematic of a Mach-Zehnder interferometer using two fibers; (b) the schematic of an in-line Mach-Zehnder interferometer using one fiber.

The Mach-Zehnder sensor interference spectrum can be expressed as [21]:

$$I = I_1 + I_2 + 2\sqrt{I_1 I_2}\cos\Delta\varnothing \tag{3}$$

where I_1 and I_2 are the irradiance of the interfering waves, $\Delta\varnothing$ is the phase difference between the core and cladding mode is defined as:

$$\Delta\varnothing = \frac{2\pi}{\lambda}\Delta n L + \Phi \tag{4}$$

where λ is the wavelength, L is the optical length of the interferometer, Δn is the difference in effective refractive index and Φ is the initial phase difference of the two waves.

PCF MZIs commonly consist of a SMF-PCF-SMF configuration; where a small section of solid core PCF is spliced between two standard SMFs. By using a fusion splicer to collapse the air holes of the PCF, the light is no longer constrained to the core and some of the light is coupled to one or multiple cladding modes and are able to propagate along the fiber [22, 23]. The splice points act as the mode couplers to form the fiber MZI, where the first splice point causes light from the core to couple to the cladding and the second splice causes the modes to recombine. MZIs composed entirely of LMA-PCF have also been realized [22] by core misalignment or by hole collapsing, as shown in **Figure 4**. By using the hole collapse technique in an all PCF MZI, alignment is less stringent as no cleaving is required. More cladding modes will also be excited and when the interference spectra were Fourier transformed, multiple spatial frequencies were seen with the number increasing with increasing interferometer length [22]. Different lengths of PCF were studied and compared for sensitivity [23–25]. It was found that the sensitivity was

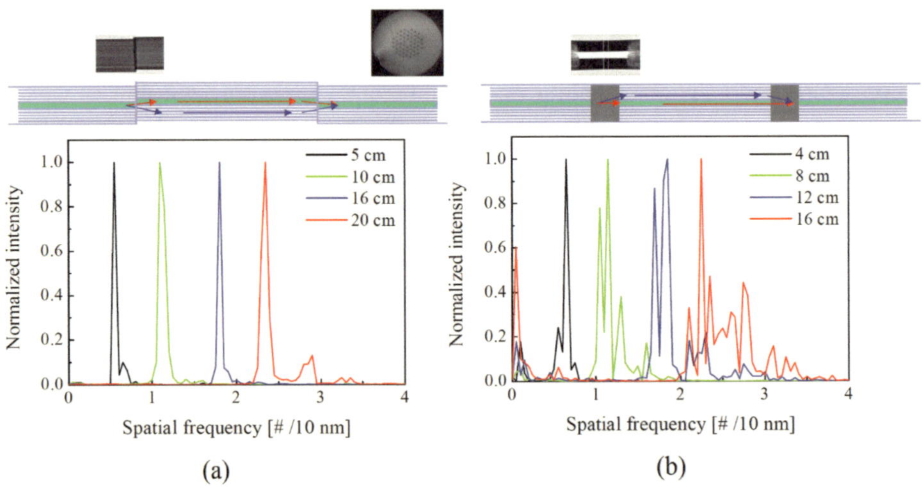

Figure 4. (a) The offset splicing method and the corresponding spatial frequencies at different interferometer lengths. Inset: Cross section of the LMA-PCF; (b) the hole collapse method and the corresponding spatial frequencies at different interferometer lengths. (a) produced one dominant spatial frequency at each length, whereas (b) induced several. Reprinted with permission from Ref. [22]. Copyright 2007 Optical Society of America.

not strongly dependent on length, though longer sections could require better handling and packaging.

These MZI configurations make useful refractometers as the effective refractive index of the cladding modes will be influenced by the surrounding environment [21, 23]. Highly sensitive refractometers have been demonstrated by means of tapering, either at just the splice points [26, 27] or the entire PCF section [28] to expose the evanescent field, making the MZI more sensitive to external RI changes. By tapering at the splice point, when compared to direct splicing, Wang et al. [27] was able to increase the sensor sensitivity from 224.2 to 260.8 nm/RIU. Wong et al. [29] used a combination of PCF-MZI and cavity ring-down technique for signal demodulation, which lead to a minimum detectable RI to be 7.8×10^{-5} RIU, almost 2.5 times greater than when compared to a PCF-MZI based on wavelength demodulation.

PCFs are generally known to be relatively temperature insensitive due to their small thermo-optic coefficient. Methods used to increase this sensitivity include, partial [30] or full [31] liquid infiltration into the holey region of the PCF, and multipath (more than two) MZIs using multicore PCFs [32] as this improves the phase sensitivity. Zhao et al. were able to achieve a temperature sensitivity of 130.6 pm/°C [32].

Different PCF configurations have been used for measuring strain [25, 33–35]. By using a multimode PCF, and careful hole collapse during splicing, a MZI was realized by coupling to two different core modes, LP_{01} and LP_{31}, allowing the light to be confined within the core and not as susceptible to ambient environment. Zheng et al. [25] were able to demonstrate a temperature and RI insensitive strain sensor with a sensitivity of 2.1 pm/με at 1550 nm with a 45 mm long PCF between two lengths of SMF. By introducing an additional collapsed region in the center of the length of PCF, two cascaded MZIs were created; the extinction ratio of the MZI induced fringes and in turn the measurement accuracy was improved [34]. The sensitivities for a normal SMF-PCF-SMF MZI and the modified MZI were 1.87 and 11.22 dB/mε, respectively. With twin core (TC) PCFs, the two cores can each act as the arm of the interferometer [36] and allow for a large strain measurement range as there are no deformations in the PCF to weaken the structure [33].

TC-PCFs have also been successfully demonstrated for use as intensity-based bend sensors [36], with a signal change found when the fiber is bent, such that both cores will experience different bend radii. Sun et al. [37] proposed a sensitive bend sensor by introducing an up- or peanut like -taper as the splitter and a down-taper as the recombiner. The up-taper improves the coupling between the PCF core and cladding modes and produces a stronger interference signal when recombined. The bend sensitivity of 50.5 nm/m^1 is one order of magnitude greater, when compared to a PCF MZI with a configuration of hole collapse and core offset (3.046 nm/m^1) [38].

The addition of a functional coating can lead to more specific and tailored sensing applications. As shown by Tao et al. [39], by coating the holey region of large mode area (LMA) and a grapefruit PCF with a polyallylamine layer with an affinity towards TNT vapor, they were able to selectively detect TNT. The LMA PCF had a lower a detection limit of 0.2 ppb_v due to a higher Q-factor. Lopez-Torres et al. demonstrated a humidity sensor capable of resolving

0.074% of relative humidity, and used a method based on the fast Fourier transform to yield a more linear device response with less noise [40]. Functionalized tip sensors have been used to detect changes in pH level [41] and changes in RI [42]. By modifying the surface of a compact PCF (~3 mm long) sensing region with biotin, Hu et al. were able to successfully demonstrate streptavidin detection [43]. Surface modified sensors can be more advantageous over air hole modification as it can be easily cleaned, reused and the analyte response is faster [43].

Long period gratings (LPGs) have also been used to fabricate in-line MZIs as they work by coupling forward propagating core light with one or more co-propagating cladding modes. This has been extended for use in all-PCF MZIs [44, 45]. Mechanically induced LPGs allow for identical, yet tuneable non-permanent LPGs to be fabricated. Yu et al. [45] demonstrated that the first LPG could be replaced by a misaligned splice point. As this is easier to manufacture than an LPG, it could reduce fabrication time and cost. The interference pattern can be tuned by adjusting the offset or the distance between the splitter and combiner [45] as well as the period and strength of the gratings [44]. It is also possible to replace the second LPG by collapsing the holes of the PCF [46]. Compact LPG based MZIs have been demonstrated, by using a CO_2 laser [35, 47] to create periodic grooves until both LPGs have coupling efficiency of around 3 dB. Compared with a standard single mode fiber MZI, Ju et al. [47] were able to obtain a higher strain sensitivity (−2.6 pm/με compared to +0.445 pm/με) and a lower temperature sensitivity (42.4 pm/°C per m compared to 1215.56 pm/°C per m). MZIs made with LPGs can be at risk of having a high insertion loss [24] due the deformation of fiber structure from inscription or from the misaligned splice point [45]. By cascading an LPG with a PCF MZI, simultaneous temperature and RI sensing was achieved as both elements in the sensor matrix responded differently to the multiple parameters [48].

Measuring low acoustic frequencies underwater can be difficult due to poor signal-to-noise ratios. An optical fiber based hydrophone using a polarization maintaining (PM) PCF sandwiched between SMFs was able to detect frequencies ranging from 5 to 200 Hz. The MZI used a two parameter detection method, namely a change in the intensity of the signal and a shift in wavelength. The change in power ranged from 0.8 to 2.32 dBm, which was much higher when compared to the ~0.1 dBm change using a SMF-MMF-SMF configuration [49].

PCF-MZIs also have potentials in communications for the demodulation of signals using differential phase shift keying (DPSK) [50] and in wavelength-division multiplexing (WDM) [51].

2.3. Michelson interferometer

Optical fiber Michelson interferometers can be realized by using two fibers or one fiber with the configurations shown in **Figure 5**. They are a similar version of MZI configurations. In the two fibers configuration, the laser light is split into two optical paths by an optical fiber coupler. The light is reflected back by the mirrors and recombined at the coupler to form the interference at the detector. In the one fiber configuration, the modes are split at a region where higher order modes or cladding modes are excited, e.g. splicing region between SMF and PCF, and are reflected by the mirrors and recombined at the splicing regions to form the interference which passes to the detector via a circulator.

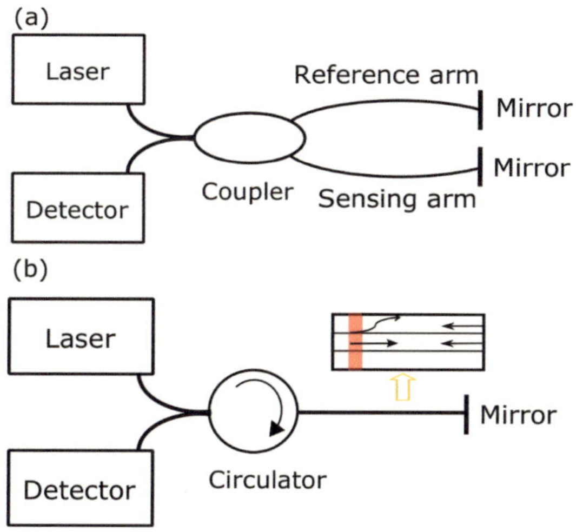

Figure 5. (a) The schematic of a Michelson interferometer using two fibers; (b) the schematic of a Michelson interferometer using one fiber.

Most PCF-based Michelson interferometers are based on one-fiber configurations. PM-PCFs have two orthogonally polarized core modes which act as two different optical paths for interference [52]. The two different optical paths can also be formed by two core modes in two-core PCFs [53, 54], or two core fibers [55]. In addition, the core mode and cladding modes that are excited at the splicing region between a SMF and a PCF due to mode mismatch [56] or a collapsed region in PCFs [57–62]. Similar configurations have also been reported in thin core fibers [63]. Due to the flexibility in the waveguide properties, PCF-based configurations can achieve high sensitivities when measuring ambient parameters, such as RI, temperature etc.

Jha et al. presented a Michelson interferometer device using a stub of a LMA PCF, with the schematic of the experimental setup and the interference spectrum in the reflected signal as shown in **Figure 6**. The PCF was fully collapsed at the splicing region between SMF and PCF, forming a multimode region for cladding mode excitation, and the end of the PCF was behaving as a reflective mirror. The dependence of the PCF length, temperature and ambient RI on the interference fringes of the device was investigated for sensing applications [58].

Enhanced temperature sensitivity was reported using a liquid-filled PCF-based Michelson interferometer [60]. The cladding holes of the PCFs were filled liquid with an RI of 1.45. The voids of the PCF were collapsed fully in the splicing process and the collapsed region between SMF and PCF was about 300 μm. The PCF end face acted as the reflective surface for the core mode and cladding modes of the PCF, which were combined and interfered in the collapsed region at the return path. The device demonstrated high temperature sensitivity with the wavelength shifts being was around 27 nm for a temperature change of 5°C [60].

Because PCF-based interferometers possess several desirable advantages including high sensitivity, linear response, and small size, they have attracted great interest in biosensing applications. Gao et al. proposed an in-line PCF Michelson interferometer for label-free, real time and sensitive detection of DNA hybridization and methylation [61]. To fabricate the interferometer,

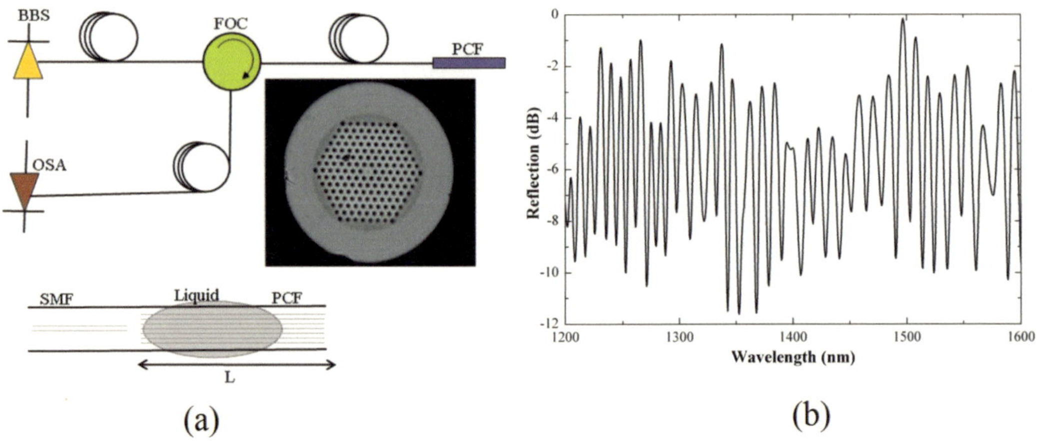

Figure 6. (a). Schematic of the experimental setup. A micrograph of the PCF used in the experiments is shown. The bottom drawing represents the interferometer, being L the length of the PCF. BBS stands for broad brand source, FOC for fiber optic circulator or coupler, and OSA for optical spectrum analyzer. (b). Reflection spectrum of a device with L = 24 mm over 400 nm. Reprinted from Ref. [58], with the permission of AIP publishing.

a section of the PCF was collapsed to excite cladding modes which possessed lower effective refractive indices than that of the core mode. The end facet of the PCF was coated with a gold film as the reflective mirror. The DNA hybridization and methylation resulted in a variation of surrounding RI, which changed the effective refractive indices of the cladding modes. The experimental results demonstrated a detection limit of 5 nM [61].

Sun et al. demonstrated a hybrid fiber interferometer by splicing a short length of PM-PCF, 177 μm, to a SMF of one output port of a 2 × 2 50:50 fiber coupler, forming a Fabry-Perot cavity in one of the optical paths of the Michelson interferometer. The spectral response of the hybrid interferometer exhibited two distinctive interference fringes and was demonstrated experimentally for simultaneous measurements of ambient RI from 1.33 to 1.38 with a resolution of 8.7×10^{-4}, and temperature in the range of 35–500°C with sensitivity of 13 pm/°C [53].

Multicore fiber (MCF) based multipath Michelson interferometers have been proposed and demonstrated for high temperature sensing recently [64]. The reflective mirror was formed via arc-fusion splicing the fiber end face. The splicing region between SMF and MCF was tapered for coupling the center core mode to surrounding cores due to reduced distances. The seven cores acted as the different optical paths in the multipath Michelson interferometer. The device demonstrated a temperature sensitivity of 165 pm/°C in the temperature range of 250–900°C [64].

Besides sensor applications, generation of logic gates such as optical add-drop multiplexers based on PCF-based Michelson interferometers has been investigated recently [54].

2.4. Sagnac interferometer

Optical fiber Sagnac interferometers (OFSI) use a Sagnac loop as the sensing element which usually uses highly birefringent (Hi-Bi) fibers or polarization-maintaining fibers (PMFs) to introduce a large optical path difference for interference between two counter-propagating waves.

The configuration of a fiber Sagnac interferometer is illustrated by **Figure 7**. The input light is split by an optical fiber coupler, usually a 3 dB coupler. Two counter-propagating waves travel in the Sagnac loop and accumulate an optical path difference due to birefringence.

Compared to conventional Hi-Bi fibers, PM-PCFs usually achieve much higher birefringence. Consequently, the required length of PM-PCF in the Sagnac loop is much shorter than that of conventional PMFs. Moreover, PM-PCFs are thermally stable due to their pure-silica material used in the fiber compared to conventional PMFs with temperature dependent birefringence. PM-PCFs also possess advantage of low bending loss due to high numerical aperture and small core diameters. As a result, PM-PCF based Sagnac interferometers have been extensively exploited and develop for many applications, such as strain, twist, pressure and curvature sensing, etc.

The temperature insensitivity of PM-PCF based Sagnac interferometers improves the accuracy of strain measurements, as the temperature crosstalk is negligible. The temperature dependence of birefringence in the PM-PCF is 35 times smaller than that of conventional PMFs [65].

Further reduced temperature sensitivities in strain measurements using PM-PCFs were reported to be 0.29 pm/K, about 3000 times lower than that of conventional PMFs, with strain sensitivity of 0.23 pm/με [66]. The experimental setup is shown in **Figure 8(a)**, consisting of a 3 dB fiber coupler to equally split the input light into two counter-propagating waves. The 86 mm long

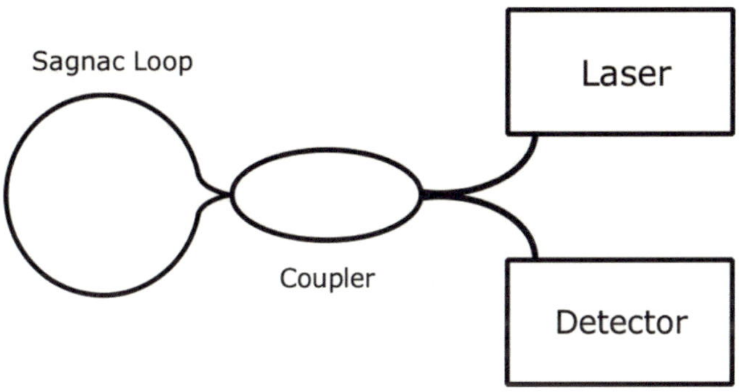

Figure 7. The schematic of a fiber Sagnac interferometer.

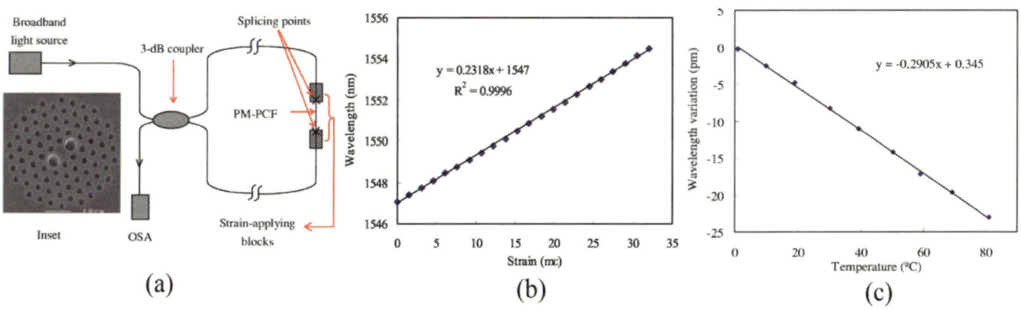

Figure 8. (a) Schematic diagram of the proposed OFSI strain sensor. Inset: SEM of the cross section of the PM-PCF. (b) wavelength shift of the transmission minimum at 1547 nm against the applied strain. (c) Wavelength variation of the transmission minimum at 1547 nm against temperature. Reprinted from Ref. [66], with the permission of AIP Publishing.

PM-PCF was spliced to the single mode fiber in the Sagnac loop. One end of the PM-PCF was fixed and the other end was stretched using a precision translation stage, for strain measurement. The scanning electron microscopic (SEM) image of the PM-PCF is shown in the inset of **Figure 8(a)**. A broadband light source is connected to the input port of the 3 dB coupler, and the transmitted light is measured by an optical spectrum analyzer (OSA) that is connected to the other input port of the 3 dB coupler. The interference occurred when two orthogonal guided modes combined at the coupler due to an accumulated phase delay in the Sagnac loop.

The transmission ratio of the optical intensity in the Sagnac loop can be described as:

$$T = [1 - \cos(\psi)]/2 \tag{5}$$

where $\psi = 2\pi LB/\lambda$ is the phase difference between the two orthogonal guided modes in PM-PCF, L is the length of the PM-PCF, B is the birefringence of the PM-PCF, and λ is the wavelength. The peak wavelength of the interference would encounter a shift due to the strain experienced by the PM-PCF, with the relationship being:

$$\Delta\lambda = \lambda\left(1 + p_e'\right)\varepsilon \tag{6}$$

where p_e' is a constant that describes the variation of strain-induced birefringence, and ε is the applied strain.

The linear relationship between strain and peak wavelength shift can be observed in **Figure 8(b)**. The temperature stability was tested by placing the PM-PCF in a temperature-controlled container, and the transmission spectrum was monitored by varying the temperature. The temperature sensitivity of the sensor was measured to be −0.29 pm/°C, as shown in **Figure 8(c)**, which is much lower than the reported value of 0.99 nm/°C of a conventional optical fiber Sagnac interferometer [67].

The influence of the coating on the fiber was investigated by Frazão et al., showing higher strain sensitivities and stronger temperature crosstalk with a nonlinear response for coated PCFs based Sagnac interferometers [68, 69]. The strain measurement sensitivities using PM-PCFs is also influenced by the ratio of the sensing PM-PCF over the entire PM-PCF length in the Sagnac loop [70]. Kim et al. develop a hollow core with an elliptical shape PBG fiber Sagnac interferometer for strain sensing with reduced temperature sensitivity when compared to conventional PMF Sagnac interferometers [71]. All-solid PCFs with Ge:SiO_2 rods and stress-induced birefringence by two Boron-doped rods have been reported to produce higher strain sensitivity of 25.6 pm/$\mu\varepsilon$ with the temperature crosstalk suppressed to −9 pm/K using a cascaded Sagnac configuration [72]. The reference signal at wavelength 1586.7 nm was used for temperature compensation of the two sensing wavelengths at 1551.5 nm and 1616.3 nm, with the wavelength difference being monitored [72]. Low-birefringence (low-Bi) PCFs with birefringence one or two orders lower than PM-PCFs were also exploited to achieve broader strain sensing range with similar strain sensitivity and a need for temperature compensation due to higher temperature sensitivity [73].

PCF-based Sagnac interferometers have been reported to develop twist or torsion sensors with potential applications in spaceflight and constructional engineering. Compared to other optical fiber twist/torsion sensors, PCF-based Sagnac interferometers exhibit higher sensitivities and reduced crosstalk due to temperature. Hi-Bi PCFs have been reported to construct twist/torsion sensors with sensitivities 0.059 and 0.057 nm/°, as measured from two interference peaks, and temperature crosstalk of −4.6 and −2.6 pm/°C, respectively [74]. Improved twist sensitivities were obtained by using a low-Bi PCF Sagnac interferometer showing sensitivity of 1.00 nm/° and temperature sensitivity of −0.5 pm/°C [75]. A Side-leakage PCF with Ge-doped core based Sagnac interferometer was reported to achieve torsion sensitivity of 0.9354 nm/°. The temperature crosstalk was around 0.054–0.178 °/°C, which could be independently determined by using matrix method [76]. Notably, matrix method was used for simultaneously multi-parameter measurement by PCF base Sagnac interferometers also. Dong et al. introduced a core-offset technique in the splicing between the PCF and standard fibers in the Sagnac loop, and measured the wavelength variation and the transmissivity difference in order to demodulate the strain and temperature [77]. More recently, Naeem et al. demonstrated a Sagnac interferometer using Hi-Bi PCF for multi-parameter measurements. The sensor consisted of hybrid interferometry; the intra-core-mode Sagnac interference and the inter-core-mode Mach-Zehnder interference. The phase shifts due to the Sagnac and Mach-Zehnder interference were measured and used to construct the sensor matrix for torsion, strain and temperature [78].

Pressure sensing using PM-PCF based Sagnac interferometers have been reported recently. Due to the high sensitivity of Sagnac interferometry, such sensors do not require modifications for sensitivity enhancement. Furthermore, the detection scheme can be wavelength shift measurement [79, 80, 81], or phase shift measurement for extended pressure measurement range up to 2.35 MPa [82]. Feng et al. demonstrated that such sensors exhibit a good linearity of the applied pressure and can accurately measure pore water pressure [81].

Gong et al. employed a low-Bi PCF in the Sagnac loop and used wavelength shift detection for curvature sensing. The achieved curvature detection resolution was 0.059 m^{-1} [83]. Comparatively, Frazão et al. demonstrated a Hi-Bi PCF Sagnac interferometer for curvature sensing with an improved detection resolution of $(1.39 \pm 0.07) \times 10^{-5}$ m^{-1}. The measurement parameter was the group birefringence β, which was defined as $\beta = \lambda^2/\Delta\lambda L$, where λ is the central wavelength in operation, $\Delta\lambda$ is the spectral width of the interferometer, and L is the length of the Hi-Bi PCF region [84].

The presence of air holes in PCF structures permits the infiltration of substances, e.g. liquid, metal etc., to introduce additional functionalities. For instance, highly sensitive temperature sensing was reported by PCF based Sagnac interferometers filled with metal [85], selectively filled with liquid [86, 87], and partially filled with alcohol [88]. The indium-filled side hole PCF was producing a change in birefringence due to the expansion of the filler metal, resulting in a high temperature sensitivity of the sensor of −9.0 nm/°C [85]. A PBG PCF was selectively filled by high index liquid, leading to temperature dependence in the bandgap properties, as well as the Sagnac interference properties. The temperature sensitivity was about −0.4 nm/°C [88]. By selectively infiltrating water at the two larger air holes adjacent to the solid core in the PM-PCF, the Sagnac interferometer showed temperature sensitivity of 0.15 nm/°C [87]. In order to realize a low cost, reusable and reliable in-line microfluidic refractometer, Wu et al. devised a device based on a C-shaped fiber and PCF based Sagnac interferometer [89]. The C-shape fiber and the PCF were fusion spliced to standard fibers in the Sagnac loop. The C-shape fibers provided openings for fluid to flow into and out of the PCF. The device was experimentally demonstrated for in-line fluid sensing, with high sensitivity of 6621 nm/RIU over RI range of 1.33–1.333 [89].

3. Conclusions and outlook

An overview of the different interferometric sensors, based on PCF, and their applications has been presented. The structure of PCFs is versatile and many different configurations can be achieved to produce interferometers with desirable properties such as high sensitivity, small sensor heads and good stability over time for sensing applications. PCF-based Fabry-Perot interferometers utilizing short sections of hollow core fibers, or forming microbubbles as the resonance cavities, or short sections of solid core PCFs, have been reported for measurements of various physical magnitudes, including RI, temperature, strain and pressure. PCF-based MZI configurations leverage on the enhanced flexibility in controlling waveguide properties in PCF, e.g. splicing between SMF and PCF can be used as an effective way to excite higher order modes and cladding modes in PCF, which exhibit greater sensitivity to ambient parameters compared to those in conventional fibers. In addition, the difference in the effective mode indices between the cladding modes and the core mode are greater in PCF, leading to much shorter device length and thus better robustness. PCF-based Michelson interferometers are a similar version of MZI configurations, except the presence of a reflective surface to reflect the modes which are combined at the same location of mode splitting. PCF-based Sagnac interferometers usually use Hi-Bi PCFs for developing compact and highly sensitive devices for measuring parameters such as strain, twist/torsion, curvature etc. PCFs can be combined with other fiber devices such as fiber Bragg grating or long period grating devices to achieve better sensor performance, e.g. higher sensitivity, minimizing cross-talk and simultaneous multiple parameter sensing. Inclusions of other substances into the holey structure of PCFs, bring additional functionalities and enhanced sensor performance such as temperature sensors. Moving forward, PCFs are expected for more exploitations and advancement of sensor development for various sensing applications.

Author details

Dora Juan Juan Hu[1]*, Rebecca Yen-Ni Wong[1] and Perry Ping Shum[2,3]

*Address all correspondence to: jjhu@i2r.a-star.edu.sg

1 Smart Energy and Environment Cluster, Institute for Infocomm Research, A*STAR, Singapore

2 Centre for Optical Fibre Technology, School of Electrical & Electronic Engineering, Nanyang Technological University, Singapore

3 CINTRA CNRS/NTU/THALES, UMI 3288, Research Techno Plaza, Singapore

References

[1] Lee BH, Kim YH, Park KS, Eom JB, Kim MJ, Rho BS, et al. Interferometric fiber optic sensors. Sensors. 2012;**12**(3):2467-2486. DOI: 10.3390/s120302467

[2] Knight JC. Photonic crystal fibres. Nature. 2003;**424**:847-851. DOI: 10.1038/nature01940

[3] Tian F, Sukhishvili S, Photonic Crystal DH. Fiber as a lab-in-fiber optofluidic platform. In: Cusano A, Consales M, Crescitelli A, Ricciardi A, editors. Lab-on-Fiber Technology. Springer International Publishing; Switzerland, 2014. p. 315-334. DOI: 10.1007/978-3-319-06998-2_15

[4] DJJ H, Ho HP. Recent advances in plasmonic photonic crystal fibers: Design, fabrication and applications. Advances in Optics and Photonics. 2017;**9**(2):257-314. DOI: 10.1364/AOP.9.000257

[5] Villatoro J, Zubia J. New perspectives in photonic crystal fibre sensors. Optics & Laser Technology. 2016;**78**(A):67-75. DOI: 10.1016/j.optlastec.2015.07.025

[6] Hu DJJ, Wang Y, Lim JL, Zhang T, Milenko KB, Chen Z, et al. Novel miniaturized Fabry–Perot refractometer based on a simplified hollow-core fiber with a hollow silica sphere tip. IEEE Sensors Journal. 2012;**12**(5):1239-1245. DOI: 10.1109/JSEN.2011.2167678

[7] Rao YJ, Zhu T, Yang XC, Duan DW. In-line fiber-optic etalon formed by hollow-core photonic crystal fiber. Optics Letters. 2007;**32**(18):2662-2664. DOI: 10.1364/OL.32.002662

[8] Villatoro J, Finazzi V, Coviello G, Pruneri V. Photonic-crystal-fiber-enables micro-Fabry-Perot interferometer. Optics Letters. 2009;**34**(16):2441-2443. DOI: 10.1364/OL.34.002441

[9] Favero FC, Bouwmans G, Finazzi V, Villatoro J, Pruneri V. Fabry–Perot interferometers built by photonic crystal fiber pressurization during fusion splicing. Optics Letters. 2011;**36**(21):4191-4193. DOI: 10.1364/OL.36.004191

[10] Wang Y, WD N, LC R, Hu T, Guo J, Wei H. Temperature-insensitive refractive index sensing by use of micro Fabry–Pérot cavity based on simplified hollow-core photonic crystal fiber. Optics Letters. 2013;**38**(3):269-271. DOI: 10.1364/OL.38.000269

[11] Deng M, Tang CP, Zhu T, Rao YJ, LC X, Han M. Refractive index measurement using photonic crystal fiber-based Fabry-Perot interferometer. Applied Optics. 2010;**49**(9):1593-1598. DOI: 10.1364/AO.49.001593

[12] Dash JN, Jha R. Fabry–Perot based strain insensitive photonic crystal fiber modal interferometer for inline sensing of refractive index and temperature. Applied Optics. 2015;**54**(35):10479-10486. DOI: 10.1364/AO.54.010479

[13] Favero FC, Araujo L, Bouwmans G, Finazzi V, Villatoro J, Pruneri V. Spheroidal Fabry-Perot microcavities in optical fibers for high-sensitivity sensing. Optics Express. 2012;**20**(7):7112-7118. DOI: 10.1364/OE.20.007112

[14] Ran ZL, Rao YJ, Deng HY, Liao X. Miniature in-line photonic crystal fber etalon fabricated by 157 nm laser micromachining. Optics Letters. 2007;**32**(21):3071-3073. DOI: 10.1364/OL.32.003071

[15] Shi Q, Wang Z, Jin L, Li Y, Zhang H, Lu F, et al. A hollow-core photonic crystal fiber cavity based multiplexed Fabry–Pérot interferometric strain sensor system. IEEE Photonics Technology Letters. 2008;**20**(15):1329-1331. DOI: 10.1109/LPT.2008.926948

[16] Choi HY, Park KS, Park SJ, Paek UC, Lee BH, Cho ES. Miniature fiber-optic high temperature sensor based on a hybrid structured Fabry–Perot interferometer. Optics Letters. 2008;**33**(21):2455-2457. DOI: 10.1364/OL.33.002455

[17] Du Y, Qiao X, Rong Q, Yang H, Feng D, Wang R, et al. A miniature Fabry–Pérot interferometer for high temperature measurement using a double-Core photonic crystal fiber. IEEE Sensors Journal. 2014;**14**(4):1069-1073. DOI: 10.1109/JSEN.2013.2286699

[18] Frazão O, Aref SH, Baptista JM, Santos JL, Latifi H, Fahari F, et al. Fabry–Pérot cavity based on a suspended-core fiber for strain and temperature measurement. IEEE Photonics Technology Letters. 2009;**21**(17):1229-1231. DOI: 10.1109/LPT.2009.2024645

[19] Wu C, HY F, Qureshi KK, Guan BO, Tam HY. High-pressure and high-temperature characteristics of a Fabry–Perot interferometer based on photonic crystal fiber. Optics Letters. 2011;**36**(3):412-414. DOI: 10.1364/OL.36.000412

[20] Chen X, Zhao D, Qiang Z, Lin G, Li H, Qiu Y, et al. Polarization-independent Fabry–Perot interferometer in a hole-type silicon photonic crystal. Applied Optics. 2010;**49**(30):5878-5881. DOI: 10.1364/AO.49.005878

[21] Lim JL, DJJ H, Shum PP, Wang Y. Cascaded photonic crystal fiber interferometers for refractive index sensing. IEEE Photonics Journal. 2012;**4**(4):1163-1169. DOI: 10.1109/JPHOT.2012.2205911

[22] Choi HY, Kim MJ, Lee BH. All-fiber Mach-Zehnder type interferometers formed in photonic crystal fiber. Optics Express. 2007;**15**(9):5711-5720. DOI: 10.1364/OE.15.005711

[23] Wang JN, Tang JL. Photonic crystal fiber Mach-Zehnder interferometer for refractive index sensing. Sensors. 2012;**12**(3):2983-2995. DOI: 10.3390/s120302983

[24] Villatoro J, Finazzi V, Badenes G, Pruneri V. Highly sensitive sensors based on photonic crystal fiber modal interferometers. Journal of Sensors. 2009;**2009**:1-11. DOI: 10.1155/2009/747803

[25] Zheng J, Yan P, Yu Y, Ou Z, Wang J, Chen X, et al. Temperature and index insensitive strain sensor based on a photonic crystal fiber inline Mach–Zehnder interferometer. Optics Communications. 2013;**297**:7-11. DOI: 10.1016/j.optcom.2013.01.063

[26] Zhao Y, Li XG, Cai L, Refractive YY. Index sensing based on photonic crystal fiber interferometer structure with up-tapered joints. Sensors and Actuators B: Chemical. 2015;**221**:406-410. DOI: 10.1016/j.snb.2015.06.148

[27] Wang Q, Kong L, Dang Y, Xia F, Zhang Y, Zhao Y, et al. High sensitivity refractive index sensor based on splicing points tapered SMF-PCF-SMF structure Mach-Zehnder mode interferometer. Sensors and Actuators B: Chemical. 2016;**225**:213-220. DOI: 10.1016/j.snb.2015.11.047

[28] Wu D, Zhao Y, Li J. PCF taper-based Mach–Zehnder interferometer for refractive index sensing in a PDMS detection cell. Sensors and Actuators B: Chemical. 2015;**213**:1-4. DOI: 10.1016/j.snb.2015.02.080

[29] Wong WC, Zhou W, Chan CC, Dong X, Leong KC. Cavity ringdown refractive index sensor using photonic crystal fiber interferometer. Sensors and Actuators B: Chemical. 2012;**161**(1):108-113. DOI: 10.1016/j.snb.2011.09.056

[30] Liang H, Zhang W, Wang H, Geng P, Zhang S, Gao S, et al. Fiber in-line Mach–Zehnder interferometer based on near-elliptical core photonic crystal fiber for temperature and strain sensing. Optics Letters. 2013;**38**(20):4019-4021. DOI: 10.1364/OL.38.004019

[31] Geng Y, Li X, Tan X, Deng Y, Hong X. Compact and ultrasensitive temperature sensor with a fully liquid-filled photonic crystal fiber Mach–Zehnder interferometer. IEEE Sensors Journal. 2014;**14**(1):167-170. DOI: 10.1109/JSEN.2013.2279537

[32] Zhao Z, Tang M, Fu S, Liu S, Wei H, Cheng Y, et al. All-solid multi-core fiber-based multipath Mach–Zehnder interferometer for temperature sensing. Applied Physics B. 2013;**112**(4):491-497. DOI: 10.1007/s00340-013-5634-8

[33] Qureshi KK, Liu Z, Tam HY, Zia MF. A strain sensor based on in-line fiber Mach–Zehnder interferometer in twin-core photonic crystal fiber. Optics Communications. 2013;**309**:68-70. DOI: 10.1016/j.optcom.2013.06.057

[34] LM H, Chan CC, Dong XY, Wang YP, Zu P, Wong WC, et al. Photonic crystal fiber strain sensor based on modified Mach–Zehnder interferometer. IEEE Photonics Journal. 2012;**4**(1):114-118. DOI: 10.1109/JPHOT.2011.2180708

[35] Shin W, Lee YL, BA Y, Noh YC, Ahn TJ. Highly sensitive strain and bending sensor based on in-line fiber Mach–Zehnder interferometer in solid core large mode area photonic crystal fiber. Optics Communications. 2010;**283**(10):2097-2101. DOI: 10.1016/j.optcom.2010.01.008

[36] Kim B, Kim TH, Cui L, Chung Y. Twin core photonic crystal fiber for in-line Mach-Zehnder interferometric sensing applications. Optics Express. 2009;**17**(18):15502-15507. DOI: 10.1364/OE.17.015502

[37] Sun B, Huang V, Liu S, Wang C, He J, Liao C, et al. Asymmetrical in-fiber Mach-Zehnder interferometer for curvature measurement. Optics Express. 2015;**23**(11):14596-14602. DOI: 10.1364/OE.23.014596

[38] Deng M, Tang CP, Zhu T, Rao YJ. Highly sensitive bend sensor based on Mach–Zehnder interferometer using photonic crystal fiber. Optics Communications. 2011;**284**(12):2849-2853. DOI: 10.1016/j.optcom.2011.02.061

[39] Tao C, Wei H, Feng W. Photonic crystal fiber in-line Mach-Zehnder interferometer for explosive detection. Optics Express. 2016;**24**(3):2806-2817. DOI: 10.1364/OE.24.002806

[40] Lopez-Torres D, Elosua C, Villatoro J, Zubia J, Rothhardt M, Schuster K, et al. Photonic crystal fiber interferometer coated with a PAH/PAA nanolayer as humidity sensor. Sensors and Actuators B: Chemical. 2017;**242**:1065-1072. DOI: 10.1016/j.snb.2016.09.144

[41] Hu P, Dong X, Wong WC, Chen LH, Ni K, CC C. Photonic crystal fiber interferometric pH sensor based on polyvinyl alcohol/polyacrylic acid hydrogel coating. Applied Optics. 2015;**54**(10):2647-2652. DOI: 10.1364/AO.54.002647

[42] Dash JN, Jha R, Temperature Insensitive PCF. Interferometer coated with graphene oxide tip sensor. IEEE Photonics Technology Letters. 2016;**28**(9):1006-1009. DOI: 10.1109/LPT.2016.2522979

[43] DJJ H, Lim JL, Park MK, Kao LTH, Wang Y, Wei H, et al. Photonic crystal fiber-based interferometric biosensor for streptavidin and biotin detection. IEEE Journal of Selected Topics in Quantum Electronics. 2012;**18**(4):1293-1297. DOI: 10.1109/JSTQE.2011.2169492

[44] Lim JH, Jang HS, Lee K, KJ C, Lee BH. Mach–Zehnder interferometer formed in a photonic crystalfiber based on a pair of long-period fiber gratings. Optics Letters. 2004;**29**(4):346-348. DOI: 10.1364/OL.29.000346

[45] Yu X, Shum P, Dong X. Photonic-crystal-fiber-based Mach–Zehnder interferometer using long-period gratings. Microwave and Optical Technology Letters. 2006;**48**(7):1379-1383. DOI: 10.1002/mop.21647

[46] Choi HY, Park KS, Lee BH. Photonic crystal fiber interferometer composed of a long period fiber grating and one point collapsing of air holes. Optics Letters. 2008;**33**(8):812-814. DOI: 10.1364/OL.33.000812

[47] Ju J, Jin W, Ho HL. Compact in-fiber interferometer formed by long-period gratings in photonic crystal fiber. IEEE Photonics Technology Letters. 2008;**20**(23):1899-1901. DOI: 10.1109/LPT.2008.2005207

[48] Hu DJJ, Lim JL, Jiang M, Wang Y, Luan F, Shum PP, et al. Long period grating cascaded to photonic crystal fiber modal interferometer for simultaneous measurement of temperature and refractive index. Optics Letters. 2012;**37**(12):2283-2285. DOI: 10.1364/OL.37.002283

[49] Pawar D, Rao CN, Choubey RK, Kale SN. Mach-Zehnder interferometric photonic crystal fiber for low acoustic frequency detections. Applied Physics Letters. 2016;**108**(4):041912-1-041912-4. DOI: 10.1063/1.4940983

[50] Du J, Dai Y, Lei GKP, Tong W, Shu C. Photonic crystal fiber based Mach-Zehnder interferometer for DPSK signal demodulation. Optics Express. 2010;**18**(8):7917-7922. DOI: 10.1364/OE.18.007917

[51] Gerosa RM, Spadoti DH, Menezes LS, de Matos CJS. In-fiber modal Mach-Zehnder interferometer based on the locally post-processed core of a photonic crystal fiber. Optics Express 2011;**19**(4):3124-3129. DOI: 10.1364/OE.19.003124

[52] Tan X, Geng Y, Li X. High-birefringence photonic crystal fiber Michelson interferometer with cascaded fiber Bragg grating for pressure and temperature discrimination. Optical Engineering. 2016;**55**(9):090508. DOI: 10.1117/1.OE.55.9.090508

[53] Sun H, Zhang J, Rong Q, Feng D, Du Y, Zhang X, et al. A hybrid fiber interferometer for simultaneous refractive index and temperature measurements based on Fabry–Perot/ Michelson interference. IEEE Sensors Journal. 2013;**13**(5):2039-2044

[54] Sousa JRR, Filho AFGF, Ferreira AC, Batista GS, Sobrinho CS, Bastos AM, et al. Generation of logic gates based on a photonic crystal fiber Michelson interferometer. Optics Communications. 2014;**322**:143-149. DOI: 10.1016/j.optcom.2014.02.023

[55] Zhou A, Li G, Zhang Y, Wang Y, Guan C, Yang J, et al. Asymmetrical twin-Core fiber based Michelson interferometer for refractive index sensing. Journal of Lightwave Technology. 2011;**29**(19):2985-2991. DOI: 10.1109/JLT.2011.2165528

[56] Chen NK, KY L, Shy JT, Lin C. Broadband micro-Michelson interferometer with multi-optical-path beating using a sphered-end hollow fiber. Optics Letters. 2011;**36**(11):2074-2076. DOI: 10.1364/OL.36.002074

[57] Hu DJJ, Lim JL, Wang Y, Shum PP. Miniaturized photonic crystal fiber tip sensor for refractive index sensing. In: Proceedings of the IEEE Sensors Conference; 28-31 October 2011; Limerick, Ireland: IEEE; 2011. p. 1488-1490

[58] Jha R, Villatoro J, Badenes G. Ultrastable in reflection photonic crystal fiber modal interferometer for accurate refractive index sensing. Applied Physics Letters. 2008;**93** (19):191106. DOI: 10.1063/1.3025576

[59] Mileńko K, Hu DJJ, Shum PP, Zhang T, Lim JL, Wang Y, et al. Photonic crystal fiber tip interferometer for refractive index sensing. Optics Letters. 2012;**37**(8):1373-1375. DOI: 10.1364/OL.37.001373

[60] Hsu JM, Horng JS, Hsu CL, Lee CL. Fiber-optic Michelson interferometer with high sensitivity based on a liquid-filled photonic crystal fiber. Optics Communications. 2014;**331**:348-352. DOI: 10.1016/j.optcom.2014.06.050

[61] Gao R, DF L, Cheng J, Jiang Y, Jiang L, JD X, et al. Fiber optofluidic biosensor for the label-free detection of DNA hybridization and methylation based on an in-line tunable mode coupler. Biosensors & Bioelectronics. 2016;**86**:321-329. DOI: 10.1016/j.bios.2016.06.060

[62] Gong H, Chan CC, Zhang YF, Wong WC, Dong X. Miniature refractometer based on modal interference in a hollow-core photonic crystal fiber with collapsed splicing. Journal of Biomedical Optics. 2011;**16**(1):017004. DOI: 10.1117/1.3527259

[63] Li Z, Wang Y, Liao C, Liu S, Zhou J, Zhong X, et al. Temperature-insensitive refractive index sensor based on in-fiber Michelson interferometer. Sensors and Actuators B: Chemical. 2014;**199**:31-35. DOI: 10.1016/j.snb.2014.03.071

[64] Duan L, Zhang P, Tang M, Wang R, Zhao Z, Fu S, et al. Heterogeneous all-solid multicore fiber based multipath Michelson interferometer for high temperature sensing. Optics Express. 2016;**24**(18):20210-20218. DOI: 10.1364/OE.24.020210

[65] Kim DH, Kang JU. Sagnac loop interferometer based on polarization maintaining photonic crystal fiber with reduced temperature sensitivity. Optics Express. 2004;**12**(19):4490-4495. DOI: 10.1364/OPEX.12.004490

[66] Dong X, Tam HY, Shum P. Temperature-insensitive strain sensor with polarization-maintaining photonic crystal fiber based Sagnac interferometer. Applied Physics Letters. 2007;**90**(15):151113. DOI: 10.1063/1.2722058

[67] Starodumov AN, Zenteno LA, Monzon D, Rosa EDL. Fiber Sagnac interferometer temperature sensor. Applied Physics Letters. 1997;**70**(1):19. DOI: 10.1063/1.119290

[68] Frazao O, Baptista JM, Santos JL. Temperature-independent strain sensor based on a Hi-Bi photonic crystal fiber loop mirror. IEEE Sensors Journal. 2007;**7**(10):1453-1455. DOI: 10.1109/JSEN.2007.904884

[69] Frazao O, Baptista JM, Santos JL, Kobelke J, Schuster K. Strain and temperature characterisation of sensing head based on suspended-core fibre in Sagnac interferometer. Electronics Letters. 2008;**44**(25):1455-1456. DOI: 10.1049/el:20081431

[70] Cui Y, Wu Z, Shum PP, Dinh XQ, Humbert G. Investigation on the impact of Hi-Bi fiber length on the sensitivity of Sagnac interferometer. IEEE Sensors Journal. 2014;**14**(6):1952-1956. DOI: 10.1109/JSEN.2014.2304521

[71] Kim G, Cho T, Hwang K, Lee K, Lee KS, Han YG, et al. Strain and temperature sensitivities of an elliptical hollow-core photonic bandgap fiber based on Sagnac interferometer. Optics Express. 2009;**17**(4):2481-2486. DOI: 10.1364/OE.17.002481

[72] Gu B, Yuan W, He S, Bang O. Temperature compensated strain sensor based on cascaded Sagnac interferometers and all-solid Birefringent hybrid photonic crystal fibers. IEEE Sensors Journal. 2012;**12**(6):1641-1646. DOI: 10.1109/JSEN.2011.2175725

[73] Gong H, Chan CC, Chen L, Dong X. Strain sensor realized by using low-birefringence photonic-crystal-fiber-based Sagnac loop. IEEE Photonics Technology Letters. 2010;**22**(16):1238-1240. DOI: 10.1109/LPT.2010.2053025

[74] Kim HM, Kim TH, Kim B, Chung Y. Temperature-insensitive torsion sensor with enhanced sensitivity by use of a highly birefringent photonic crystal fiber. IEEE Photonics Technology Letters. 2010;**22**(20):1539-1541. DOI: 10.1109/LPT.2010.2068043

[75] Zu P, Chan CC, Jin Y, Gong T, Zhang Y, Chen LH, et al. A temperature-insensitive twist sensor by using low-birefringence photonic-crystal-fiber-based Sagnac interferometer. IEEE Photonics Technology Letters. 2011;**23**(13):920-922. DOI: 10.1109/LPT.2011.2143400

[76] Chen W, Lou S, Wang L, Zou H, Lu W, Jian S. Highly sensitive torsion sensor based on Sagnac interferometer using side-leakage photonic crystal fiber. IEEE Photonics Technology Letters. 2011;**23**(21):1639-1641. DOI: 10.1109/LPT.2011.2166062

[77] Dong B, Hao J, Chin-Yi L, Xu Z. Cladding-mode resonance in polarization-maintaining photonic-crystal-fiber-based Sagnac interferometer and its application for fiber sensor. Journal of Lightwave Technology. 2011;**29**(12):1759-1763. DOI: 10.1109/JLT.2011.2140313

[78] Naeem K, Kim BH, Kim B, Chung Y. Simultaneous multi-parameter measurement using Sagnac loop hybrid interferometer based on a highly birefringent photonic crystal fiber with two asymmetric cores. Optics Express. 2015;**23**(3):3589-3601. DOI: 10.1364/OE.23.003589

[79] HY F, Tam HY, Shao LY, Dong X, Wai PKA, Lu C, et al. Pressure sensor realized with polarization-maintaining photonic crystal fiber-based Sagnac interferometer. Applied Optics. 2005;**47**(15):2835-2839. DOI: 10.1364/AO.47.002835

[80] HY F, Wu C, Tse MLV, Zhang L, Cheng KCD, Tam HY, et al. High pressure sensor based on photonic crystal fiber for downhole application. Applied Optics. 2010;**49**(14):2639-2643. DOI: 10.1364/AO.49.002639

[81] Feng WQ, Liu ZY, Tam HY, Yin JH. The pore water pressure sensor based on Sagnac interferometer with polarization-maintaining photonic crystal fiber for the geotechnical engineering. Measurement. 2016;**90**:208-214. DOI: 10.1016/j.measurement.2016.04.067

[82] Cho LH, Wu C, Lu C, Tam HY. A highly sensitive and low-cost Sagnac loop based pressure sensor. IEEE Sensors Journal. 2013;**13**(8):3073-3078. DOI: 10.1109/JSEN.2013.2261291

[83] Gong HP, Chan CC, Zu P, Chen LH, Dong XY. Curvature measurement by using low-birefringence photonic crystal fiber based Sagnac loop. Optics Communications. 2010;**283**(16):3142-3144. DOI: 10.1016/j.optcom.2010.04.023

[84] Frazão O, Baptista JM, Santos JL, Roy P. Curvature sensor using a highly birefringent photonic crystal fiber with two asymmetric hole regions in a Sagnac interferometer. Applied Optics. 2008;**47**(13):2520-2523. DOI: 10.1364/AO.47.002520

[85] Reyes-Vera E, Cordeiro CMB, Torres P. Highly sensitive temperature sensor using a Sagnac loop interferometer based on a side-hole photonic crystal fiber filled with metal. Applied Optics. 2017;**56**(2):156-162. DOI: 10.1364/AO.56.000156

[86] Zheng X, Yg L, Wang Z, Han T, Wei C, Chen J. Transmission and temperature sensing characteristics of a selectively liquid-filled photonic-bandgap-fiber-based Sagnac interferometer. Applied Physics Letters. 2012;**100**(14):141104. DOI: 10.1063/1.3699026

[87] Cui Y, Shum PP, Hu DJJ, Wang G, Humbert G, Dinh XQ. Temperature sensor by using selectively filled photonic crystal fiber Sagnac interferometer. IEEE Photonics Journal. 2012;**4**(5):1801-1808. DOI: 10.1109/JPHOT.2012.2217945

[88] Zhao CL, Wang Z, Zhang S, Li Q, Zhong C, Zhang Z, et al. Phenomenon in an alcohol not full-filled temperature sensor based on an optical fiber Sagnac interferometer. Optics Letters. 2012;**37**(22):4789-4791. DOI: 10.1364/OL.37.004789

[89] Wu C, Tse MLV, Liu Z, Guan BO, Lu C, Tam HY. In-line microfluidic refractometer based on C-shaped fiber assisted photonic crystal fiber Sagnac interferometer. Optics Letters. 2013;**38**(17):3283-3286. DOI: 10.1364/OL.38.003283

Chapter 3

Dual-Core Transversally Chirped Microstructured Optical Fiber for Mode-Converter Device and Sensing Application

Erick Reyes Vera, Juan Úsuga Restrepo,
Margarita Varon and Pedro Torres

Additional information is available at the end of the chapter

http://dx.doi.org/10.5772/intechopen.70989

Abstract

We propose and demonstrate the concept of transversally chirped microstructured optical fiber and its application for the development of new platforms for sensing and telecommunications devices. First, the feasibility of the structure is demonstrated through two different techniques of manufacture. Based on the proposed structure, a novel mode-converter device is numerically studied. It is found that the mode conversion between LP_{01} and LP_{11} modes can be continuously tuned by temperature changes from 25 to 75°C. And that, the coupling efficiency in the wavelength range between 1.2 μm and 1.7 μm is always higher than 65%. Consequently, the proposed mode converter can operate in the E + S + C + L + U bands. Finally, a similar structure was used to design a new sensing architecture, which consisting of a dual-core transversally chirped microstructured optical fiber for refractive index sensing of fluids. We show that by introducing a chirp in the hole size, the microstructured optical fiber can be a structure with decoupled cores, forming a Mach–Zehnder interferometer in which the analyte directly modulates the device transmittance by its differential influence on the effective refractive index of each core mode. We show that by filling all fiber holes with analyte, the sensing structure achieves high sensitivity (transmittance changes of 302.8 per RIU at 1.42) and has the potential for use over a wide range of analyte refractive index.

Keywords: microstructured optical fibers, fiber optics sensors, interferometry, mode converter, space, division multiplexing, refractive index sensor, mode conversion, Mach–Zehnder interferometer

1. Introduction

In the last two decades, several technological breakthroughs were needed to increase and satisfy the capacities of the optical links. Some important advances allow the connection between users through the implementation of optical fibers. One of the most important advances, related to these technological breakthroughs, is the use of broadband optical amplifiers to increase the length of optical links. On the other hand, different multiplexing techniques such as optical time division multiplexing (OTDM), wavelength division multiplexing (WDM) and polarization division multiplexing (PDM) have been implemented in transmission channels to increase the optical transmission system capacity. However, due the high growth in the demand, the transmission capacity of this technology has reached the limits imposed by the nonlinear effects in optical fiber [1, 2]. In order to keep up the growth of current optical communication networks, it is necessary to implement new technological breakthroughs. One possibility is the implementation of independent spatial channels to send information. This technique is known as spatial division multiplexing (SDM) and can be implemented in two different schemes. Intuitively, in the multi-core fiber (MCF) scheme, each core acts as an independent channel for sending the information [1], while in the modal division multiplexing (MDM) scheme, each mode is considered an independent transmission channel as in single-mode fiber [3, 4]; hence, the key is to convert the fundamental fiber modes to higher order modes. As the processing systems are not prepared to work with hundreds of modes, in the MDM scheme it is preferred to work with few modes fibers (FMFs) [3, 4]. As with any new technology, emerging SDM systems require the development of new components such as optical fibers that support multiple spatial modes and integrated mode converters to control propagating modes, spatial mode multiplexers (SMUXs) and demultiplexers (SDEMUXs).

To address these needs, several works have reported different mechanisms to control the propagation modes in FMFs, such as long period fiber grating (LPFG), Fiber Bragg Gratings (FBG), tapers and phase mask [5–10]. Another interesting alternative is to use microstructured optical fibers (MOFs), also called photonic crystal fibers (PCFs), which offer flexibility in its design and the possibility of manipulating the optical properties of the device because its characteristics—dispersion, effective area, birefringence and nonlinearity, among others—depends on the diameter of holes, the separation between them and the shape of the microstructure [11–17]. Owing these characteristics, these type of optical fibers have been employed for the development of different devices such as polarization beam splitters [18], dispersion compensators [19, 20] and mode converters [5] to name a few.

Mode selective couplers (MSC) based on microstructured optical fiber (MOF) represent one of the best approaches to achieve mode conversion, avoiding the problems of other techniques—bulky free-space optics, polished- and fused-type MSC—since the devices are compact, robust and efficient, and allows the possibility of manipulating its behavior based on the MOF geometrical parameters [3, 21–24]. The principle of MSC is to phase match the fundamental mode in a single-mode fiber with a high-order mode in FMF.

In [23], Cai et al. proposed a mode converter based on a hybrid dual-core MOF, which contains an index-guided core and a photonic bandgap core. The air holes of the first ring around one of the cores are replaced with high-index rods, then mode conversions can be continuously achieved by varying the refractive index of high-index rods The all-solid bandgap structure requires two suitable materials that are also compatible for drawing and splicing with few-mode fibers. In [24], the authors proposed a tuneable MSC based on a fully liquid-filled dual-core MOF with non-identical cores. The tuning of the wavelength in the S + C + L + U bands is performed by changing the refractive index (RI) of the filling fluid.

Simultaneously, the implementation of MOFs has allowed the development of a new family of optical fiber sensors, which present higher sensitivity and compact sizes compared to sensors based in standard optical fibers and other technologies [25]. Some novel configurations of these sensors have been implemented in the measurement of refractive index (RI) changes [26, 27], temperature [28, 29] and force [30–32], among others. The refractive index sensors are the most studied and applied in recent years in biological, medical and chemical applications [33–38]. Two general configurations for the interaction between the light and analyte in MOFs may be identified. In the first option, the analyte is located in the evanescent field of the waveguide [39, 40]. In the second option, the analyte can be inserted into the fiber holes and experience long range interaction with the guided light while maintaining the waveguide, thereby ensuring a robust device [26, 41]. In addition, optical fiber sensors based on the dual-core MOF configuration are able to achieve improved sensitivity for RI measurements. In these structures, the fiber holes are filled with the analyte. Then, the refractive index of the sample modulates the device transmittance by its influence on the coupling between the cores. In [37], Markos et al. presented an experimentally feasible design of a dual-core microstructured polymer optical fiber (mPOF), which can act as a label-free selective biosensor. Numerical results indicate a sensitivity of 20.3 nm/nm—wavelength shift per nm thickness of biolayer—achieved with a 15-cm-long device at visible region where the mPOF has the lowest absorbance. Recently, Wu et al. proposed and demonstrated a novel configuration with a sensitivity of 30,100 nm per refractive index unit (nm/RIU). This configuration is based on a directional coupler architecture using a solid- core PCF [42]. Yuan et al. demonstrated the design of an all-solid dual-core photonic bandgap fiber, in which a single hole between the cores acts as microfluidic channel for the analyte [43]. The predicted sensitivity was 70,000 nm/RIU. In 2011 [44], Sun et al. proposed and demonstrated a refractive index sensor based on the selectively resonant coupling between a conventional solid core and a microstructured core. Numerical results shown that this configuration could achieve a sensitivity of 8500 nm/RIU. However, these configurations have also some drawbacks, for instance, have complex design for the fiber cores or require selective filling. For this reason, the implementation of interferometric schemes in combination with these specialty fibers has emerged as a new alternative. In [26], we introduced the concept of transversally chirped solid-core MOF and reported a dual-core chirped MOF that could act as a structure with decoupled cores, thus forming a Mach-Zehnder interferometer in which the analyte directly modulated the device transmittance by its differential influence on the effective RI of each core mode, achieving a sensitivity of 300 per RIU for a 12-mm-long device and analyte RI of 1.42. A year later, we designed a label-free biosensor by immobilization of an antigen sensor layer onto the walls of the air holes of rings surrounding one core of the fiber. A sensitivity of around 3.7 nm/nm was achieved for a 10 mm long device at near IR wavelengths [38]. Then, we studied some refractometric properties of these configuration using numerical models to improve the performance of this device [41].

The chirp concept has been widely used in one-dimensional structures such as chirped mirrors [45] and chirped fiber Bragg gratings where different spectral components are localized at different positions inside the chirped structure [46]. In both cases, the chirp was implemented on the propagation direction. Chirping also has already been applied for designing hollow core fibers with a radially chirped microstructured cladding [47]. By introducing a radial chirp into the photonic crystal structure, it was demonstrated a novel concept that breaks with the paradigm of lattice homogeneity and enables a new degree of freedom in the design of MOFs. Another important variation was reported by Ghosh et al. [48]. The authors proposed a novel chirped cladding as a novel tailoring tool to attain wider transmission window and reduced temporal dispersion in an all-solid Bragg-like MOF.

In this chapter, dual-core transversally chirped MOFs for active mode conversion in telecommunications and sensing applications are presented. In the first part of this chapter, we explain the fabrication process of this novel structure. Next, we demonstrate that this type of MOF can be used to design a novel and tuneable mode-converter to improve the performance of the modern optical systems. Finally, a dual-core transversally chirped MOF is proposed to create a compact highly sensitivity optical fiber sensor.

2. Fabrication methodology

Two techniques can be used to fabricate dual-core transversally chirped MOFs. The first alternative consists in the implementation of the standard stack and draw technique [49, 50]. The first step is producing the preform, which is based on stacking of capillaries and rods. In our case, the diameter of the capillary should have a slight reduction in its diameter along the cross-section. Then, several slightly chirped preforms of about 1 mm were obtained by pulling a ~1 cm preform with a small transversal temperature gradient. The temperature gradient was produced by pulling the preform off-center [26]. After this process the fiber shown in **Figure 1(a)** was obtained, which is characterized by a slight transverse chirp in the hole distribution.

The second alternative consists in tapering the MOF from the previous step, in such a way that fiber structures with a larger transverse chirp can be achieved. In our case, MOF tapers were produced by using the flame brushing technique. The MOF was mounted on a motorized stage. The fiber was heated using a butane flame, which was mounted on a second motorized stage. The butane flame was moved back and forth along the fiber axis as the taper was pulled simultaneously. In order to ensure that the holes do not collapse, it was applied pressure within the holes. **Figures 1(b)** and **(c)** show the cross-section of two tapered MOFs obtained with an applied pressure of 6 and 7 bars, respectively. From these results, it is evident that the pressure applied inside of fiber holes can control the transversal chirp of the pristine MOF. For example, the MOF with an applied pressure of 6 bars has a structure in which none

http://dx.doi.org/10.5772/intechopen.70989

Figure 1. SEM images of (a) a dual-core transversally chirped MOF obtained with the standard stack and draw technique. (b) and (c) Dual-core transversally chirped MOF tapers obtained through the flame brushing technique at 6 and 7 bars, respectively, within the fiber holes.

of the holes collapsed during the tapering process and the transverse chirping slope is smaller than the MOF obtained when the pressure was 7 bars.

3. Mode converter device

The cross-section of the proposed dual-core transversally chirped MOF MSC is shown in **Figure 2**. As we can see, the cladding holes are arranged in a hexagonal lattice with constant pitch $\Lambda = 6$ μm. The diameter of the circular air holes decreases linearly from $d_{max} = 6$ μm on the left side of the fiber to $d_{min} = 0.9$ μm on the right side. The considered MOF has two solid

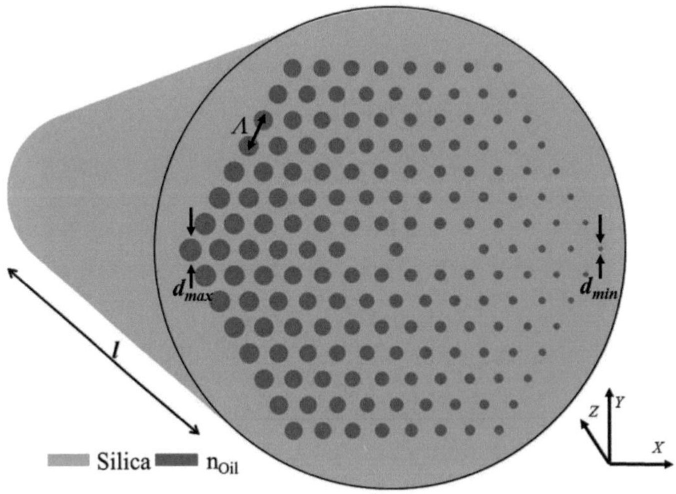

Figure 2. Structure of the mode selective coupler based on a dual-core transversally chirped MOF.

cores, which are separated by only one hole in the microstructure, that is, by 2Λ. In this case, we employ a small separation between both cores in order to guarantee power transfer. In addition, background material is silica, and its dispersion is given by Sellmeier's equation [51]. Note that by the transverse chirping of the microstructure, the fiber cores obviously do not have the same shape. In this case, the right core is almost 1.8 times wider than the left core.

In our device, a RI-matching oil (Cargille Labs., $n_0 = 1.42$ at room temperature) is chosen to be filled onto the MOF, and its thermo-optic coefficient is $\alpha = 3.94 \times 10^{-4}/C°$, and the relationship between refractive index n and temperature T is to be $n = n_0 - \alpha(T - T_0)$ where T_0 is the room temperature. A variety of functional fiber devices have been fabricated based on MOF fully infiltrated by fluid such as optical switches [52, 53], all-optical modulator [54], tunable optical filters [55] and fiber polarimeters [56].

According to our design, the fundamental mode LP_{01} in the left core is converted to the LP_{11} mode in the right core. Because the right core supports two modes (LP_{01} and LP_{11}), supermode analysis [57] was used to investigate the behavior of the temperature-controlled mode converter. We performed finite element simulations under different temperature conditions at the particular free-space wavelength $\lambda = 1.55$ μm.

Figure 3 shows the effective index curves of symmetric (supermode 1) and antisymmetric (supermode 2) modes. From these results, it is evident that LP_{01} mode in the left core interacts with LP_{11} mode in the right core. As expected, the effective refractive index n_{eff} of both supermodes decrease with increasing temperature. In addition, the effective index of symmetric mode is always slightly larger than that of antisymmetric mode, indicating that these supermodes have different propagation constants and therefore there is a beating between these two modes and, thus, the power fluctuates back and forth between the two cores.

Once it was determined which modes exchange energy, the coupled mode theory was applied to find the phase-matching conditions [57, 58]. Here, each core is analyzed as an independent

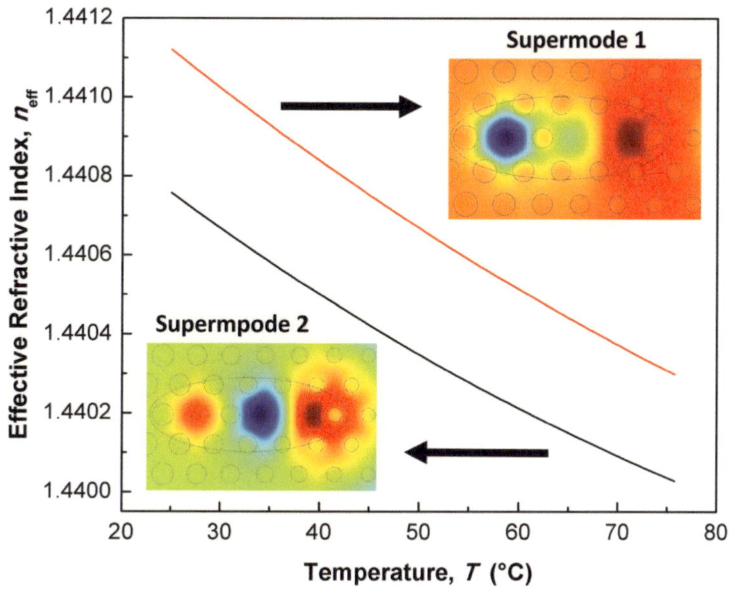

Figure 3. Supermode analysis of the dual core transversally chirped MOF when temperature varies from 25 to 75°C at an operating wavelength λ=1.55 μm.

waveguide to avoid the perturbations caused by the presence of the other core. When the mode of the left core interacts with the mode of the right core, crossing occurs and propagation constants of the two modes are matched. Then, maximum power transfer can be achieved at the phase-matching wavelength.

Figure 4(a) presents the modal dispersion curves of LP_{01} mode in the left core and of LP_{11} mode in the right core with different temperature values in the wavelength range 1.2–1.8 μm. Colored points in this figure represent the phase-matching wavelengths. **Figure 4(b)** shows the dependence of the operating wavelength on temperature. Therefore, the liquid-filling method is an easy method to tune the behavior of this device. According to this result, this mode converter could operate in the E + S + C + L + U bands.

To test the mode-conversion performance of the coupler, **Figure 5** presents the normalized power as a function of fiber length at a wavelength of 1.55 μm. It is observed that almost 100% of the power is coupled between the cores with the beating length L_c = 2 mm. This result shows that the proposed device is compact compared with other MOF-based mode converters [5, 7, 23–25].

Finally, the mode coupling efficiency of the device was evaluated. The results are depicted in **Figure 6**. From these results, it is evident that the mode coupling efficiency obtained with this structure presents a good performance in the E + S + C + L + U bands. As expected, the behavior of this device can be thermally controlled. It is observed that the phase-matching wavelength varies with the temperature change. It is found that when T increases, the operating wavelength also increases. In addition, coupling efficiencies above 65% were obtained in this study.

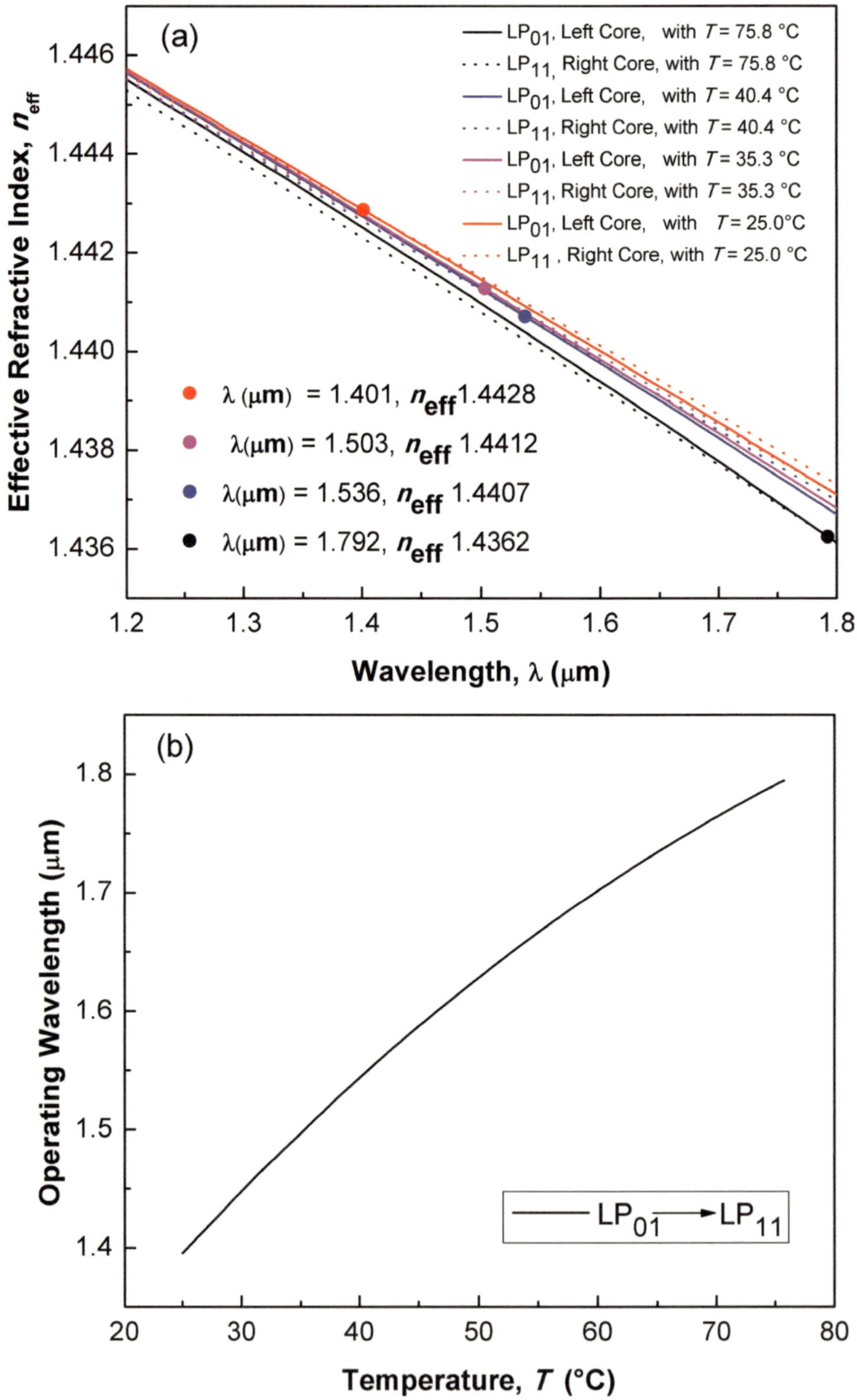

Figure 4. (a) Modal dispersion curves for LP_{01} mode in the left core and LP_{11} in the right core with temperature. (b) Operating wavelength as function of applied temperature on mode converter based on dual-core transversally chirped MOF.

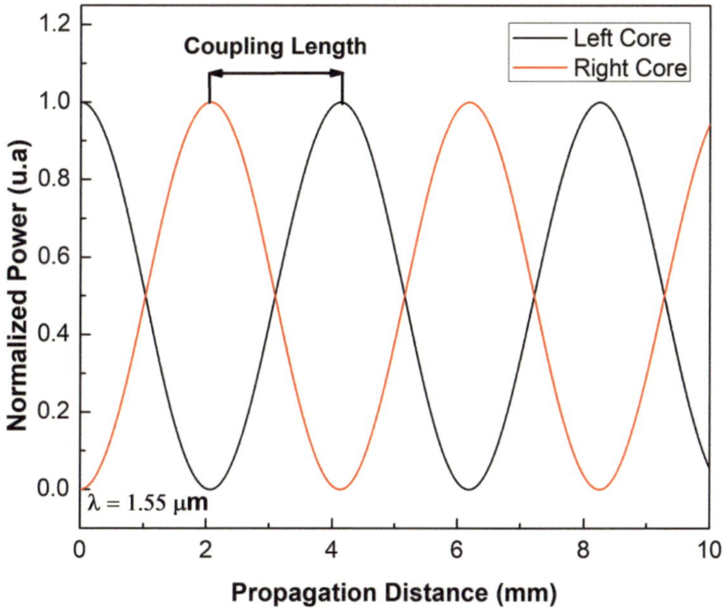

Figure 5. Normalized power transfer for the proposed mode-converter device with $T = 25°C$.

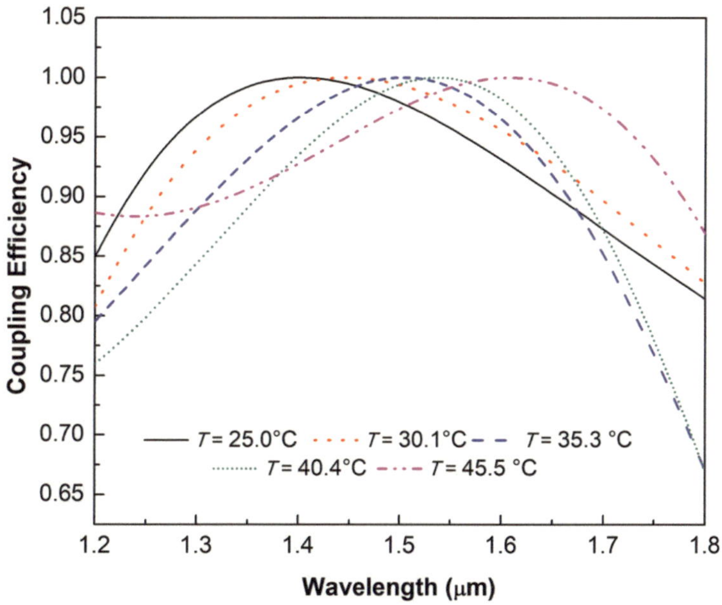

Figure 6. Comparison of mode coupling efficiency for different temperature values in the telecommunication windows.

4. Refractive index sensor

The dual-core transversally chirped MOF that is considered for refractive index sensing of fluids has a similar structure to the fibers in **Figures 1(b)** and **(c)**. Now, it is necessary to guarantee that the cores are uncoupled to exploit the fiber as a Mach-Zehnder interferometer (MZI) [27]. Then, the distance between the cores is increased to 4Λ, where in this new design Λ = 4.33 μm. This separation was considered because small fluctuations in fiber diameter due to fabrication tolerances may affect the performance of the sensor. In this structure, the diameter of the air holes decreases linearly from d_{max} = 2.6 μm to d_{min} = 0.6 μm, so the relative sizes of holes (d/Λ) range from 0.6 to 0.13 μm. As expected, the cores are non-identical because the holes around the right core are smaller than the holes around the left core. However, both cores are single-mode waveguides due to the slight chirp.

In addition, we consider as an example label-free antibody detection using the highly selective antigen-antibody binding based on our previous experience as another important variation [39]. Then, the first ring of air holes around the right core are functionalized for antibody detection by immobilization of an antigen sensor layer onto the walls of the holes as is shown in **Figure 7**. This layer can consist of a functionalization layer of a certain thickness in addition to the antigen layer. Then, we consider a layer with a thickness t_s equal to 40 nm for sensing the antibody α-streptavidin with thickness t_a = 5 nm. The refractive index of the sensor layer and α-streptavidin is 1.45 (we neglect the dispersion of the biomolecule layer).

From **Figure 8** the operation of the sensor can be clearly understood. The refractometric sensor gains its sensitivity from the fact that only the mode of the right core has substantial overlap with the analyte. This arises because of the low fraction ratio (d/Λ) of holes that surrounding this core. Then the light is not well confined. Now, the RI of the analyte directly modulates the device transmittance by its differential influence on the effective refractive index of each core mode, resulting in a variable phase difference between the optical path lengths of the interferometer arms. Therefore, the proposed configuration was classified as a modal interferometer in the sense that two modes of the dual-core structure are interfering among them. Here, the performance of the sensor is compared with an interesting variation. It consists in

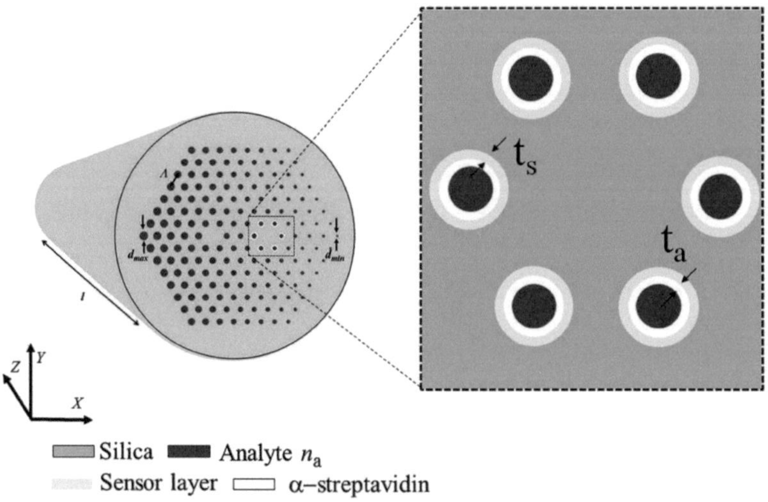

Figure 7. Dual-core transversally chirped MOF biosensor with Λ = 4.33 μm and the hole diameter vary from d_{max} = 2.6 μm to d_{min} = 0.6 μm. In this design, the fiber cores are uncoupled.

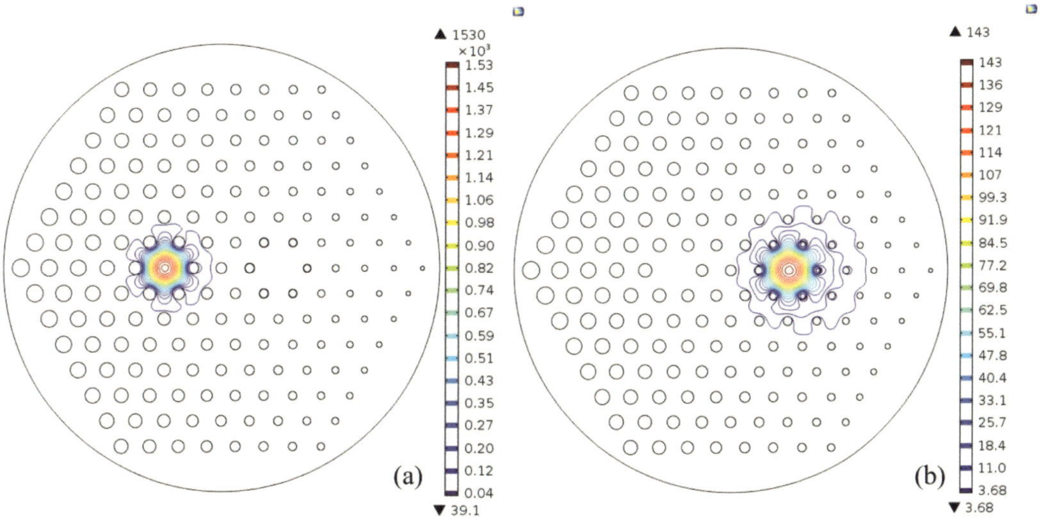

Figure 8. Fundamental mode distribution in left and right core at $\lambda = 633$ nm: Pitch $\Lambda = 4.33$ μm, $d_{max} = 2.6$ μm, $d_{min} = 0.6$ μm; all fiber holes are filled with an analyte of refractive index 1.32.

the inclusion of biomolecule layers onto the walls of the holes, as already explained. We only apply this variation on the first ring of air holes around the right core in order to determine the impact on sensitivity of the proposed sensor.

Figure 9 shows the effective refractive index of each mode when the analyte RI changes from 1.32 to 1.44 at $\lambda = 633$ nm. In this figure, we compared the obtained results of the MZI with and without biomolecule layers onto the walls of the holes. From these results, it is clear that in all cases the effective refractive index increases with the analyte RI. As expected, the behavior of the left core is the same in both configurations, due that this core has good confinement fraction and no biomolecule layer. On the other hand, the results of the right core are different, the results with biomolecule layers presents bigger values in the effective refractive index in the whole range, which indicate that the presence of biomolecules can affect the behavior of our configuration. Although the asymmetric nature of the proposed schematic, the chirped MOF-based interferometer is insensitive to the polarization state of the input beam, as we can see from results illustrated in **Figure 9**. This is a great advantage because our sensor does not require controls of polarization. Then, its implementation could be easier than other configurations.

Figure 10(a) shows the effective refractive index difference between the two fundamental modes—for the two orthogonal polarizations and the two studied configurations—that propagate through the fiber cores as a function of analyte RI. Now, from the effective refractive index difference, it is possible to determine the phase difference per unit length, which is given by Eq. (1). In this equation, Δn_{eff} is the effective refractive index difference between the two fundamental modes that propagate through the fiber cores.

$$\delta = \frac{2\pi\Delta n_{eff}}{\lambda} \quad (1)$$

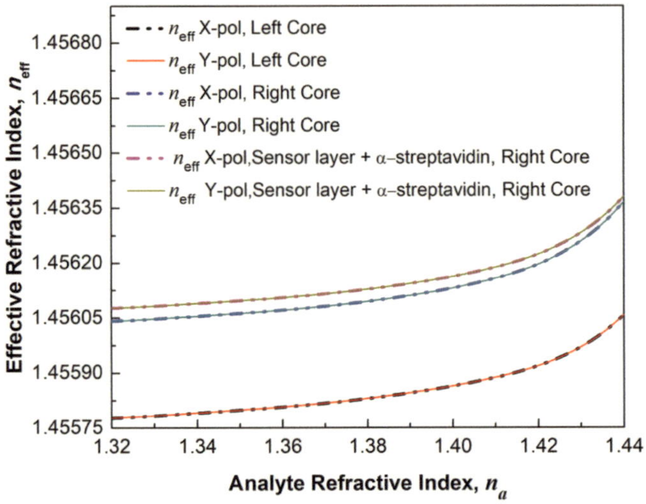

Figure 9. Effective refractive index of the fundamental modes for both polarizations as a function of analyte refractive index. This figure compares obtained results for configurations with and without biomolecules layers.

Figure 10(b) shows the phase difference per unit length as a function of analyte RI. Here, we only present the results for x-polarization. As mentioned before, this sensor is polarization-insensitive. As we can see, even though the analyte RI is low, the large differential overlap of the mode cores with the analyte results in a significant amount of phase difference. In both configurations, the phase difference increases exponentially. The configuration with biomolecules layers present higher δ in the whole range compared with the configuration without biomolecule layers. The region from 1.32 to 1.40 presents a phase difference per unit length almost constant in both cases, while for analyte RI higher than 1.40 the phase difference per unit length increases strongly with analyte RI. These results show a better behavior for the

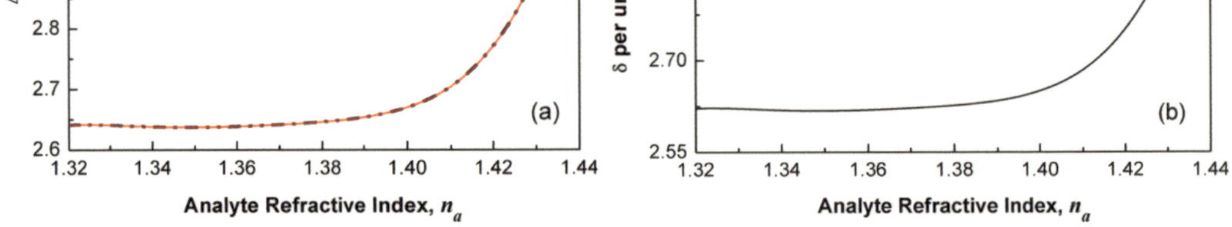

Figure 10. (a) Effective refractive index difference between the two fundamental modes that propagate through the fiber cores as a function of analyte refractive index. (b) Phase difference between the two fundamental modes that propagate through the fiber cores as a function of analyte refractive index (x-polarization).

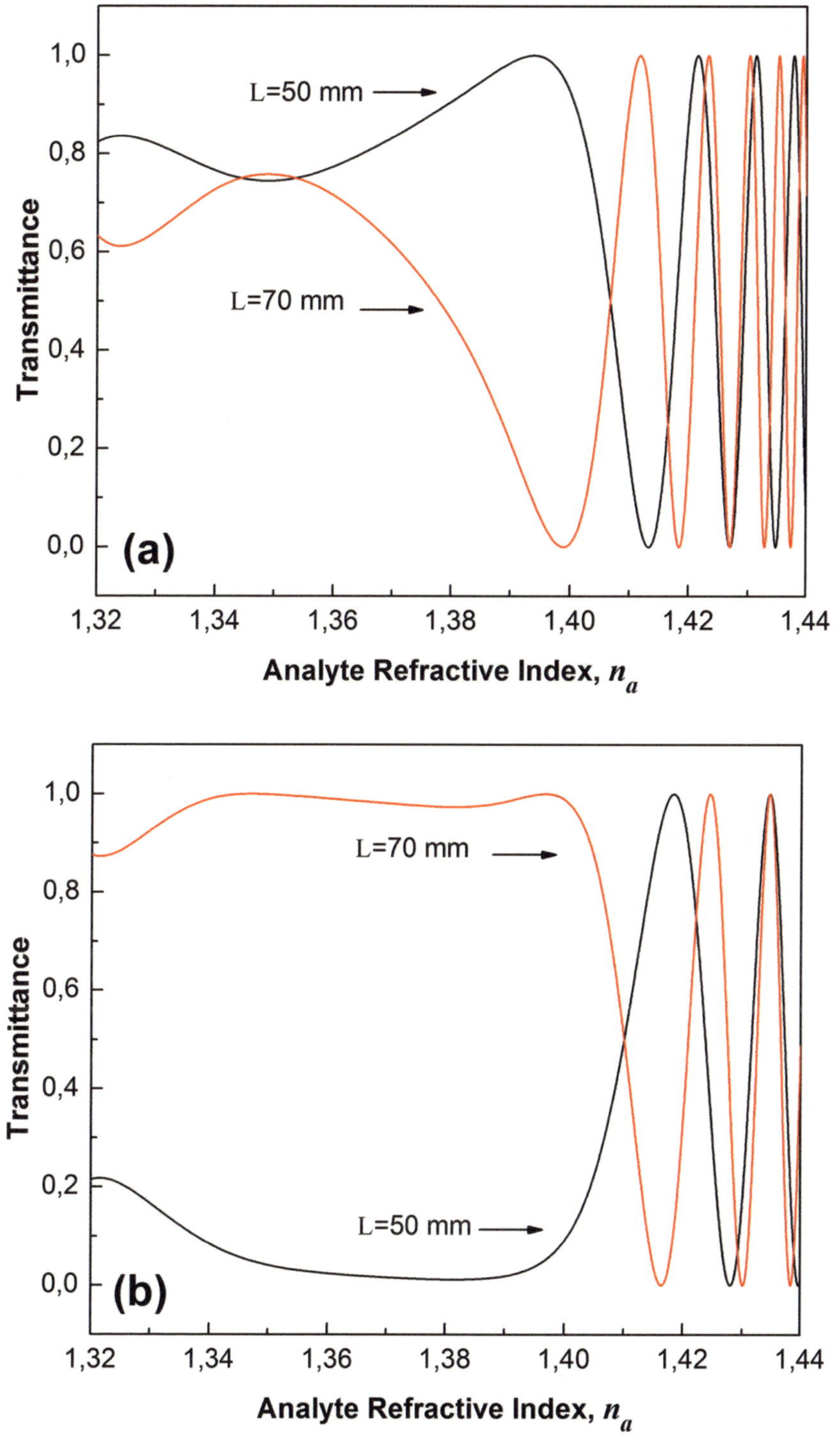

Figure 11. Transmittance of the dual-core transversally chirped MOF for L = 50 mm and L = 70 mm as a function of analyte refractive index: (a) RI sensor without biomolecules layers; (b) RI sensor with biomolecules layers into the first ring of air holes around the right core.

configuration without biomolecules layers. We believe that it is due to that the layers help to confine the light within the right core.

The phase difference per unit length obtained in **Figure 10(b)** was analytically approximated by an exponential function, which was used to obtain the transmittance of the both sensing configurations. For a balanced interferometer, the normalized transmittance can be calculated by using the following expression

$$T = \cos^2\left(\frac{\delta L}{2}\right) \quad (2)$$

where L is the total length of the sensor. Note that as the analyte refractive index increases, the transmittance passes through a series of nulls and peaks as the phase difference increases. In our case, the fact that the periodic variation of transmittance is reducing indicates that the phase difference between the optical paths of the chirped MOF interferometer changes increasingly rapidly with increasing analyte RI. In addition, these results show that the best region to implement the proposed interferometric schemes is for analyte RI from 1.40 to 1.44.

For best sensitivity, the sensor must be biased to operate at 50% transmittance around a given value of refractive index. In practice, this condition may be achieved by fabricating the device with a length so that $\delta = \pi/2$ (plus any multiple of π radians), or by temperature or wavelength tuning. In **Figure 11(a)**, we can see that the sensitivity of the dual-core transversally chirped MOF structure scales with device length. For example, the sensitivity around $n_a = 1.42$ is 3.028×10^2 RIU^{-1} for a 70-mm-long device, which gives a detection limit of 3.303×10^{-6} RIU assuming that we can detect transmittance variations of 10^{-3}. On the other hand, from the **Figure 11(b)** the configuration with biomolecules layers in the first ring of air holes around the right core present a sensitivity equal to 1.83×10^2 RIU^{-1} around $n_a = 1.42$. In this case, the detection limit is 5.464×10^{-6} RIU. Based on the obtained results, it is clear that the best configuration to measure refractive index changes is the scheme without biomolecules layers. In addition, other works presents the same order of sensitivity [44, 59] but using selective filling of some holes of the microstructure in order to improve the sensor performance.

5. Conclusion

In this chapter, we have presented the concept of dual-core transversally chirped microstructured optical fiber and how this structure can be used in different applications such as mode-converter devices and refractive index sensors. We have shown two simple methods to manufacture this specialty fiber. The effect of pressure inside of fiber holes was demonstrated and the transversal chirp of the MOF can be controlled.

Based on this novel concept, a mode selective coupler was designed and analyzed. We demonstrated a promising platform to manufacture compact and highly efficient mode converters. Through the fluid-filling post-processing technique the operating wavelength and the coupling efficiency can be can be continuously tuned by varying the temperature. The coupling efficiency over the entire wavelength range between 1.2 μm and 1.7 μm was greater than 65%.

Consequently, the proposed mode converter can operate in the E + S + C + L + U bands. In general, this kind of mode selective coupler has potential applications in MDM optical fiber communications since it can increase the channel capacity.

Finally, sensing possibilities enabled by the concept of transversally chirped microstructure have been proved, which can be exploited for refractive index sensing in an interferometric arrangement. We have also identified some features of this sensor, including high sensitivity and resolution and scalability of the sensitivity with sensor length. The sensor can be operated over a wide range of analyte refractive index values with a higher sensitivity compared to other selectively filled MOF sensors.

Acknowledgements

This work was supported in part by the Universidad Nacional de Colombia and the Instituto Tecnológico Metropolitano (projects P15108). Erick Reyes-Vera was supported in part by a grant from SPIE.

Author details

Erick Reyes Vera[1,2]*, Juan Úsuga Restrepo[2], Margarita Varon[1] and Pedro Torres[3]

*Address all correspondence to: erickreyes@itm.edu.co

1 Department of Electrical and Electronic Engineering, Universidad Nacional de Colombia, Bogota, Colombia

2 Department of Electronic and Telecommunications, Instituto Tecnológico Metropolitano, Medellín, Colombia

3 Escuela de Física, Universidad Nacional de Colombia, Medellín, Colombia

References

[1] Richardson DJ, Fini JM, Nelson LE. Space-division multiplexing in optical fibres. Nature Photonics. 2013;**7**:354-362

[2] Tkach RW. Scaling optical communications for the next decade and beyond. Bell Labs Technical Journal. 2010;**14**:3-9

[3] Cai S, Yu S, Lan M, Gao L, Nie S, Gu W. Broadband mode converter based on photonic crystal fiber. IEEE Photonics Technology Letters. 2015;**27**:474-477

[4] Schulze C, Bruning R, Schroter S, Duparre M. Mode coupling in few-mode fibers induced by mechanical stress. Journal of Lightwave Technology. 2015;**33**:4488-4496

[5] Zhang Y, Wang Y, Cai S, Lan M, Yu S, Gu W. Mode converter based on dual-core all-solid photonic bandgap fiber. Photonics Research. 2015;**3**:220-223

[6] Yunhe Zhao, Y. Liu, Jianxiang Wen, and Tingyun Wang, Mode converter based on the long period fiber gratings written in two mode fiber, in 2015 Opto-Electronics and Communications Conference (OECC) (IEEE, 2015), Vol. 24, pp. 1-3

[7] Martelli P, Gatto A, Boffi P, Martinelli M. Free-space optical transmission with orbital angular momentum division multiplexing. Electronics Letters. 2011;**47**:972

[8] Weng Y, He X, Wang J, Pan Z. All-optical ultrafast wavelength and mode converter based on inter-modal four-wave mixing in few-mode fibers. Optics Communication. 2015;**348**:7-12

[9] Hellwig T, Walbaum T, Fallnich C. Optically induced mode conversion in graded-index fibers using ultra-short laser pulses. Applied Physics B: Lasers and Optics. 2013;**112**:499-505

[10] Shi CX, Okoshi T. Mode conversion based on the periodic coupling by a reflective fiber grating. Optics Letters. 1992;**17**:1655-1657

[11] Usuga J, Amariles D, Correa N, Reyes-Vera E, Gomez-Cardona N. Analysis of chromatic dispersion compensator using a PCF with elliptical holes. Revista Colombiana de Física. 2016;**33**:38-41

[12] Birks TA, Knight JC, Russell PSJ. Endlessly single-mode photonic crystal fiber. Optics Letters. 1997;**22**:961

[13] Ortigosa-Blanch A, Knight JC, Wadsworth WJ, Arriaga J, Mangan BJ, Birks TA, Russell PS. Highly birefringent photonic crystal fibers. Optics Letters. 2000;**25**:1325-1327

[14] Eggleton B, Kerbage C, Westbrook P, Windeler R, Hale A. Microstructured optical fiber devices. Optics Express. 2001;**9**:698

[15] Reyes-Vera E, Gonzalez-Valencia E, Torres P. Understanding the birefringence effects in an all-fiber device based on photonic crystal fibers with integrated electrodes. Photonics Letters of Poland. 2010;**2**:168-170

[16] Reyes-Vera E, Gonzalez-Valencia E, Torres P. Transverse stress response of FBGs in large-mode-area microstructured Panda-type fibers. In: Hernández-Cordero J, Torres-Gómez I, Méndez A, editors. Proceedings of SPIE. Vol. 7839. 2010. p. 78391G

[17] Reyes-Vera E, Torres P. Influence of filler metal on birefringent optical properties of photonic crystal fiber with integrated electrodes. Journal of Optics. 2016;**18**:85804

[18] Khaleque A., Franco MAR, and Hattori H, Ultra-broadband and compact polarization splitter for sensing applications, in Photonics and Fiber Technology 2016 (ACOFT, BGPP, NP) (OSA, 2016), Vol. 2, p. JM6A.2

[19] Franco MAR, Serrão VA, Sircilli F. Microstructured optical fiber for residual dispersion compensation over S+C+L+U Wavelength Bands. IEEE Photonics Technology Letters. 2008;**20**:751-753

[20] Betancur-Pérez AF, Botero-Cadavid JF, Reyes-Vera E, Gómez-Cardona N. Hexagonal photonic crystal fiber behaviour as a chromatic dispersion compensator of a 40 Gbps link. International Journal of Electronics and Telecommunications. 2017;**63**:93-98

[21] Chen MY, Chiang KS. Mode-selective characteristics of an optical fiber with a high-index core and a photonic bandgap cladding. IEEE Journal of Selected Topics in Quantum Electronics. 2016;**22**:251-257

[22] Huang W, Liu Y, Wang Z, Zhang W, Luo M, Liu X, Guo J, Liu B, Lin L. Generation and excitation of different orbital angular momentum states in a tunable microstructure optical fiber. Optics Express. 2015;**23**:33741-33752

[23] Cai S, Yu S, Wang Y, Lan M, Gao L, Gu W. Hybrid dual-core photonic crystal fiber for spatial mode conversion. IEEE Photonics Technology Letters. 2016;**28**:339-342

[24] Reyes Vera EE, Usuga Restrepo JE, Gómez Cardona NE, and Varón M. Mode selective coupler based in a dual-core photonic crystal fiber with non-identical cores for spatial mode conversion, in Latin America Optics and Photonics Conference (OSA, 2016), p. LTu3C.1

[25] Frazão O, Santos JL, Araújo FM, Ferreira LA. Optical sensing with photonic crystal fibers. Laser & Photonics Reviews. 2008;**2**:449-459

[26] Torres P, Reyes-Vera E, Díez A, Andrés MV. Two-core transversally chirped microstructured optical fiber refractive index sensor. Optics Letters. 2014;**39**:1593-1596

[27] Rindorf L, Bang O. Sensitivity of photonic crystal fiber grating sensors: biosensing, refractive index, strain, and temperature sensing. Journal of the Optical Society of America B: Optical Physics. 2008;**25**:310

[28] Zhang X, Peng W. Fiber Bragg grating inscribed in dual-core photonic crystal fiber. IEEE Photonics Technology Letters. 2015;**27**:391-394

[29] Reyes-Vera E, Cordeiro CMB, Torres P. Highly sensitive temperature sensor using a Sagnac loop interferometer based on a side-hole photonic crystal fiber filled with metal. Applied Optics. 2017;**56**:156

[30] Villatoro J, Minkovich VP, Zubia J. Photonic crystal fiber interferometric force sensor. IEEE Photonics Technology Letters. 2015;**27**:1181-1184

[31] Villatoro J, Minkovich VP, Zubia J. Photonic crystal fiber interferometric vector bending sensor. Optics Letters. 2015;**40**:3113

[32] Osório JH, Hayashi JG, Espinel YAV, Franco MAR, Andrés MV, Cordeiro CMB. Photonic-crystal fiber-based pressure sensor for dual environment monitoring. Applied Optics. 2014;**53**(3668)

[33] Wang J-N, Tang J-L. Photonic crystal fiber Mach-Zehnder interferometer for refractive index sensing. Sensors (Basel). 2012;**12**:2983-2995

[34] Silva S, Roriz P, Frazão O. Refractive index measurement of liquids based on microstructured optical fibers. Photonics. 2014;**1**:516-529

[35] Oliveira R, Osório JH, Aristilde S, Bilro L, Nogueira RN, Cordeiro CMB. Simultaneous measurement of strain, temperature and refractive index based on multimode interference, fiber tapering and fiber Bragg gratings. Measurement Science and Technology. 2016;**27**:75107

[36] Osório JH, Oliveira R, Aristilde S, Chesini G, Franco MAR, Nogueira RN, Cordeiro CMB. Bragg gratings in surface-core fibers: Refractive index and directional curvature sensing. Optical Fiber Technology. 2017;**34**:86-90

[37] Markos C, Yuan W, Vlachos K, Town GE, Bang O. Label-free biosensing with high sensitivity in dual-core microstructured polymer optical fibers. Optics Express. 2011; **19**(7790-8)

[38] Reyes-Vera E, Gómez-Cardona N, Torres P. Label-free biosensor based on a dual-core transversally chirped microstructured optical fiber. In: López-Higuera JM, Jones JDC, López-Amo M, Santos JL, editors. Proceedings of SPIE. Vol. 9157. 2014 91578J

[39] Monro TM, Belardi W, Furusawa K, Baggett JC, Broderick NGR, Richardson DJ. Sensing with microstructured optical fibres. Measurement Science and Technology. 2001;**12**: 854-858

[40] Mägi E, Steinvurzel P, Eggleton B. Tapered photonic crystal fibers. Optics Express. 2004; **12**(776-84)

[41] Velasquez-Botero F, Reyes-Vera E, Torres P. Some refractometric features of dual-core chirped microstructured optical fibers. In: Kalinowski HJ, Fabris JL, Bock WJ, editors. Proceedings of SPIE. Vol. 9634. 2015 963450

[42] Wu DKC, Kuhlmey BT, Eggleton BJ. Ultrasensitive photonic crystal fiber refractive index sensor. Optics Letters. 2009;**34**:322

[43] Kuhlmey BT, Coen S, Mahmoodian S. Coated photonic bandgap fibres for low-index sensing applications: cutoff analysis. Optics Express. 2009;**17**:16306-16321

[44] Sun B, Chen M-Y, Zhang Y-K, Yang J-C, Yao J, Cui H-X. Microstructured-core photonic-crystal fiber for ultra-sensitive refractive index sensing. Optics Express. 2011; **19**(4091-4100)

[45] Matuschek N, Kartner FX, Keller U. Analytical design of double-chirped mirrors with custom-tailored dispersion characteristics. IEEE Journal of Quantum Electronics. 1999;**35**:129-137

[46] Riant I, Gurib S, Gourhant J, Sansonetti P, Bungarzeanu C, Kashyap R. Chirped fiber Bragg gratings for WDM chromatic dispersion compensation in multispan 10-Gb/s transmission. IEEE Journal of Selected Topics in Quantum Electronics. 1999;**5**:1312-1324

[47] Skibina JS, Iliew R, Bethge J, Bock M, Fischer D, Beloglasov VI, Wedell R, Steinmeyer G. A chirped photonic-crystal fibre. Nature Photonics. 2008;**2**:679-683

[48] Ghosh S, Varshney RK, Pal BP, Monnom G. A Bragg-like chirped clad all-solid microstructured optical fiber with ultra-wide bandwidth for short pulse delivery and pulse reshaping. Optical and Quantum Electronics. 2010;**42**:1-14

[49] Russell PSJ. Photonic-crystal fibers. Journal of Lightwave Technology. 2006;**24**:4729-4749

[50] Méndez A, Morse TF. Specialty Optical Fibers Handbook. London, Academic Press; 2007

[51] Bansal NP. Handbook of Glass Properties. London, Academic Press; 1986

[52] Du F, Lu Y-Q, Wu S-T. Electrically tunable liquid-crystal photonic crystal fiber. Applied Physics Letters. 2004;**85**:2181-2183

[53] Larsen T, Bjarklev A, Hermann D, Broeng J. Optical devices based on liquid crystal photonic bandgap fibres. Optics Express. 2003;**11**(2589-96)

[54] Alkeskjold TT, Lægsgaard J, Bjarklev A, Hermann DS, Anawati A, Broeng J, Li J, Wu S-T. All-optical modulation in dye-doped nematic liquid crystal photonic bandgap fibers. Optics Express. 2004;**12**:5857

[55] Steinvurzel P, Eggleton BJ, de Sterke CM, Steel MJ. Continuously tunable bandpass filtering using high-index inclusion microstructured optical fibre. Electronics Letters 2005;**41**:463

[56] Alkeskjold TT, Bjarklev A. Electrically controlled broadband liquid crystal photonic bandgap fiber polarimeter. Optics Letters. 2007;**32**:1707

[57] Snyder AW, Love JD. Optical Waveguide Theory. New York, Springer US; 1983

[58] Lee DL. Electromagnetic Principles of Integrated Optics. 1st ed. New Jersey, Wiley; 1986

[59] Wu Y, Town GE, Bang O. Refractive index sensing in an all-solid twin-core photonic bandgap fiber. IEEE Sensors Journal. 2010;**10**:1192-1199

Chapter 4

Multi-Core Optical Fibers: Theory, Applications and Opportunities

Andrés Macho Ortiz and Roberto Llorente Sáez

Additional information is available at the end of the chapter

http://dx.doi.org/10.5772/intechopen.72458

Abstract

Multi-core fibers (MCFs) have sparked a new paradigm in optical communications, as they can significantly increase the Shannon capacity of optical networks based on single-core fibers. In addition, MCFs constitute a useful platform for testing different physical phenomena, such as quantum or relativistic effects, as well as to develop interesting applications in various fields, such as biological and medical imaging. Motivated by the potential applications of these new fibers, we will perform a detailed review of the MCF technology including a theoretical analysis of the main physical impairments and new dispersive effects of these fibers, and we will discuss their emerging applications and opportunities in different branches of science.

Keywords: multi-core fiber, inter-core crosstalk, birefringent effects, intermodal dispersion, microwave photonics, optical sensors, medical imaging

1. Introduction

Data traffic demand in access and backbone networks has been increased exponentially in the last three decades [1, 2]. Remarkably, in the last decade, the development of streaming transmissions and cloud computing has accelerated this growth [3]. Nowadays, in spite of the fact that this data traffic demand is easily covered by wavelength-division multiplexed (WDM) systems based on single-mode single-core fibers (SM-SCFs),[1] recent works show that the WDM systems are rapidly approaching their Shannon capacity limit [4].

Aimed to overcome the Shannon capacity limit of WDM networks using SM-SCFs, space-division multiplexing (SDM) has been extensively investigated in recent years [5–7].

[1]SM-SCFs are also termed in the literature as single-mode fibers (SMFs).

Remarkably, the SDM concept within the context of optical communications was proposed for the first time in the decade of 1980 [8–10]. Unfortunately, the technology underneath SDM was immature and extremely expensive. Nevertheless, the fabrication methods of the SDM fibers and optical devices have been extensively developed in the last decade reducing their manufacturing cost [11]. In this scenario, new types of optical fibers based on the SDM concept have been proposed [5–12]: fiber bundle based on SM-SCFs, multi-mode single-core fibers (MM-SCFs),[2] single/multi-mode multi-core fibers (SM/MM-MCFs) and photonic crystal fibers.

In contrast with the other aforementioned SDM fibers, MCFs allow us to increase the channel capacity limit of SM-SCFs by exploiting the six signal dimensions (time, wavelength, amplitude, phase, polarization and space) through spatial multi-dimensional modulation formats and digital signal processing at the receiver [13–15]. Interestingly, SCFs have also been used as an experimental platform for testing different phenomena related to diverse branches of physics, such as fluid dynamics, quantum mechanics, general relativity and condensed matter physics, as well as to develop applications in other fields [16–23]. Along this line, MCFs are potential laboratories that could extend the possibilities offered by SCFs. As an example, disordered MCFs exhibiting transverse Anderson localization have been proposed with potential applications in biological and medical imaging [22].

Inspired by the potential applications of these new fibers, we perform a detailed review of the MCF technology including a theoretical analysis of the main physical impairments and new dispersive effects of these fibers, and we discuss their applications and opportunities in different branches of physics, engineering and medicine. The chapter is organized as follows. In Section II, the different MCF types are revisited. In Section III, we include some fundamental aspects of light propagation in the linear and nonlinear fiber regime. Specifically, we focus on the theoretical description of the physical impairments observed in these fibers in the single-mode regime: the linear and nonlinear inter-core crosstalk, the intra- and inter-core birefringent effects, the intermodal dispersion and higher-order coupling and nonlinear effects. In Section IV we discuss the main applications and opportunities of MCFs in photonics, medicine and experimental physics. Finally, in Section V the main conclusions of the chapter and the open research lines in the topic are highlighted.

2. Multi-core fiber types and fabrication

MCF designs can be classified in different categories attending to diverse fiber parameters and characteristics. **Table 1** shows the usual MCF designs employed for SDM transmissions and MCF laser and sensing applications:

1. The refractive index profile of each core allows us to differentiate between step-index (SI-MCF) and gradual-index MCFs (GI-MCF). In the former case, the refractive index profile of all cores has a step between two constant values in the core and cladding interface. However, in the latter case, a MCF is referred to as GI-MCF if at least one core

[2] MM-SCFs are also referred to as multi-mode fibers (MMFs).

has a continuous refractive index profile. Along this line, we can make a distinction with a third type of MCF: a trench- or hole-assisted MCF (TA-MCF or HA-MCF). In general, a TA- and HA-MCF present a multi-step refractive index profile in the cladding to reduce the mode-coupling (inter-core crosstalk) between the linearly polarized (LP) modes of adjacent cores [7, 11, 12, 24]. Specifically, in a HA-MCF an additional step is included in the cladding by performing holes around the cores [7].

MCF classification	Type 1	Type 2	Figure/comments
Refractive index profile	Step-index SI-MCF	Gradual-index GI-MCF	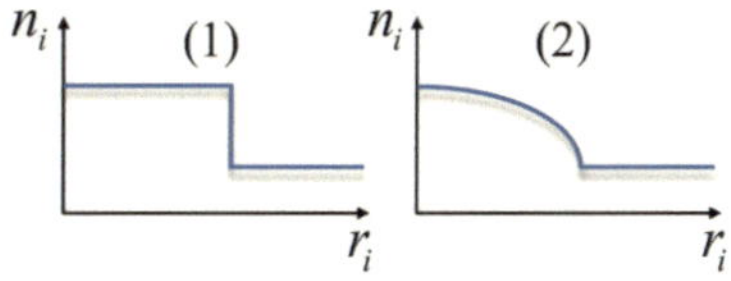
Modal regime	Single-mode SM-MCF	Multi-mode MM-MCF	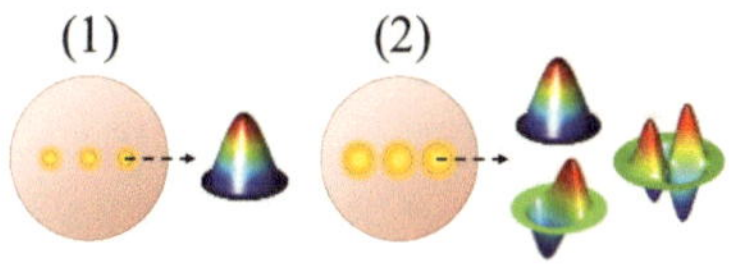
Spatial homogeneity	Homogeneous cores HO-MCF	Heterogeneous cores HE-MCF	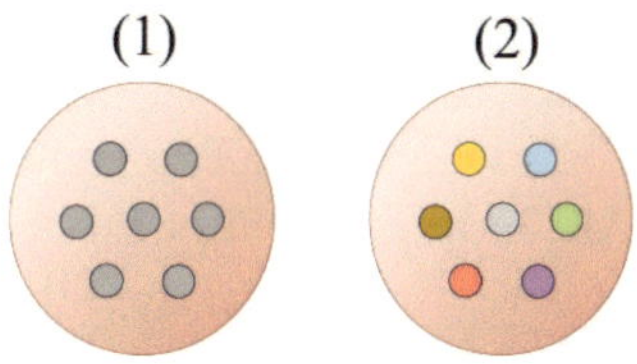
Core-to-core distance (d_{ab})	Uncoupled cores ($d_{ab} \geq 7 \cdot R_0$) UC-MCF	Coupled cores ($d_{ab} < 7 \cdot R_0$) CC-MCF	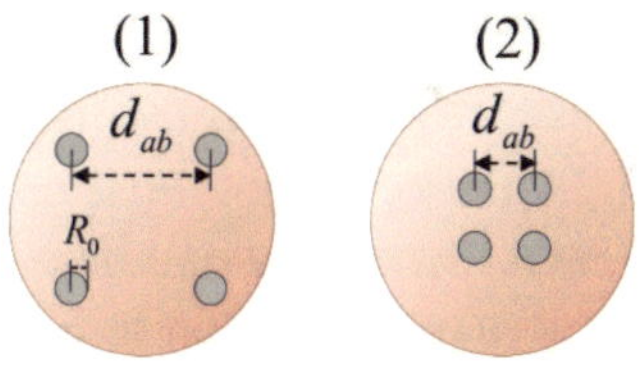
Intrinsic linear birefringence	Lowly birefringent cores LB-MCF	Highly birefringent cores HB-MCF	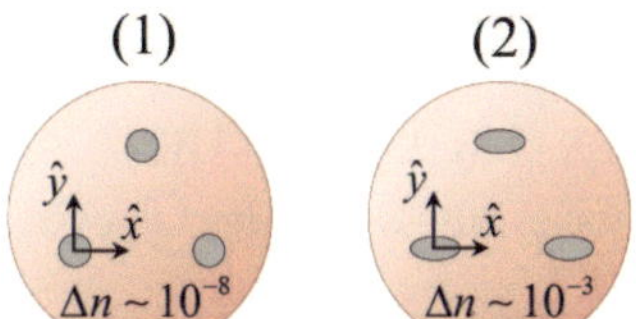
Others	(1) Trench-assisted MCF (TA-MCF), (2) Hole-assisted MCF (HA-MCF), (3) MCF with coupled and uncoupled cores, (4) Dispersion-shifted MCF (DS-MCF), (5) MCF Bragg gratings, (6) Hexagonal shaped cores… 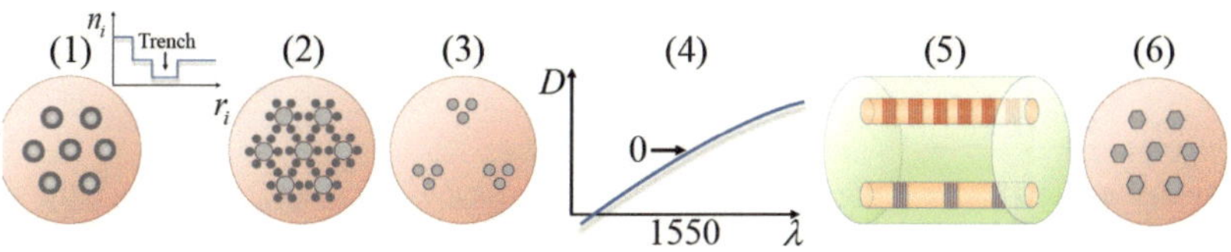		

Table 1. Classification of multi-core fiber types.

2. A single-mode MCF (SM-MCF) supports only the LP_{01} mode in each core. In contrast, if a given core guides several LP modes, the fiber is known as a multi-mode MCF (MM-MCF). Moreover, a MCF supporting only the first three or four LP mode groups (LP_{01}, LP_{11}, LP_{21}, LP_{02}) is usually termed as a few-mode MCF (FM-MCF) [7].

3. Attending to the spatial homogeneity of the MCF structure, we can make a distinction between a homogeneous MCF (HO-MCF) or a heterogeneous MCF (HE-MCF). In the former case, all cores present the same refractive index profile, and in the latter case, the MCF comprises at least one core with a different refractive index profile[3].

4. The core-to-core distance (or core pitch) is the main fiber parameter which determinates the inter-core crosstalk level among the LP modes of each core. Usually, if the core pitch between two homogeneous cores a and b (d_{ab}) is lower than seven times[4] the core radius R_0, the LP modes of the cores are found to be degenerated and the MCF supports supermodes [25]. In such a case, the MCF is referred to as a coupled-core MCF (CC-MCF). On the contrary, if the LP modes of the cores a and b are non-degenerated, the supermodes cannot be generated and each core is considered as an individual light path. This fiber design is termed as uncoupled-core MCF (UC-MCF). Recent works have reported a mixed design using coupled and uncoupled cores [26, 27].

5. If the intrinsic linear birefringence of each core $\Delta n = |n_x - n_y|$ is lower than 10^{-7}, the MCF is referred to as lowly birefringent MCF (LB-MCF). Nonetheless, if a given core has a $\Delta n > 10^{-7}$ the MCF is known as a highly birefringent MCF (HB-MCF). In general, a HB-MCF comprises elliptical or panda cores for polarization-maintaining applications [28, 29].

6. Other designs of MCFs involve: dispersion-shifted cores (DS-MCFs) [30], selective-inscribed Bragg gratings [31] and hexagonal shaped cores [32].

Once the MCF cross-section design is established, the specific MCF fabrication method is of key importance for the final optical transmission characteristics. MCF fabrication processes have been refined and optimized in the last years with an intensive research work [5, 7, 33–37]. In this scenario, the main technological challenge in the design and fabrication of uncoupled MCFs is to minimize the crosstalk providing the maximum core isolation. Usually, the design work is performed numerically using commercial simulation packages. The simulation analysis targets to determine the cross-section modal distribution, the spectral power density of the LP modes, the associated power losses and the resulting inter-core crosstalk.

MCF fabrication can be addressed by microstructured stack-and-draw technology [36], a flexible technology which allows us to fabricate very different fibers on the same machinery. Unfortunately, MCF manufacturing is a complex process with nonlinear results on the process parameters. In particular, some rods or capillaries configurations may be technically difficult to draw into the designed form, which results in a MCF with higher crosstalk levels than in the

[3]Two cores a and b have a different refractive index profile if $n_a(r_a) \neq n_b(r_b)$, where $r_{a(b)}$ is the local radial coordinate of each core. Hence, two step-index cores have a different refractive index profile if $n_a(r_a = 0) \neq n_b(r_b = 0)$.

[4]The condition $d_{ab} < 7R_0$ is only an approximation in the third transmission window and in single-mode regime of each core, with $R_0 \sim 4$ μm. In general, the criteria to achieve the supermode regime in the MCF structure depend on additional fiber parameters such as the refractive index profile and the wavelength of the optical carrier.

original design. In this scenario, it is necessary to investigate the linear and nonlinear MCF propagation taking into account not only the MCF manufacturing imperfections, but also additional external fiber perturbations (see below). This would be of great benefit for investigating multi-dimensional modulation formats, spatial encoding techniques and sharing of receiver resources in MCF systems [38–41].

3. Linear and nonlinear propagation in real MCFs

In general, the electromagnetic analysis of a MCF should be performed by solving the macroscopic Maxwell equations (MMEs) in each dielectric region of the fiber (cores + cladding) and applying the boundary conditions between the cores-cladding interfaces. However, the calculation of the exact MCF eigenmodes from the MMEs presents a high degree of complexity, and usually, they should be calculated numerically. Therefore, in order to analyze theoretically the electromagnetic phenomena in MCF media, the perturbation theory is usually employed. **Figure 1** shows a flowchart of this approach.

The goal is to derive a set of coupled equations from the MMEs in terms of the complex envelopes of the electric field strength in each core performing the next steps:

1. First, we should propose the ansatz of the global electric field strength ($\mathcal{E}$) of the MCF structure following the assumption of the classical perturbation theory [42]: the exact electric field strength is approximated by a linear combination of the *mi* polarized core modes[5] considering isolated cores, that is, assuming that the geometry of each core m is not perturbed by the presence of adjacent cores [$\mathcal{E} \approx \sum \mathcal{E}_{mi}$]. At the same time, we should decouple the rapid- and the slowly varying temporal and longitudinal changes of $\mathcal{E}_{mi}$. The

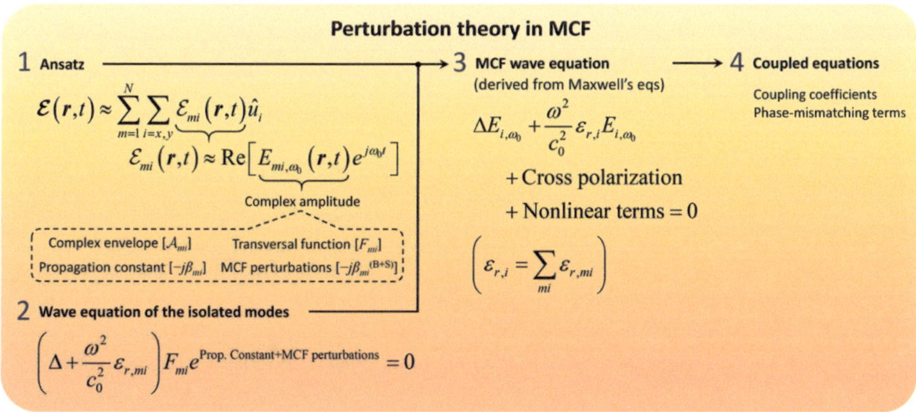

Figure 1. Flowchart of the perturbation theory in MCF media to derive the coupled equations from the macroscopic Maxwell equations.

[5]The polarized core mode *mi* refers to the $LP_{01,mi}$ mode associated with core m and polarization axis i.

rapidly varying temporal changes are decoupled by using the slowly varying complex amplitude approximation with $\boldsymbol{\mathcal{E}}_{mi}(\boldsymbol{r},t) \approx f(E_{mi,\omega0}(\boldsymbol{r},t))$, where $E_{mi,\omega0}$ is the complex amplitude and ω_0 is the angular frequency of the optical carrier[6]. In a similar way, the rapidly varying longitudinal variations are decoupled by writing $E_{mi,\omega0}$ as a function of the complex envelopes $\mathcal{A}_{mi}(z,t)$, that is, $E_{mi,\omega0}(\boldsymbol{r},t) \approx f(\mathcal{A}_{mi}(z,t))$. Although ε_{mi} is written assuming isolated conditions, the complex envelopes should be assumed longitudinal dependent to describe not only the usual longitudinal distortion[7] of the optical pulses in SM-SCFs, but also the longitudinal fluctuations induced by the mode-coupling. Moreover, $E_{mi,\omega0}$ also involves fundamental information such as the ideal propagation constant $(-j\beta_{mi})$, the transversal eigenfunction (F_{mi}) and the MCF perturbations [bending, twisting and additional fiber birefringent fluctuations modeled by the longitudinal and temporal dependent phase function $\beta_{mi}{}^{B+S}(z,t)$][8].

2. Each polarized core mode $\boldsymbol{\mathcal{E}}_{mi}$ is written by assuming isolated conditions of each core. Therefore, the transversal eigenfunction F_{mi} and the ideal phase constant β_{mi} can be expressed as indicated in [44] for the LP_{01} mode of a single-core fiber. Moreover, taking into account that the nonlinear effects are not included in the modal solution of [44] and in the MCF perturbations $\beta_{mi}{}^{B+S}$, thus the eigenfunction $F_{mi}\cdot\exp(-j(\beta_{mi} + \beta_{mi}{}^{B+S})z)$ should satisfy the linear wave equation in each core and polarization in $\delta z \sim \lambda_0$, where λ_0 is the maximum value of the wavelength of the optical carrier in the multi-dielectric medium[9].

3. In the third step, the wave equation of the MCF should be derived for the complex amplitudes $E_{mi,\omega0}$ from the MMEs by taking into account the cross- and nonlinear polarization.

4. Finally, using the results of the first and second step in the MCF wave equation, we finally derive the coupled equations of the complex envelopes by assuming the slowly varying complex envelope approximation (SVEA), that is, $\delta_z\mathcal{A}_{mi} << |\mathcal{A}_{mi}|$ in $\delta z \sim \lambda_0$ and $\delta_t\mathcal{A}_{mi} << |\mathcal{A}_{mi}|$ in $\delta t \sim 2\pi/\omega_0$. More specifically[10]:

$$\left|\frac{\partial^2 \mathcal{A}_{mi}}{\partial z^2}\right| \ll k_0 \left|\frac{\partial \mathcal{A}_{mi}}{\partial z}\right| \ll k_0^2 |\mathcal{A}_{mi}|; \quad \left|\frac{\partial^2 \mathcal{A}_{mi}}{\partial t^2}\right| \ll \omega_0 \left|\frac{\partial \mathcal{A}_{mi}}{\partial t}\right| \ll \omega_0^2 |\mathcal{A}_{mi}|, \tag{1}$$

where $k_0 = 2\pi/\lambda_0$. Thus, we can approximate $\partial_z^2\mathcal{A}_{mi} \approx 0$.

In the following subsections, we will review the new physical impairments observed in SM-MCFs using the aforementioned perturbation theory. First, we will describe the inter-core

---

[6] In general, we cannot consider a single-optical carrier in SDM-WDM systems using SM-MCFs. However, the interchannel nonlinearities should only be taken into account for optical pulses higher than 50 ps [43]. Therefore, the assumption of a single-optical carrier allows us to investigate the major physical impairments in SM-MCFs.

[7] Chromatic dispersion, polarization-mode dispersion and additional distortions induced by the intra-core nonlinear effects.

[8] In Section 3.1, we will be more specific with the description of the MCF perturbations.

[9] The symbol λ_0 is commonly used in the literature to describe the wavelength of the optical carrier at the vacuum. The context should avoid any confusion.

[10] Note that $\delta_z\mathcal{A}_{mi}$ and $\delta_t\mathcal{A}_{mi}$ are defined as $\delta_z\mathcal{A}_{mi} := |\mathcal{A}_{mi}(z,t) - \mathcal{A}_{mi}(z+\delta z,t)|$ and $\delta_t\mathcal{A}_{mi} := |\mathcal{A}_{mi}(z,t) - \mathcal{A}_{mi}(z,t+\delta t)|$.

crosstalk among cores when assuming a single polarization. Second, we will discuss the intra- and inter-core birefringent effects by including two polarizations per core. Later, we will analyze the intermodal dispersion and its impact on Gaussian pulses and optical solitons. And finally, higher-order coupling and nonlinear effects will be investigated when propagating optical pulses in the femtosecond regime.

3.1. Inter-core crosstalk

In multi-dielectric media, we can observe mode-coupling among adjacent dielectric regions. The continuity of the electromagnetic field in such media is the physical origin of the mode-coupling, referred to as the inter-core crosstalk (IC-XT) in MCFs.

The IC-XT behavior is induced by the longitudinal and temporal deterministic and random MCF perturbations. The longitudinal perturbations include the macrobending, microbending, fiber twisting and the intrinsic manufacturing imperfections of the fiber. The temporal perturbations are induced by external environmental factors, such as temperature variations and floor vibrations modifying the propagation constant of each polarized core mode, the bending radius and the twist rate of the optical medium. In spite of the fact that the deterministic nature of the intrinsic manufacturing imperfections, the remaining perturbations present a random nature, and therefore, the IC-XT will have a stochastic evolution in the time and space domain [33, 45–50].

Figure 2 shows the temporal evolution of the IC-XT measured during 26 hours in a homogeneous 4-core MCF [Fibercore SM-4C1500(8.0/125)] between two adjacent cores in the linear and nonlinear regimes (power launch levels of 0 and 17 dBm, respectively). Although the bending radius and the twist rate present a constant value in the experimental set-up (see [49] for more details), the slight longitudinal and temporal local variations of both fiber parameters induce a longitudinal and temporal random evolution of the IC-XT in both power regimes. In addition, in the nonlinear regime, the Kerr effect is stimulated in the illuminated core 3 reducing the index-matching between the measured cores 1 and 3. In this scenario, the homogeneous cores 1 and 3 become heterogeneous when high power launch levels are injected in a

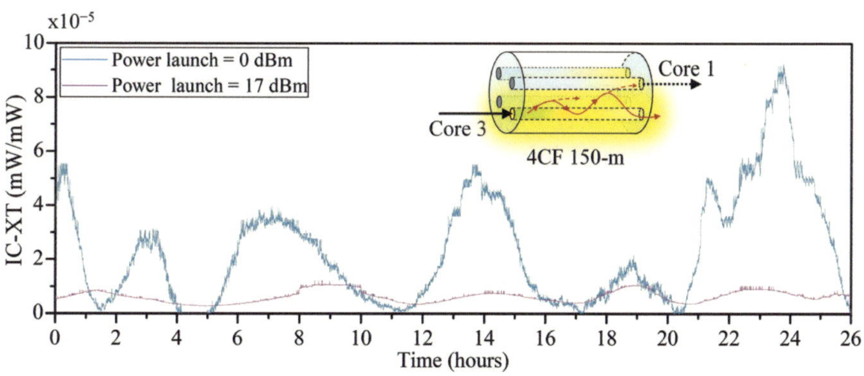

Figure 2. Measured temporal profile of the linear and nonlinear IC-XT between adjacent cores in a homogeneous 4-core MCF (results based on [49]).

given core. As a result, the MCF can be modeled in the nonlinear regime as an asymmetric optical coupler with random behavior, and consequently, the IC-XT mean and variance is reduced.

From these results, we conclude that the IC-XT has a random nature in both linear and nonlinear power regimes. Hence, at this point it is natural to ask how the probability distribution is. In a first investigation of this impairment, the answer can be easily found from the perturbation theory by assuming several initial simplifications: two cores a and b, a single polarization x and monochromatic electric fields. The last two approximations allow us to reduce the mathematical discussion of the IC-XT. In the next sections, these approximations will be revisited. Therefore, following a similar and simple approach as in the first works of the IC-XT [33, 45–50], the global electric field strength of the weakly guiding MCF (only two cores a and b and a single polarization x) is expressed as:

$$\begin{aligned} \mathcal{E}(\mathbf{r},t) &\approx \sum_{m=a,\,b} \mathcal{E}_m(\mathbf{r},t)\hat{x} \approx \sum_{m=a,\,b} \mathrm{Re}\left[E_{m,\omega_0}(\mathbf{r})\exp(j\omega_0 t)\right]\hat{x} \\ &= \sum_{m=a,\,b} \mathrm{Re}\left[A_m(z)F_m(x,y,\omega_0)\exp(-j\beta_m(\omega_0)z)\exp(j\omega_0 t)\right]\hat{x}, \end{aligned} \tag{2}$$

where A_m is the complex envelope[11] of the continuous wave in the core $m = a,b$ satisfying the SVEA; F_m is the transversal eigenfunction of the LP_{01} mode in the core m; and β_m is the phase constant at the angular frequency ω_0 of the optical carrier. Note that the eigenmodes are written in Eq. (2) assuming isolated cores. Nonetheless, as mentioned before, A_m should be assumed z-dependent to model the longitudinal variations induced by the IC-XT in the core modes. Moreover, it can be noted that the MCF perturbations have been omitted in our ansatz (first step of **Figure 1**). In these previous works [33, 45–50], the MCF perturbations will be inserted heuristically[12] in the fourth step with fortunate final results. Now, using Eq. (2) in the MMEs and following the steps detailed in **Figure 1** omitting the MCF perturbations, we obtain the coupled equations of the classical coupled-mode theory (CMT) of an asymmetric and nonlinear optical coupler:

$$j\frac{\mathrm{d}A_a(z)}{\mathrm{d}z} = k_{ab}\exp(-j\Delta\beta_{ba}z)A_b(z) + q_{1a}|A_a(z)|^2 A_a(z), \tag{3}$$

and a similar expression is found for $\mathrm{d}A_b(z)/\mathrm{d}z$. In Eq. (3), k_{ab} is the linear coupling coefficient, q_{1a} is the self-coupling nonlinear coefficient and $\Delta\beta_{ba}:= \beta_b(\omega_0)-\beta_a(\omega_0)$. In general, additional linear and nonlinear coupling coefficients appear in Eq. (3) [50]. However, considering that the core pitch is usually higher than four times de core radius in CC- and UC-MCFs, these

[11] Note that in Eq. (2) we have employed a different function for the complex envelope as in Eq. (1). We will use $\mathcal{A}(z,t)$ to describe the complex envelope in the non-monochromatic regime (optical pulses) and $A(z)$ in the monochromatic regime (continuous waves). Both functions are related as indicated in Section 3.3.

[12] This strategy is mathematically questionable. Note that the propagation constants are assumed invariant in the ansatz [Eq. (2)], but once we derive the coupled equations, we will assume that the MCF perturbations modify the propagation constants. Although the final estimation of the IC-XT is in line with the experimental results in [47-51], in Section 3.2 we will solve this mathematical inconsistence.

coupling coefficients can be neglected [50]. The predominant coupling coefficients describe the linear and nonlinear mode-coupling: k_{ab} models the linear mode overlapping between the transversal eigenfunctions F_a and F_b, and q_{1a} allows us to investigate the self-coupling effect in the nonlinear regime (analog to the self-phase modulation which can be observed when propagating optical pulses in a single-core fiber). At this point, the MCF perturbations can be described by modifying heuristically the exponential term in Eq. (3) as follows:

$$\exp\left(-j\Delta\beta_{ba}z\right) \rightarrow \exp(-j\Delta\phi_{ba}(z)) = \exp\left(-j\int_0^z \Delta\beta_{ba}^{(eq)}(\tau)\mathrm{d}\tau\right), \tag{4}$$

with $\Delta\beta_{ba}^{(eq)}(z) = \Delta\beta_{ba} + \Delta\beta_{ba}^{(B+S)}(z)$, where $\Delta\beta_{ba}^{(B+S)}$ describes the phase fluctuation induced by the MCF perturbations. As can be seen, the temporal fluctuations of the propagation constants are also omitted in Eq. (4) to simplify the first analysis of the IC-XT. The temporal fluctuations of the crosstalk will be discussed later. Hence, the coupled-mode equation in a real MCF is finally found as:

$$j\frac{\mathrm{d}A_a(z)}{\mathrm{d}z} = k_{ab}\exp(-j\Delta\phi_{ba}(z))A_b(z) + q_{1a}|A_a(z)|^2 A_a(z). \tag{5}$$

Remarkably, the revisited CMT constitutes a fundamental tool to estimate numerically the IC-XT in SM-MCFs using the Monte Carlo method [50]. The numerical calculation can also be performed in HA- and TA-MCFs by using Eq. (5) along with the corresponding closed-form expression of the linear coupling coefficient k_{ab} detailed in [51]. Furthermore, the revisited CMT allows us to derive the closed-form expressions of the IC-XT cumulative distribution function (cdf), probability density function (pdf), mean and variance in the linear and nonlinear regimes. Although the details of the mathematical discussion can be found in [33] for the linear regime and in [50] for the nonlinear regime, let us summarize the main results of these works.

The starting point is to consider a constant or quasi-constant bending and twisting conditions, i.e. their average value much higher than their longitudinal random fluctuations. In such a case, the phase-mismatching function of Eq. (4) can be expressed as [52]:

$$\Delta\phi_{ba}(z) \approx \Delta\beta_{ba}z - \frac{\beta_a d_{ab}}{2\pi f_T(z) R_B(z)}\sin\left(2\pi f_T(z)z\right), \tag{6}$$

where $R_B(z)$ and $f_T(z)$ are the bending radius and the twist rate along the MCF length, respectively[13]. The previous expression is the same as Eq. (2) of [52], but assuming to be null the offset of the twist angle of the core a at $z = 0$. It should be noted that the power exchanged between the cores a and b is maximized at the z-points where phase-mismatching function becomes null. These points are referred to as the phase-matching points (denoted as N_L and N_{NL} in the linear

---

[13] Eq. (6) is valid if and only if we can assume that $R_B >> \delta R_B$ and $f_T >> \delta f_T$ along the MCF length. In other case, Eq. (4) must be solved numerically using the refractive index model of [33].

and nonlinear regime, respectively). In general, in homogeneous SM-MCFs $N_{L(NL)} \neq 0$, but in the heterogeneous case, the phase-matching points can only be observed for a bending radius with an average value[14] R_B lower than the threshold R_{pk} (phase-matching region[15]) [33]:

$$R_B < R_{pk} = d_{ab}\beta_a / |\Delta\beta_{ba}|. \tag{7}$$

In the phase-matching region, we can use a first-order solution of Eq. (5) to perform the statistical analysis of the IC-XT [33, 50]. **Table 2** shows the analytical expressions of the linear and nonlinear IC-XT distribution and its statistical parameters derived from the CMT. As can be seen, the measured IC-XT pdf fits correctly to a chi-squared distribution with 4 degrees of freedom. In the linear regime, the mean, variance and N_L are constant with the optical power launch level (P_L). However, in the nonlinear regime, the Kerr effect detunes the phase constant of the core modes as P_L increases in the excited core 3 and, therefore, the homogeneous MCF becomes heterogeneous. As a result, the statistical IC-XT parameters are reduced with $P_L > 2$ dBm, the critical optical power in silica MCFs [49].

Furthermore, note that these statistical parameters can be estimated from the mean of the linear crosstalk $\mu_{L,ab}$. Hayashi, Koshiba and co-workers reported in [47, 48] the closed-form expressions to estimate $\mu_{L,ab}$ in different MCF designs with different bending twisting conditions. For small bending radius with $R_B < R_{pk}$, $\mu_{L,ab}$ can be estimated using Eq. (27) of [48], and for large bending radius with $R_B > R_{pk}$, $\mu_{L,ab}$ can be estimated from Eq. (21)[16] of [47]. In **Table 2** we also include the evolution of $\mu_{L,ab}$ with the average value of the bending radius in a heterogeneous SM-MCF [47]. In the phase-matching region, the mean of the linear IC-XT increases with R_B. However, in the phase-mismatching region, the mean of the linear IC-XT is reduced when R_B increases.

Finally, it should be noted that the statistical analysis previously described is only focused on the random longitudinal evolution of the IC-XT along the MCF considering a single polarization and temporal invariant conditions of the optical medium[17]. Hence, the following natural step is to consider temporal varying conditions of the dielectric medium and two polarizations per core.

3.2. Birefringent effects

Now, let us assume a 2-core SM-MCF operating in the monochromatic regime as in the previous section, but considering two polarizations per core and both longitudinal and time-varying random perturbations. These initial assumptions will allow us to predict the different

[14] In the following equations, we denote the average value of the bending radius and the twist rate without the usual brackets < > to simplify the mathematical expressions, that is, $R_B(z) \equiv R_B + \delta R_B(z)$ and $f_T(z) = f_T + \delta f_T(z)$.

[15] Note that $R_{pk} = \infty$ when considering homogeneous cores.

[16] In [48], Eq. (21) is given as a function of the correlation length of MCF structural fluctuations. The MCF structural fluctuations are all the medium perturbations except the macrobends. Microbends, fiber twisting or floor vibrations are specific examples of the MCF structural fluctuations.

[17] That is, the electrical permittivity is assumed to be temporally invariant.

http://dx.doi.org/10.5772/intechopen.72458

IC-XT	Equations	Figure/Comments
Probability density function (pdf)	$f_X(x) \approx \frac{(16L^2 q_{1a}^2 P_L^2 x^3 + 4x)}{N_{NL}^2 \lvert K_{ab}\rvert^4} \times \exp\left(-\frac{L^2 q_{1a}^2 P_L^2 x^3 + x}{N_{NL}\lvert K_{ab}\rvert^2/2}\right) u(x)$	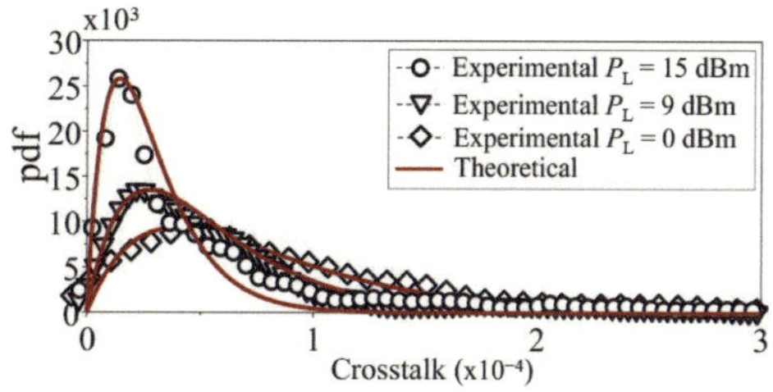
Mean	$\mu_{NL,ab} \approx \frac{\mu_{L,ab}}{1+bLP_L}$ $\mu_{L,ab} \approx \frac{2K_{ab}^2 R_B L}{\sqrt{\left(\frac{\beta_a d_{ab}}{R_B}\right)^2 - \Delta\beta_{ba}^2}}; R_B < R_{pk}$ $\mu_{L,ab} \approx \frac{2K_{ab}^2 l_c L}{1+(\Delta\beta_{ba} l_c)^2}; \quad R_B > R_{pk}$	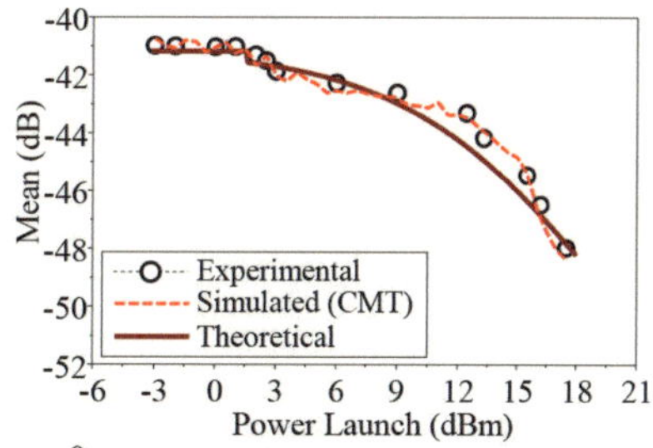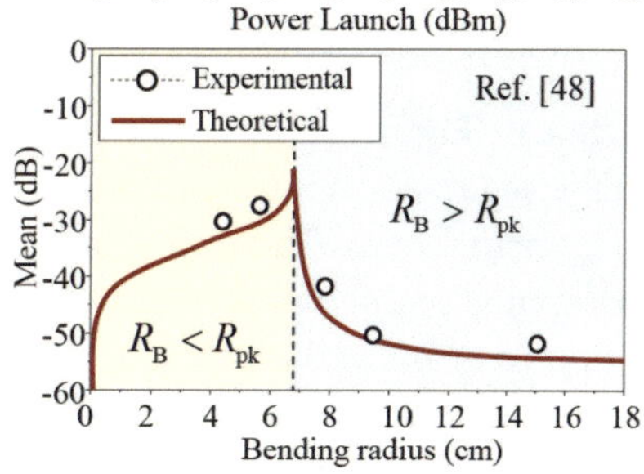
Variance	$\sigma_{NL,ab}^2 \approx \frac{\mu_{NL,ab}^2}{2} \equiv \frac{\sigma_{L,ab}^2}{(1+bLP_L)^2}$	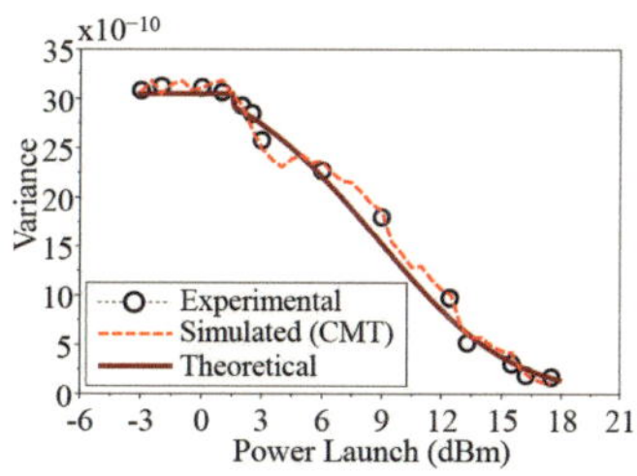
Phase-matching points	$N_{NL} \approx \frac{\mu_{NL,ab}}{\lvert K_{ab}\rvert^2} \equiv \frac{N_L}{1+bLP_L}$	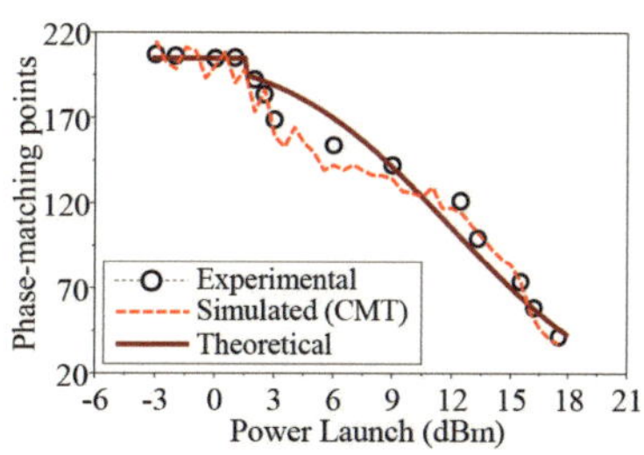

Table 2. Statistical distribution and parameters of the linear and nonlinear IC-XT. L is the MCF length, u is the Heaviside step function, P_L is the power launch level in the excited core b, K_{ab} is the discrete coupling coefficient calculated from Eq. (12) of [33], q_{1a} is the nonlinear coupling coefficient, b is a constant which depends on additional MCF parameters [50] [b = 0.5 in the Fibercore SM-4C1500(8.0/125)] and l_c is the correlation length of the autocorrelation function of the MCF structural fluctuations [47].

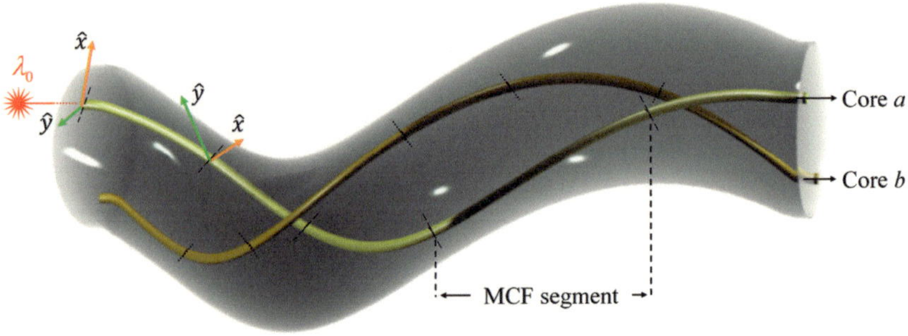

Figure 3. MCF comprising different birefringent segments in cores a and b with longitudinal and temporal varying fluctuations in the first-order electrical permittivity tensor.

crosstalk types between the polarized core modes (PCMs) in a SM-MCF: (i) the intra-core crosstalk (iC-XT) which describes the mode-coupling between orthogonal polarizations in a given core; (ii) the direct inter-core crosstalk (DIC-XT) modeling the mode coupling between the same polarization axis in different cores; and (iii) the cross inter-core crosstalk (XIC-XT) involving mode coupling between orthogonal polarizations in different cores.

As depicted in **Figure 3**, in a real MCF, each core $m = a,b$ can be modeled as a series of birefringent segments with a different time-varying retardation and random orientation of the local principal axes. Therefore, the first-order electrical susceptibility tensor $\chi_{ij}^{(1)}(\mathbf{r};t)$ will have both spatial and temporal dependence. As a result, in each segment of a given core m, the propagation constant of the polarized core modes (PCMs) LP_{01x} and LP_{01y} presents a different value due to the mentioned slight changes of $\chi_{ij}^{(1)}$, and therefore, the transversal function $F_{\underline{mi}}$ of each PCM "mi" ($i = x,y$) is also found to be spatial and temporal dependent.

In order to model theoretically this scenario, the concept of local mode is included in the perturbation theory. A local mode can be defined as an eigenfunction in a short core segment where the equivalent phase constant $\beta_{mi}^{(eq)}$ and the transversal function F_{mi} are approximately invariant. Hence, each core can be separated in different segments and local modes where the longitudinal and temporal birefringence conditions are approximately constant. In this way, in contrast with the previous section, now the MCF perturbations are considered from the ansatz inserted in the Maxwell equations. Specifically, the ansatz of the global electric field strength of the MCF structure is now written as [53]:

$$\begin{aligned}\mathcal{E}(\mathbf{r},t) &\approx \sum_{m=a,b}\sum_{i=x,y}\mathcal{E}_{mi}(\mathbf{r},t)\hat{u}_i \approx \sum_{m=a,b}\sum_{i=x,y}\mathrm{Re}\left[E_{mi,\omega_0}(\mathbf{r};t)\exp(j\omega_0 t)\right]\hat{u}_i \\ &= \sum_{m=a,b}\sum_{i=x,y}\mathrm{Re}\left[A_{mi}(z;t)F_{mi}(x,y,\omega_0;z,t)\exp(-j\Phi_{mi}(z,\omega_0;t))\exp(j\omega_0 t)\right]\hat{u}_i,\end{aligned} \tag{8}$$

where the semicolon symbol is used to separate explicitly longitudinal and temporal changes induced by the slowly varying MCF perturbations. Thus, note that the complex amplitude $E_{mi,\omega 0}$ is only a phasor with temporal changes much lower than the temporal oscillation of the optical carrier ($T_0 = 2\pi/\omega_0$). The MCF perturbations and the optical attenuation are described by the complex phase function Φ_{mi} defined as:

$$\Phi_{mi}(z,\omega_0;t):=\phi_{mi}(z,\omega_0;t)-j\frac{1}{2}\alpha(\omega_0)z=\int_0^z\beta_{mi}^{(\mathrm{eq})}(\delta,\omega_0;t)\mathrm{d}\delta-j\frac{1}{2}\alpha(\omega_0)z$$
$$=\beta_{mi}(\omega_0)z+\int_0^z\beta_{mi}^{(\mathrm{B+S})}(\delta,\omega_0;t)\mathrm{d}\delta-j\frac{1}{2}\alpha(\omega_0)z, \tag{9}$$

with α modeling the power attenuation coefficient of the MCF (assumed similar in each PCM), and the real phase function ϕ_{mi} involving the ideal phase constant of the PCM and the longitudinal and temporal MCF perturbations. Now, using Eq. (8) and performing the derivation of the perturbation theory as depicted in **Figure 1**, the following coupled local-mode equation is found [53]:

$$j\left(\frac{\partial}{\partial z}+\frac{\alpha}{2}\right)A_{ax}(z;t)=m_{ax,ay}(z;t)\exp\left(-j\Delta\phi_{ay,ax}(z;t)\right)A_{ay}(z;t)$$
$$+k_{ax,bx}(z;t)\exp(-j\Delta\phi_{bx,ax}(z;t))A_{bx}(z;t)$$
$$+\left(q_{ax}(z;t)|A_{ax}(z;t)|^2+g_{ax}(z;t)\left|A_{ay}(z;t)\right|^2\right)A_{ax}(z;t)$$
$$+\frac{1}{2}g_{ax}(z;t)\exp\left(-j2\Delta\phi_{ay,ax}(z;t)\right)A_{ax}^*(z;t)A_{ay}^2(z;t), \tag{10}$$

where $m_{ax,ay}$, $k_{ax,bx}$, q_{ax} and g_{ax} are the coupling coefficients defined in [53]; and the phase-mismatching functions are defined as $\Delta\phi_{ay,ax}:=\phi_{ay}-\phi_{ax}$, $\Delta\phi_{bx,ax}:=\phi_{bx}-\phi_{ax}$ and $\Delta\phi_{by,ax}:=\phi_{by}-\phi_{ax}$. From the above equation, the following considerations are in order:

- In contrast with the previous section, the longitudinal and temporal MCF perturbations are now modeled, not only by the phase-mismatching functions $\Delta\phi(z;t)$, but also by space- and time-varying coupling coefficients[18]. This coupled local-mode theory (CLMT) inherently incorporates these stochastic perturbations in both functions, as they were directly included in the Maxwell equations using Eq. (8).
- The CLMT is completed by three additional coupled local-mode equations for the *ay*, *bx* and *by* PCMs, which can be obtained just by exchanging the corresponding subindexes in Eq. (10). The herein presented theory is a general model which can be applied to SM-MCFs comprising: coupled or uncoupled cores, lowly or highly birefringent cores, trench-assisted, hole-assisted, and with gradual-index or step-index profile. In SM-CC-MCFs with a core pitch value (d_{ab}) lower than three times the maximum core radius (R_0 = max $\{R_{0a}, R_{0b}\}$), additional nonlinear terms modeling cross-coupling effects should be included in Eq. (10). However, if we assume a MCF with $d_{ab} >> 3R_0$, the self-coupling effect is the predominant nonlinear coupling effect and the additional nonlinear terms can be neglected [53].

[18]Note that the explicit dependence with ω_0 has been omitted in the phase-mismatching functions and in the coupling coefficients for the sake of simplicity.

- The monochromatic equivalent refractive index model (ERIM) reported in [53] must be used to calculate numerically the coupling coefficients and the phase-mismatching functions. Thanks to the CLMT and the ERIM, we will observe that the temporal birefringence fluctuation of each core modifies the average value of the iC-, DIC- and XIC-XT.

First, in order to analyze the longitudinal MCF random perturbations induced by MCF bending and twisting, a Monte Carlo simulation was performed using the CLMT along with the ERIM considering a 2-m SI-SM-HO-UC-LB-2CF with cores *a* and *b* comprising a single birefringent segment with the same birefringence average value of $<\Delta n_{aj}> = <\Delta n_{bj}> = 10^{-7}$ (see the specific fiber parameters in [53]). The temporal birefringence fluctuation of the 2CF was omitted in this simulation. The numerical results are shown in **Figure 4**, where we can observe the behavior of the mean of the iC-XT *ay-ax* DIC-XT *bx-ax* and XIC-XT *by-ax* when changing the bending radius R_B and fiber twisting conditions f_T.

As it can be noticed from **Figure 4(a)**, we cannot observe intra-core mode-coupling between *ax-ay* with $f_T = 0$ in 2 m of the MCF. Macrobending increases the phase-mismatching between the PCMs *ax* and *ay* without inducing iC-XT due to the photo-elastic effect [54]. As a result, significant XIC-XT cannot be observed for short MCF lengths when $f_T = 0$, as in the case of **Figure 4(c)**. Nevertheless, an average level of DIC-XT between −100 and −50 dB can be noted from **Figure 4(b)** in non-twisting conditions depending on the R_B value. In addition, the higher the twist rate and the bending radius, the higher the iC-, DIC- and XIC-XT mean due to the reduction of the phase-mismatching between the different PCMs of the 2CF. Note that DIC- and XIC-XT means are balanced when the iC-XT mean achieves the value of 0 dB in **Figure 4(a)**. Therefore, MCF twisting can be proposed as a potential strategy for birefringence management to balance the inter-core crosstalk between the different PCMs for short MCF distances. For MCF distances of several kilometers, the iC-XT mean will be increased and the difference between the mean of the DIC- and XIC-XT will be reduced.

In addition, experimental measurements were performed on a 4CF [Fibercore SM-4C1500 (8.0/125)] analyzing the temporal birefringence of the optical media and its impact on the mean of the crosstalk between the PCMs of the cores 1 and 3. **Figure 5** shows the temporal fluctuation of the linear birefringence and the crosstalk mean behavior between the PCMs of cores 1

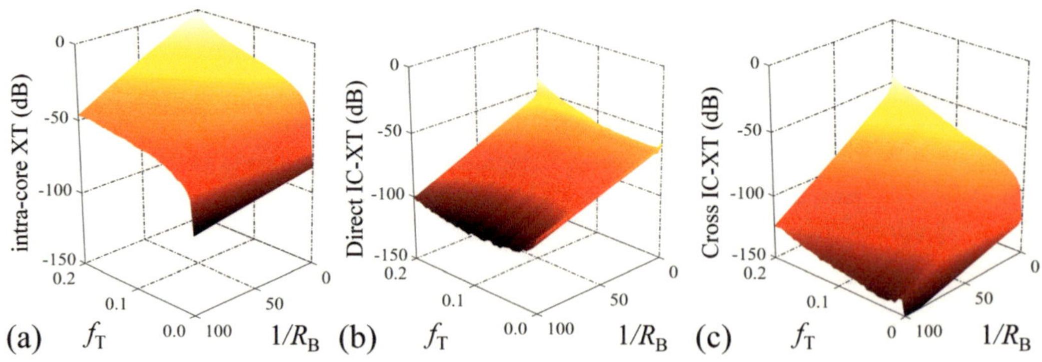

Figure 4. Numerical simulation of the crosstalk between PCMs varying the bending radius and the twist rate in a 2-m SI-SM-HO-UC-LB-2CF: (a) iC-XT mean *ay-ax*, (b) DIC-XT mean *bx-ax*, and (c) XIC-XT mean *by-ax*.

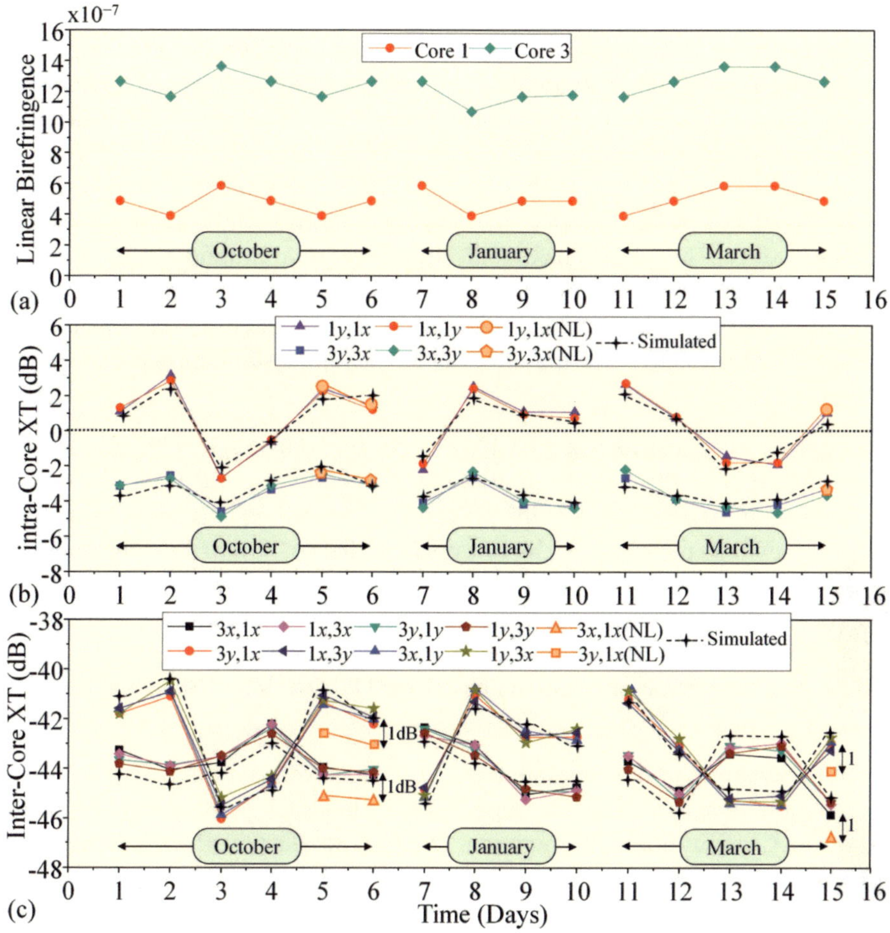

Figure 5. Experimental results of the temporal linear birefringence fluctuation over different days and months of a 150-m 4CF, and corresponding intra- and inter-core crosstalk mean between cores 1 and 3 (NL: nonlinear regime). (a) Linear birefringence of the cores 1 and 3, (b) iC-XT and (c) DIC- and XIC-XT. Results based on [53].

and 3 measured in different days and months[19]. As shown in **Figure 5(a)**, cores 1 and 3 present a different average value of the linear birefringence estimated to be $<\Delta n_{1,j}> = 4.9{\cdot}10^{-7}$ and $<\Delta n_{3,j}> = 1.2{\cdot}10^{-6}$, respectively ($j = 1,\ldots,15$). It can be noted that the average value of the linear birefringence is found to be constant in each core the three different measured months. Moreover, although the average value of the linear birefringence is different in each core, the temporal evolution of the linear birefringence presents a similar shape in both cores. As can be seen from **Figure 5(a)** and **(b)**, the higher the linear birefringence in a given core, the lower the mean of the iC-XT. Furthermore, it should be noted that iC-XT in core 3 is lower than in core 1 due to a higher index-mismatching between the orthogonal polarizations. Remarkably,

[19]As the linear birefringence of each core remains unchanged during more than 10 hours in the laboratory room, we analyzed the temporal birefringence fluctuation in different days and months, with similar temperature conditions in the laboratory.

the iC-XT mean presents a lower temporal fluctuation in the more birefringent core (core 3), which occurs when the average value of the birefringence is higher than the temporal random birefringence fluctuation (~10^{-7}). In addition, we can observe from **Figure 5(c)** that the temporal evolution of the XIC-XT mean presents the same behavior as the iC-XT mean, indicating that XIC-XT depends directly on the iC-XT of both cores. As a result, DIC-XT is higher than XIC-XT when iC-XT is lower than 0 dB in both cores. The nonlinear crosstalk between de PCMs $1y$-$1x$, $3y$-$3x$, $3x$-$1x$ and $3y$-$1x$ was also measured the 5th, 6th and 15th days with a power launch level of 6 dBm. The DIC- and XIC-XT mean is reduced around 1 dB keeping constant the difference between both inter-core crosstalk types as a direct consequence of the constant behavior of the iC-XT mean in nonlinear regime. Finally, note that the experimental measurements of **Figure 5** fit correctly with the CLMT when using the simulation parameters detailed in [53].

Additional numerical calculations of the CLMT can be found in [53] involving both LB and HB cores. Interesting, it is worth mentioning that the CLMT and the ERIM can be used to design HB-MCFs with random orientation of the principal axes between adjacent cores to reduce the mode-coupling between their PCMs. The concept is similar to the crosstalk behavior which can be found in disordered MCFs exhibiting transverse Anderson localization [21]. Along this line, a TA- and HA-cladding can also be considered in these fibers to obtain low DIC- and XIC-XT levels. In all these scenarios, the CLMT can be used in the design work, with a lower computational time than numerical simulations based on FDTD (Finite-Difference Time Domain) calculations.

The temporal fluctuation of the crosstalk has also been investigated in [55], but considering a single polarization per core and inserting heuristically the MCF perturbations in the exponential terms of the CMT, in line with the initial crosstalk works [33, 45–50]. Specifically, in ref. [55], the crosstalk transfer function has been discussed at the MCF output considering small modulated signals, i.e. non-monochromatic electric fields. However, the comprehension of the MCF propagation and the IC-XT in the non-monochromatic regime is not as straightforward as initially foreseen. Hence, at this point, let us discuss the non-monochromatic regime with a similar rigorous formalism as in [53] for the monochromatic case.

3.3. Intermodal dispersion and higher-order coupling and nonlinear effects

The theoretical study of the non-monochromatic regime will allow us to describe the propagation of optical pulses through a MCF. Focusing our efforts on SM-MCFs, additional physical impairments should be included in Eq. (10), such as the group-velocity dispersion (GVD), polarization-mode dispersion (PMD), intermodal dispersion and additional nonlinear effects. Moreover, if we also consider the propagation of ultra-short optical pulses in the femtosecond regime, the analysis of higher-order coupling and nonlinear effects should also be incorporated to the coupled equations.

Although in the picosecond regime MCF propagation models have been proposed in [56, 57] including polarization effects and the random longitudinal fiber perturbations (but omitting the temporal fluctuations), in the femtosecond regime, existing MCF propagation models exclude polarization effects and omit both temporal and longitudinal random perturbations of the fiber [58–60]. In order to include these realistic fiber conditions in the mathematical description of the propagation of femtosecond optical pulses through a MCF, a theoretical model is proposed in [61] based on the concept of local modes. As can be seen later, the intermodal dispersion induced by the MCF random perturbations can become one of the major physical impairment in the single-mode regime of the fiber. Specifically, the intermodal dispersion, also referred to as the mode-coupling dispersion (MCD) in this work, is induced in the femtosecond regime not only by the mismatching between the propagation constants of the PCMs, but also by the frequency dependence of their mode overlapping.

Our initial goal is to revisit the CLMT of the previous section but now assuming non-monochromatic fields. In such a case, the ansatz of the global electric field strength of a SM-MCF should be written as:

$$\mathcal{E}(\boldsymbol{r},t) \approx \sum_{m=a,b}\sum_{i=x,y} \mathcal{E}_{mi}(\boldsymbol{r},t)\hat{u}_i \approx \sum_{m=a,b}\sum_{i=x,y} \mathrm{Re}\left[E_{mi,\omega_0}(\boldsymbol{r},t)\exp(j\omega_0 t)\right]\hat{u}_i, \tag{11}$$

where the complex amplitude $E_{mi,\omega 0}$ is now a bandpass signal with the slowly varying temporal changes of the electric field strength, given by the expression:

$$\begin{aligned} E_{mi,\omega_0}(\boldsymbol{r},t) = \frac{1}{2\pi}\int \tilde{A}_{mi}(z,\omega-\omega_0;t)F_{mi}(x,y,\omega;z,t)\times \\ \times \exp[-j\Phi_{mi}(z,\omega;t)]\exp[j(\omega-\omega_0)t]\mathrm{d}\omega. \end{aligned} \tag{12}$$

The functions involved in the previous equations are the same as in Eq. (8), but now expressed in the frequency domain. Nevertheless, a fundamental remark of the complex envelope should be taken into account at this point. As previously discussed in **Figure 1**, the slowly varying longitudinal changes should also be decoupled from the rapidly varying longitudinal fluctuations via the complex envelope. However, in Eq. (12) the rapidly and slowly varying longitudinal changes are coupled in the first exponential term via the function $\phi_{mi}(z,\omega;t)$. Therefore, in order to decouple them and model the analytic function of the optical pulses, the complex envelope should be rewritten as:

$$\tilde{\mathcal{A}}_{mi}(z,\omega-\omega_0;t) = \tilde{A}_{mi}(z,\omega-\omega_0;t)\exp[-j(\phi_{mi}(z,\omega;t)-\phi_{mi}(z,\omega_0;t))]. \tag{13}$$

Once we have written our ansatz of the global electric field strength, the following step is to propose the wave equation of the PCMs (second step) and the wave equation of the MCF (third step). In particular, in the third step, we will able to incorporate the higher-order nonlinear effects via the constitutive relation between the global electric field strength and the nonlinear polarization. Note that in the femtosecond regime, the aforementioned constitutive relation should include the delay response of the electronic and nuclei structure of silica atoms [62]. For optical frequencies well below the electronic transitions, the electronic contribution to the nonlinear polarization can be considered instantaneous. However, since nucleons (protons and neutrons) are considerably heavier than electrons, the nuclei motions have resonant frequencies much lower than the electronic resonances and, consequently, they should be retained in the constitutive relation as indicated in Eq. (S36) of [61]. Specifically, Raman scattering is a well-known effect arising from the nuclear contribution to the nonlinear polarization. All in all, the coupled local-mode equations can be derived to describe the propagation of ultra-short pulses in SM-MCFs. In particular, the coupled local-mode equation modeling the propagation of the PCM *ax* is found to be [61]:

$$
\begin{aligned}
j\left(\frac{\partial}{\partial z}+\widehat{\mathrm{D}}_{ax}^{(\mathrm{eq})}+\frac{1}{2}\widehat{\alpha}\right)\mathcal{A}_{ax}(z,t) &= \widehat{\mathrm{M}}_{ax,ay}^{(\mathrm{eq})}\mathcal{A}_{ay}(z,t)+\sum_{m=b}^{N}\widehat{\mathrm{K}}_{ax,mx}^{(\mathrm{eq})}\mathcal{A}_{mx}(z,t)\\
&+\widehat{\mathrm{q}}_{ax}^{(\mathrm{I})}\left(|\mathcal{A}_{ax}(z,t)|^2\mathcal{A}_{ax}(z,t)\right)+\frac{2}{3}\widehat{\mathrm{g}}_{ax,ay}^{(\mathrm{I})}\left(|\mathcal{A}_{ay}(z,t)|^2\mathcal{A}_{ax}(z,t)\right)\\
&+\frac{1}{3}\exp\left(-j2\Delta\phi_{ay,ax}^{(0)}(z;t)\right)\widehat{\mathrm{g}}_{ax,ay}^{(\mathrm{I})}\left(\mathcal{A}_{ax}^{*}(z,t)\mathcal{A}_{ay}^{2}(z,t)\right)\\
&+\widehat{\mathrm{q}}_{ax}^{(\mathrm{R})}\left[\left(f(t)*|\mathcal{A}_{ax}(z,t)|^2\right)\mathcal{A}_{ax}(z,t)\right]+\widehat{\mathrm{g}}_{ax,ay}^{(\mathrm{R})}\left[\left(h(t)*|\mathcal{A}_{ay}(z,t)|^2\right)\mathcal{A}_{ax}(z,t)\right]\\
&+\frac{1}{2}\widehat{\mathrm{g}}_{ax,ay}^{(\mathrm{R})}\left\{\left[u(t)*\left(\mathcal{A}_{ax}(z,t)\mathcal{A}_{ay}^{*}(z,t)\right)\right]\mathcal{A}_{ay}(z,t)\right\}\\
&+\frac{1}{2}\exp\left(-j2\Delta\phi_{ay,ax}^{(0)}(z;t)\right)\widehat{\mathrm{g}}_{ax,ay}^{(\mathrm{R})}\left\{\left[u(t)*\left(\mathcal{A}_{ax}^{*}(z,t)\mathcal{A}_{ay}(z,t)\right)\right]\mathcal{A}_{ay}(z,t)\right\}.
\end{aligned}
\tag{14}
$$

where $\widehat{\mathrm{D}}_{ax}^{(\mathrm{eq})}$ is the equivalent dispersion operator in the PCM *ax* including the frequency dependence of the MCF perturbations in the time domain; $\widehat{\alpha}$ is the attenuation operator; the h and u functions describe the isotropic and anisotropic Raman response, respectively; the f function is $f:= h + u$; the phase-mismatching term $\Delta\phi_{ay,ax}^{(0)}(z;t):=\phi_{ay}(z,\omega_0;t)-\phi_{ax}(z,\omega_0;t)$ describes the phase-mismatching between the PCMs *ax* and *ay* at ω_0; $\widehat{\mathrm{M}}_{ax,ay}^{(\mathrm{eq})}$ and $\widehat{\mathrm{K}}_{ax,mx}^{(\mathrm{eq})}$ are, respectively, the equivalent intra- and inter-core mode-coupling dispersion operators between the PCMs *ax-ay* and *ax-mx*; $\widehat{\mathrm{q}}_{ax}^{(\mathrm{I})}$ and $\widehat{\mathrm{g}}_{ax,ay}^{(\mathrm{I})}$ are the nonlinear mode-coupling dispersion operators associated with the instantaneous response of the nonlinear polarization and accounting for the nonlinear mode overlapping between the PCMs *ax-ax* and *ax-ay*; and $\widehat{\mathrm{q}}_{ax}^{(\mathrm{R})}$ and $\widehat{\mathrm{g}}_{ax,ay}^{(\mathrm{R})}$ are analogous to $\widehat{\mathrm{q}}_{ax}^{(\mathrm{I})}$ and $\widehat{\mathrm{g}}_{ax,ay}^{(\mathrm{I})}$, but associated with the nonlinear polarization induced by the delay response of the nuclei motion of silica atoms (Raman effect). A comprehensive description of the main parameters of the model can be found in [61].

It should be remarked that the linear operators of Eq. (14) are found to be longitudinal and temporal dependent, instead of constant coupling coefficients and unperturbed propagation constants. Thanks to these linear operators, Eq. (14) is able to describe accurately the linear and nonlinear propagation of each PCM and the linear and nonlinear MCD including the longitudinal and temporal MCF perturbations. Furthermore, it is worthy to note that the MCD is induced in each birefringent segment by two different dispersive effects when propagating femtosecond optical pulse through a MCF: (i) the frequency dependence of the local mismatching between the phase functions $\phi_{mi}(z,\omega;t)$ of the PCMs, referred to as the phase-mismatching dispersion (PhMD); and (ii) the frequency dependence of the mode overlapping between the PCMs, modeled by the coupling coefficients and referred to as the coupling coefficient dispersion (CCD). As an example, the PhMD between the PCMs *ax* and *mx* is given by the phase-mismatching $\Delta\phi_{mx,ax}(z,\omega;t)$ and the CCD by the coupling coefficients

$\tilde{k}_{ax,mx}(z,\omega;t)$ and $\tilde{k}_{mx,ax}(z,\omega;t)$, both dispersive effects modeled by the operators $\widehat{D}_{ax}^{(eq)}$, $\widehat{D}_{mx}^{(eq)}$, $\widehat{K}_{ax,mx}^{(eq)}$ and $\widehat{K}_{mx,ax}^{(eq)}$. Note that the equivalent dispersion operators $\widehat{D}_{ax}^{(eq)}$ and $\widehat{D}_{mx}^{(eq)}$ describe not only the linear propagation of the PCMs *ax* and *mx*, but also the exact phase-mismatching $\Delta\phi_{mx,ax}(z,\omega;t)$ at each angular frequency ω at a given *z* point. The MCD can be observed in a SM-MCF between the PCMs of different cores (inter-core MCD) and between the PCMs of a single core (intra-core MCD). Note that the intra-core MCD is the well-known linear and nonlinear polarization-mode dispersion (PMD). Hence, we will discuss in this subsection the inter-core MCD (IMCD) involving the mode-coupling between the PCMs of different cores.

Although the proposed model allows us to investigate a wide range of propagation phenomena in MCFs, our efforts are mainly focused on a deeper understanding of the IMCD induced by the fiber perturbations. In order to clarify the impact of the MCF birefringence on this physical impairment when propagating femtosecond optical pulses, Eq. (16) is solved numerically in the linear and nonlinear regime of the fiber. In all the analyzed cases, we considered a MCF comprising a fiber length of $L = 40$ m and two cores *a* and *b* distributed in a square lattice as in the Fibercore SM 4C1500(8.0/125) but with a core-to-core distance d_{ab} = 26 μm. The wavelength of the optical carrier λ_0 was selected to be in the third transmission window with λ_0 = 1550 nm. The time variable was normalized using the group delay of the PCM *ax* as a reference with $t_N = (t - \beta_{ax}^{(1)}z)/T_P$, where T_P is defined in this work as the full width at 1/2*e* (~18%) of the peak power.

As a first simple example, we considered an ideal homogeneous MCF, with $R_B = \infty$ and $f_T = 0$ turns/m. **Figure 6(a)** shows the simulation results of the CLMT when a 350-fs Gaussian optical pulse is launched into the PCM *ax* at *z* = 0. In this example, the GVD and the PMD (induced by the intrinsic random fiber birefringence) were omitted to isolate the effects of the first-order

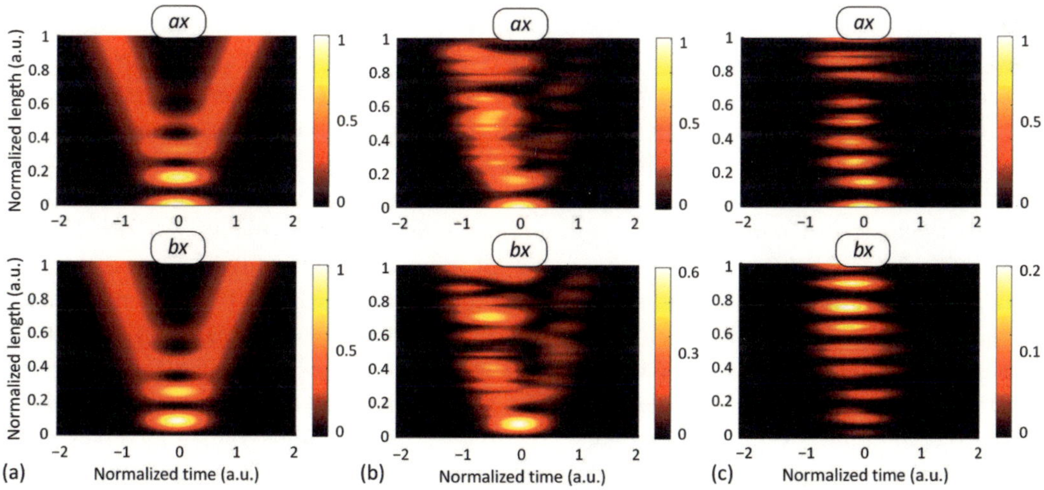

Figure 6. IMCD impact on Gaussian pulses and optical solitons propagating through a SM-MCF. (a) 350-fs Gaussian optical pulse propagation under ideal conditions. (b) 250-fs Gaussian optical pulse propagation with random bending conditions. (c) 600-fs fundamental bright soliton with random bending and twisting conditions. The numerical results for the PCMs *ay* and *by* can be found in [61].

IMCD. In this way, the pulse is only propagated by the PCMs *ax* and *bx*. Remarkably, the pulse splitting predicted by Chiang et al. in [58] appears induced by the first-order CCD: each spectral component of the pulse presents a different coupling length as a direct consequence of the linear frequency dependence of the power confinement ratio of the LP01 mode in each core. As a result, the pulse propagation can be modeled in this case by two linear and time-invariant systems with the impulse response proportional to the Dirac delta functions $\delta(t \pm Kz)$, where K is the first-order frequency derivative of the coupling coefficient between the PCMs *ax* and *bx*.

Another interesting effect of the first-order IMCD is related to the random birefringence that arises from a randomly varying fiber bending radius. In this case, the effect of the first-order PhMD along with the CCD can also be observed when considering a high number of MCF birefringent segments where the bending radius fluctuates with a Normal distribution between adjacent segments. We simulate the MCF of the first example considering a 250-fs Gaussian optical pulse and 50 birefringent segments with a bending radius Normal distribution $R_B = N(\mu = 100, \sigma^2 = 40)$ cm. The numerical results are shown in **Figure 6(b)**. It can be seen that the group delay and the pulse splitting present a random evolution in each core due to the stochastic nature of the MCF perturbations inducing a random differential group delay between the PCMs *ax* and *bx*. In particular, this result can be employed for pulse shaping and dispersion engineering applications.

In the third example, the IMCD effects are also investigated in the nonlinear fiber regime along with the PMD (intra-core MCD). Specifically, the impact of such perturbations on a bright soliton is analyzed. A 600-fs fundamental soliton (~350 fs full width at half maximum) was launched into the PCM *ax* of a dispersion-shifted homogeneous 2-core MCF with usual GVD parameters of $\beta^{(2)} = -1$ ps^2/km and $\beta^{(3)} = 0.1$ ps^3/km. The peak power (P_0) required to support the fundamental soliton is found to be around ~40 dBm considering a nonlinear refractive index of $n_{NL} = 2.6 \cdot 10^{-20}$ m^2/W at 1550 nm. In order to simulate realistic MCF conditions, we assume $\Delta\beta_{bx,ax}{}^{(1)} = 0.2$ ps/km and $\Delta\beta_{bx,ax}{}^{(2)} = 0.1$ ps^2/km induced by manufacturing imperfections (similar values for the *y*-polarization). In this case, we also include the intrinsic linear birefringence of the medium along with the linear and circular birefringence induced by the fiber bending and twisting conditions. We consider 50 birefringent segments along the MCF length, where the linear and circular birefringence fluctuate between adjacent segments. The circular birefringence is induced by a random twist rate f_T given by the Normal distribution $f_T = N(\mu = 0.1, \sigma^2 = 0.01)$ turns/m. The linear birefringence is induced by: (i) the random bending conditions with $R_B = N(\mu = 100, \sigma^2 = 40)$ cm; and (ii) the intrinsic linear birefringence of each core, fixed to $2 \cdot 10^{-7}$ in both cores *a* and *b*. According to **Figure 6(c)**, we can observe that the soliton condition is broken along the MCF propagation. The second-order PhMD becomes the main physical impairment when $\Delta\beta_{bx,ax}{}^{(2)} \neq 0$ in dispersion-shifted coupled-core MCFs, with a reduced second-order GVD coefficient and core-to-core distance. Therefore, in the first propagation meters, the additional chirp induced by the second-order PhMD along with the first-order CCD increases the pulse width and reduces the peak power. As a result of the peak power reduction, the pulse width is increased along the MCF length and the soliton peak is shifted from its original position due to the first-order PhMD and the third-order GVD. In this case, note that the effects of the Raman-induced frequency shift (RIFS) [63] and the self-steepening are difficult to observe with $T_P = 600$ fs, $L = 40$ m, $\beta^{(2)} = -1$ ps^2/km, and $P_0 \approx 40$ dBm. Nevertheless, in optical pulses of few femtoseconds and in MCFs with a higher second-order GVD coefficient, the soliton distortion will be increased not only by the IMCD and the third-order GVD, but also by the RIFS and the self-steepening nonlinear effects.

Although we have only discussed the main effects of the longitudinal birefringence of the MCF, the analysis of the temporal perturbations of the medium can be found in [61]. It should be noted that the IMCD can also fluctuate in time due to the temporal fluctuation of the MCF birefringence modifying the value of the phase functions $\phi_{mi}(z,\omega;t)$ for the PCM mi. Therefore, the random group delay induced by the first-order PhMD in each MCF segment may present a time-varying evolution.

For completeness, we investigate the fiber length scales over which the dispersive effects of the IMCD should be considered in the pulse propagation phenomena when comparing this physical impairment with the GVD. To this end, we compare the GVD, CCD and PhMD lengths considering a MCF without random perturbations, given by the expressions for the PCMs ax and bx [61]:

$$L_{\mathrm{GVD}}{:=}T_{\mathrm{P}}^2/\left|\beta_{ax}^{(2)}\right|;\quad L_{\mathrm{CCD}}{:=}T_{\mathrm{P}}/2\left|\tilde{k}_{ax,bx}^{(1)}\right|;\quad L_{\mathrm{PhMD}}{:=}T_{\mathrm{P}}^2/\left|\Delta\beta_{bx,ax}^{(2)}\right|. \tag{15}$$

Figure 7 depicts the comparison of the GVD, CCD and PhMD dispersion lengths. As can be seen, the IMCD induced by the CCD becomes the predominant impairment in dispersion-shifted coupled-core MCFs with a reduced core-to-core distance and $\Delta\beta_{bx,ax}{}^{(2)} = 0$. On the other hand, the GVD is expected to become the major physical impairment in homogeneous uncoupled-core MCFs, with a core-to-core distance d_{ab} higher than 30 μm and $\Delta\beta_{bx,ax}{}^{(2)} \approx 0$, or in heterogeneous MCFs with inter-core crosstalk levels lower than −30 dB. Nevertheless, the GVD along with the IMCD induced by the second-order PhMD will be the predominant physical impairments in homogeneous coupled-core MCFs with $\Delta\beta_{bx,ax}{}^{(2)} \neq 0$.

Finally, it should be noted that the extension of Eq. (14) to the multi-mode regime is straightforward when including additional LP mode groups in the complex amplitude of the global electric field strength $E_{i,\omega 0}$. Inserting $E_{i,\omega 0}$ in the Maxwell equations, the CLMT can be extended

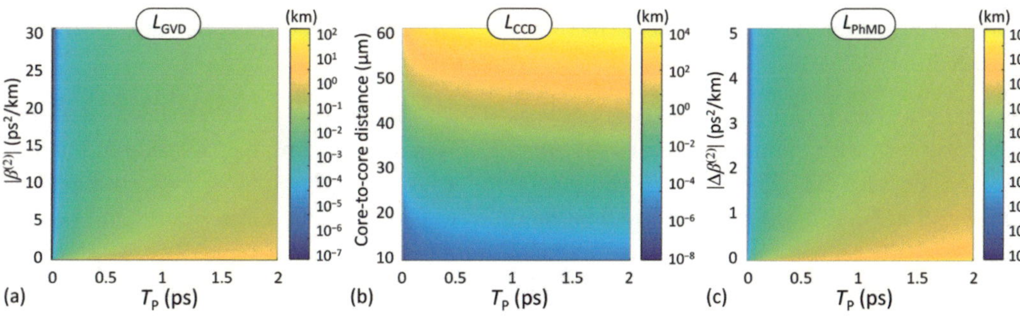

Figure 7. Comparison of the dispersion lengths. (a) Group-velocity dispersion (GVD) length, (b) coupling coefficient dispersion (CCD) length, and (c) phase-mismatching dispersion (PhMD) length.

to the multi-mode regime performing a similar mathematical discussion as in [61] in the single-mode regime.

4. Current and emerging applications

Once we have reviewed the fundamental aspects of the linear and nonlinear propagation in MCF media, we will discuss in this section the main applications and opportunities of the MCF technology in photonics and diverse branches of sciences.

4.1. Backbone and access optical networks using multi-core fibers

SDM systems using MCFs have been extensively investigated in recent years targeting to overcome the exponential growth of data traffic in the backbone and in the access network [4–7].

The first laboratory MCF transmission was demonstrated in May 2010 [64]. Zhu and co-workers used a SI-SM-HO-UC-LB-7CF with a hexagonal lattice. A novel network configuration was proposed for passive optical network (PON) based on a bidirectional parallel transmission at 1310 nm and 1490 nm and using a tapered MCF connector (TMC) for injecting and extracting the optical signals in the MCF.

A set of MCF experiments were reported since 2011. Scaling in capacity demonstrations, [65–67] should be mentioned. In [65] the authors demonstrated a 210 Tb/s self-homodyne transmission system using distributed feedback (DFB) lasers and a 19-core TA-SM-HO-UC--LB-MCF. Sakaguchi et al. reported in [66] a record capacity of 305 Tb/s over 10.1 km using the same MCF as in [65], with an IC-XT mean of −32 dB between adjacent cores at 1550 nm. The authors also fabricated a 19-channel SDM multiplexer/demultiplexer using free-space optics with low insertion losses and low additional crosstalk. As another interesting example, Takara et al. reported in [67] 1.01 Pb/s transmission over 52 km with the highest aggregate spectral efficiency of 91.4 b/s/Hz by using a one-ring-structured 12-core TA-SM-HO-UC-LB-MCF. They generated 222-channel WDM signals of 456-Gb/s PDM-32QAM-SC-FDM signals[20] with 50-GHz spacing in the C and L bands. Following significant efforts on the design and fabrication of MCFs, demonstrations of SDM transmissions using MCF media for long-haul applications have shown impressive progress in terms of capacity, reach, and spectral efficiency, as detailed in **Table 3**.

On the other hand, cloud radio-access network (C-RAN) systems should also deal with this huge future capacity demand in the next-generation wireless systems, e.g. 5G cellular technology and Beyond-5G [75–77]. According to some telecom equipment manufacturers, it is expected that 5G cellular networks will be required to provide 1000 times higher mobile data traffic in 2025 as compared with 2013, including flexibility and adaptability solutions to maximize the energy efficiency of the network [78, 79]. A new radio-access model supporting

[20] PDM: Polarization-Division Multiplexing, QAM: Quadrature Amplitude Modulation, SC: Single Carrier, FDM: Frequency-Division Multiplexing.

Year	Ref.	Fiber type	Cores × modes	Distance (km)	Channel rate (Gb/s)	WDM channels per core	S/E (b/s/Hz)	Total capacity (Tb/s)
2011	[68]	SM-MCF	7 × 1	2688	128	10	15	7
2012	[67]	SM-MCF	12 × 1	52	456	222	91.40	1012
2012	[66]	SM-MCF	19 × 1	10.1	172	100	30.50	305
2013	[65]	SM-MCF	19 × 1	10.1	100	125	33.60	210
2014	[7]	FM-MCF	7 × 2	1	4000	50	102	200
2015	[69]	SM-MCF	7 × 1	2520	100	73	16	51
2015	[70]	FM-MCF	36 × 2	5.5	107	40	108	432
2015	[71]	FM-MCF	12 × 2	527	80	20	90.28	45
2015	[72]	FM-MCF	19 × 4	9.8	40	8	345	29
2016	[73]	FM-MCF	19 × 4	9.8	60	360	456	2050
2017	[74]	SM-MCF	32 × 1	205.6	768	46	217.6	1001

Table 3. Summary of progress in MCF transmissions in recent years. The MCF type indicates only the modal regime (additional characteristics of the MCF involving the index profile, the spatial homogeneity, the core pitch and the birefringence can be found in the corresponding reference). The number of modes indicate the number of LP mode groups supported by the MCF transmission. The channel rate includes PDM and the overhead for forward-error-correction (FEC). The spectral efficiency and total capacity exclude the FEC overhead.

massive data uploading will be required considering additional transport facilities provided by the physical layer [78–80].

Fronthaul connectivity performed by radio-over-fiber (RoF) transmission using single-input single-output (SISO), multiple-input multiple-output (MIMO) configuration [81], sub-Nyquist sampling [82], and ultra-wideband signals exceeding 400 MHz bandwidth has been proposed for the 5G cellular generation [76, 77, 83]. The required channel capacity is further extended in the case of Beyond-5G systems, where a massive number of antennas operating in MIMO configuration, should be connected using RoF. To overcome the massive increment in the data capacity demand, MCF has been recently proposed as a suitable medium for LTE-Advanced (LTE-A) MIMO fronthaul systems [52, 83, 84].

MCFs open up attractive possibilities in RoF systems as different wireless signals can be transmitted simultaneously over the same optical wavelengths and electrical frequencies in different cores of the optical waveguide to provide multi-wireless service using a single laser at the transmitter. Thus, MCF can also be proposed as an alternative to the classical SM-SCF [also termed in the literature as the standard single-mode fiber (SSMF)] providing fronthaul connectivity using multiple wavelength channels with multiple lasers. Additionally, MCFs with high core density are suitable for connecting large phase array antennas performing multi-user MIMO (MU-MIMO) processing [85]. Furthermore, network operators can offer a dynamic and scalable capacity in the next cellular generation due to the aggregated channel capacity provided by the MCF technology [86]. Moreover, the possibility of combining MCF-RoF transmissions with additional multiplexing techniques such as time-division multiplexing (TDM), WDM, PDM and mode-division multiplexing (MDM) [12] should be considered.

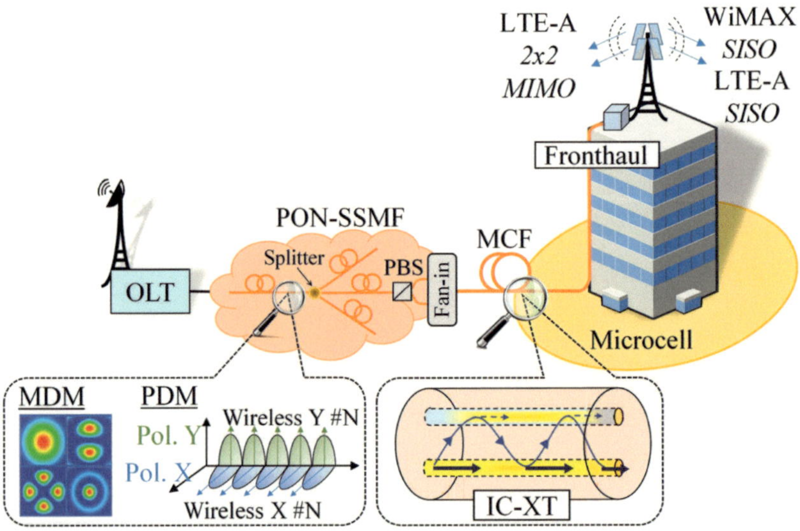

Figure 8. Next-generation optical fronthaul system using MCF medium operating with a converged fiber-wireless PON including optical polarization-division multiplexing (PDM) and mode-division multiplexing (MDM) transmissions.

Figure 8 depicts the proposed fronthaul provision applied to converged fiber-wireless PON including PDM to provide connectivity between the SSMF and MCF media.

Remarkably, the use of MCFs in the next-generation RoF fronthaul systems is proposed for the first time in [52, 87]. In these works, it is investigated the performance of fully standard LTE-A signals in MIMO and SISO configurations with the random IC-XT fluctuations and the demonstration of fronthaul provision of both LTE-A and WiMAX signals using a 150-m SI-SM-HO-UC-LB-4CF. In order to reduce the random fluctuations of the error vector magnitude (EVM) induced by the IC-XT, the core interleaving nonlinear stimulation (CINLS) was proposed to mismatch the phase constant of adjacent core modes reducing the temporal and spectral EVM fluctuations of the MCF-RoF transmissions.

4.2. Signal processing

The potential application of MCFs is not only restricted to SDM transmissions. The inherent capability of a MCF to modify the propagated signals allows us to investigate a vast scenario of new applications for ultra-high capacity SDM transmissions and microwave photonics (MWP) based on signal processing techniques. As we will see, the basic concept of the signal processing using MCFs is a far richer scope than initially foreseen.

In particular, the use of MCFs for MWP applications based on signal processing was firstly proposed by Gasulla and Capmany in [88]. In this work the authors investigate the suitability of these new fibers to perform true-time delay lines (TTDLs), optical beamforming, optical filtering and arbitrary waveform generation using heterogeneous cores. These applications have been extensively researched in [30, 31, 89–93] with different MCF designs and experimental setups. As an attractive example, it should be remarked the proposal reported in [31, 90], where the inscription of selective Bragg gratings in a homogeneous MCF it was introduced in [90] and experimentally verified in [31] to achieve compact fiber-based TTDL without using heterogeneous cores. Along this line, other MWP applications such as optical beamforming can also be performed by using homogeneous cores as described in [94]. In this work, Llorente and co-workers propose a compact all-fiber beamformer based on a *N*-core homogeneous MCF.

On the other hand, the MCF signal processing also involves additional applications and functionalities such as pulse shaping, dispersion engineering, modal conversion and modal filtering applications. Remarkably, the engineering of the refractive index profile allows us to implement these fashion features in MCF media. In this scenario, a fascinating proposal recovered from the string and quantum field theory was firstly introduced in [95] within the framework of photonics and further developed in [19, 96] to design SCFs and MCFs: the supersymmetry (SUSY). Specifically, one-dimensional SUSY allows us to perform the aforementioned MWP applications. The specific details can be found in [96] for cylindrical potentials with axial symmetry. As an interesting example (among other applications detailed in this work), we include here the description of a true modal (de)multiplexer (M-MUX/DEMUX) using a 3-core MCF. **Figure 9** shows the optical device and its functionality.

The device is designed using a 60-cm MCF comprising three cores a, b and c with a core-to-core distance $d_{ab} = d_{ac} = 55$ μm, $R_0 = 25$ μm, and $\lambda_0 = 1550$ nm [**Figure 9(a)**]. The index profiles of the cores a and c are calculated by using the Darboux procedure. The index profile of the core b is taken to be the step-index profile, with $n_b = 1.45$ when $r < R_0$. A 10-ps Gaussian optical pulse is launched to the central core b, first in the LP_{01} mode, and later in the LP_{11} and LP_{21} modes with a peak power of 0 dBm. The numerical simulation was performed using a beam propagation method at $\lambda_0 = 1550$ nm. **Figure 9(b)** shows the numerical results of the optical pulse propagating through each LP mode in the M-DEMUX. It is worth noting that, in contrast with other mode (de)multiplexing strategies [97–100], a true mode demultiplexing is achieved for each LP mode. At the device output, the pulse launched into the LP_{11} mode of the core b is found in the LP_{11} mode of the core a, the pulse launched into the LP_{01} mode of the core b is found in the same mode and core, and the pulse launched into the LP_{21} mode of the core b can be observed

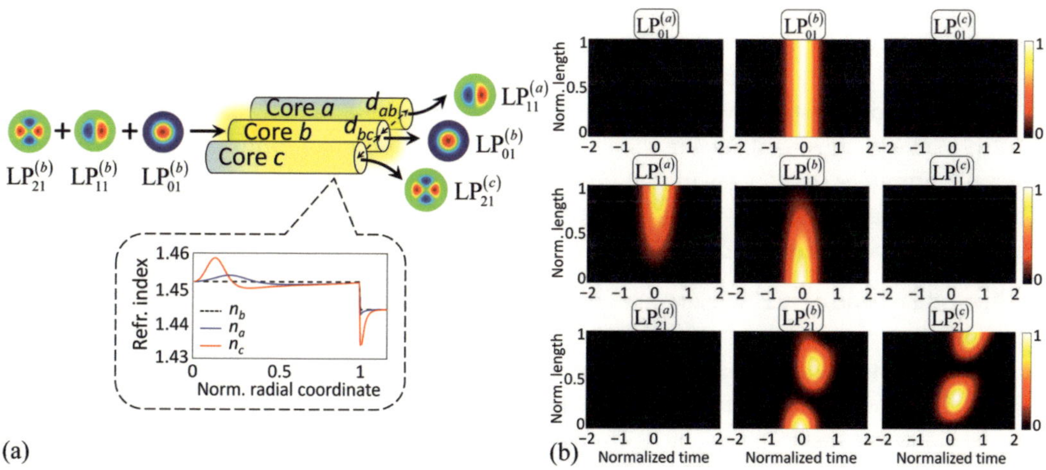

Figure 9. Modal (de)multiplexer based on a 60-cm 3-core MCF [96]. (a) Schematic structure of the optical device. (b) A 10-ps Gaussian pulse propagating through the: LP_{01}, LP_{11} and LP_{21} modes of the cores a, b, and c.

in the LP_{21} mode of the core *c*. Moreover, pulse shaping and dispersion engineering functionalities can be incorporated in the proposed device as indicated in [96]. On the other hand, it should be noted that the SUSY transformations presented in [96] can also be applied to axially symmetric quantum and acoustic potentials as discussed in this work.

4.3. Multi-core fiber lasers, amplifiers and optical sensors

All-fiber designs of optical lasers, amplifiers and sensors using MCFs have been extensively investigated in recent years [32, 101–111]. In particular, the multi-mode interference (MMI) which can be observed through a chain SMF-MCF-SMF is widely employed in lasers, amplifiers and optical sensors to improve the performance of classical designs based on SCFs [32, 104].

As one can expect, the basic concept of an active MCF is the natural evolution for the cladding pumped rare-earth-doped fibers. The classical design using a single core offers an excellent combination of high efficiency and beam quality. However, high output powers are limited by the stimulation of nonlinear effects. In that case, the increment of the mode field area is the obvious solution to decrease the nonlinear effects. In this scenario, active MCFs offer the possibility of reducing the nonlinear effects using a coupled-core design to generate supermodes with large mode field area [101, 105]. Moreover, note that the gain medium is split at discrete regions (cores) inside the cladding, and therefore, the thermal dissipation is higher than in the classical single-core design. As a result, higher output powers can be achieved in MCF media [105]. On the other hand, in contrast with a SCF bundle, a *N*-core MCF laser/amplifier only requires a single pumped laser for *N* optical paths, with the corresponding energy cost reduction for the network operators [5, 106, 107].

In this topic, an intense research work has been developed in the last decade [5, 11, 12]. To date, most CC-MCF lasers/amplifiers operate in the in-phase supermode combining high brightness and near-diffraction limited far field profile. The selection of the in-phase supermode can be performed by using diverse methods such as phase-locking and Talbot cavities [102]. As an example, a monolithic fiber laser using a CC-MCF with highly and lowly reflective fiber Bragg gratings (HR/LR-FBG) is shown in **Figure 10(a)** [104]. The MCF segment is located between the HR-FBG and the LR-FBG creating an active cavity, where the MMI allows us to obtain a high-contrast spectral modulation. In addition, the uniform illumination of the cores is achieved by performing a cladding pumping scheme. Remarkably, this MCF laser design demonstrates the direct correlation between the MMI in few-mode SCF systems and in the laser operation when multiple supermodes oscillate simultaneously. Following a similar approach, additional MCF laser and amplifier designs have been proposed in [103, 105]. Nevertheless, in long-haul SDM transmissions the usual design is the multi-core erbium-doped-fiber-amplifier based on a cladding pumped scheme [106, 107].

On the other hand, MCF sensors are also based on a similar concept as in the laser of the previous example [see **Figure 10(b)**]. The sensor comprises two SSMFs spliced to a short MCF segment with hexagonal shaped cores. The operating principle within the MCF segment is the MMI, which induces a deep peak in the transmission spectrum. An external environmental change shifts the spectral position of the minimum. As an specific example, let us consider a temperature change. When increasing the temperature, the thermal expansion of the MCF medium will increase the refractive index of the silica cores, and consequently, the peak will be shifted to a longer wavelength [32].

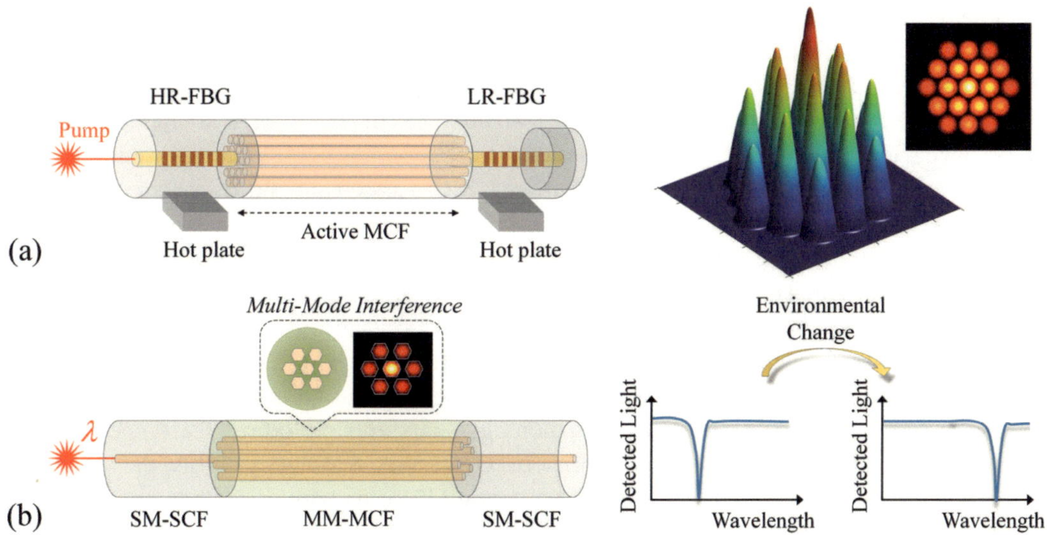

Figure 10. MCF laser and optical sensor operating on the principle of multi-mode interference (MMI). (a) MCF laser comprising a highly and lowly reflective fiber Bragg grating (HR/LR-FBG). (b) MCF optical sensor with hexagonal shaped cores. Results based on [32, 104].

In the past, fiber optic sensors using SCFs have been widely discussed for sensing in a broad range of industrial and scientific applications including temperature, force, liquid level, pressure and acoustic waves, among other. Nowadays, the MCF technology allows us to design and fabricate new optical sensors providing accuracy, high resolution, compactness, stability, reproducibility and reliability [32, 108–111].

4.4. Multi-core fibers for medical applications

Multi-core optical fibers have also been studied in recent years within the context of medicine for biomedical sensing and imaging applications [112–121]. Basically, biomedical sensors using MCFs are based on the MMI technique previously described. Thus, let us now focus our attention on biomedical imaging applications in the next paragraphs.

Nowadays, the main challenge in biomedical imaging is the study of cells in biological tissues. In this scenario, the multiphoton microscopy and adaptive optics become fundamental technologies because of their benefits in cellular resolution, high sensitivity, and high imaging rate [121]. In particular, the two-photon excited fluorescence (TPEF) microscopy requires the use of adaptive optics to increase the imaging depth, in practice limited to 1 mm [122]. Remarkably, the so-called *lensless endoscope* is based on the TPEF microscopy and adaptive optics adding at the same time the use of an optical waveguide [121]. The waveguide should be capable of acquiring a multiphoton image of an object located at its tip. To this end, MCFs have been proposed as a necessary technology for the realization of ultrathin lensless endoscopes [112–121]. **Figure 11** depicts different MCF types proposed for biomedical imaging along with a basic scheme of adaptive optics using a spatial light modulator (SLM).

In general, MCFs used for image transport require a high number of cores (>100) with low IC-XT levels and low intermodal dispersion among cores. Therefore, the preferred design is a SI-SM-HO-CC-LB-MCF, in line with the MCF shown in **Figure 11(a)**. Examples of this MCF type fabricated for medical imaging purposes can be found in [113–115], with $d_{ab} < 20$ μm and IC-XT levels lower than −20 dB/m. In spite of the fact that the intermodal dispersion can be reduced with a homogeneous design, disordered MCFs based on the transverse Anderson localization have been reported in [22] to improve the image transport quality [see **Figure 11(b)**]. Specifically, the transverse Anderson localization of light allows localized optical-beam-transport through a transversely disordered medium. Interesting, in disordered multi-dielectric media, the resultant image quality can also be understood with the perturbation theory. In general, disordered arranged non-homogeneous cores exhibit a high phase-mismatching between their LP modes. As a result, the IC-XT level between adjacent core modes is found to be of the same order or lower than in a homogeneous and periodically arranged design [**Figure 11(a)**]. In a similar way and from our viewpoint, additional highly density MCF designs could be investigated from the CLMT using HB cores with a random orientation of the principal axes to minimize the IC-XT.

On the other hand, adaptive optics is required in the TPEF microscopy to recover the initial imaging of the biological tissue [**Figure 11(c)**]. The advance on wave front shapers composed by 2-D SLMs and deformable mirrors have spurred the main evolution in ultrathin endoscopes [121]. Thompson et al. were the first to report imaging with a lensless endoscope based on a waveguide with multiple cores [112]. Later, in 2013, Andresen and co-workers realized a lensless endoscope employing a MCF similar to **Figure 11(a)** with extremely low IC-XT between adjacent cores [113]. In the same line, additional works have been reported combining MCF and MM-SCFs with adaptive optics in [114–120]. At present, the major aim in lensless endoscopy using MCF media is to increase the core density with a reduced IC-XT and intermodal dispersion between neighboring cores [121].

4.5. Multi-core fiber opportunities in experimental physics

In the past, fiber-optical analogies have been investigated to use optical fibers as an experimental platform for testing different physical phenomena in various fields, such as in quantum

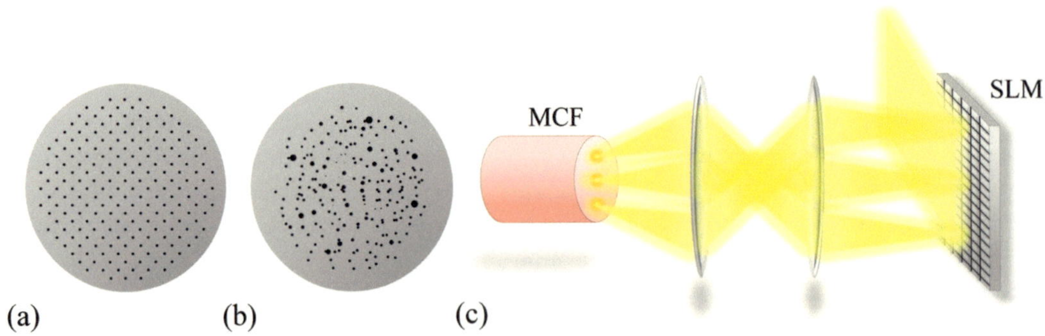

Figure 11. MCFs and adaptive optics for medical imaging. (a) MCF with low IC-XT and periodically arranged cores [113], (b) disorder MCF based on transverse Anderson localization, and (c) wavefront shaping with a single spatial light modulator (SLM) for a MCF based lensless endoscope. Results based on [22, 114].

http://dx.doi.org/10.5772/intechopen.72458

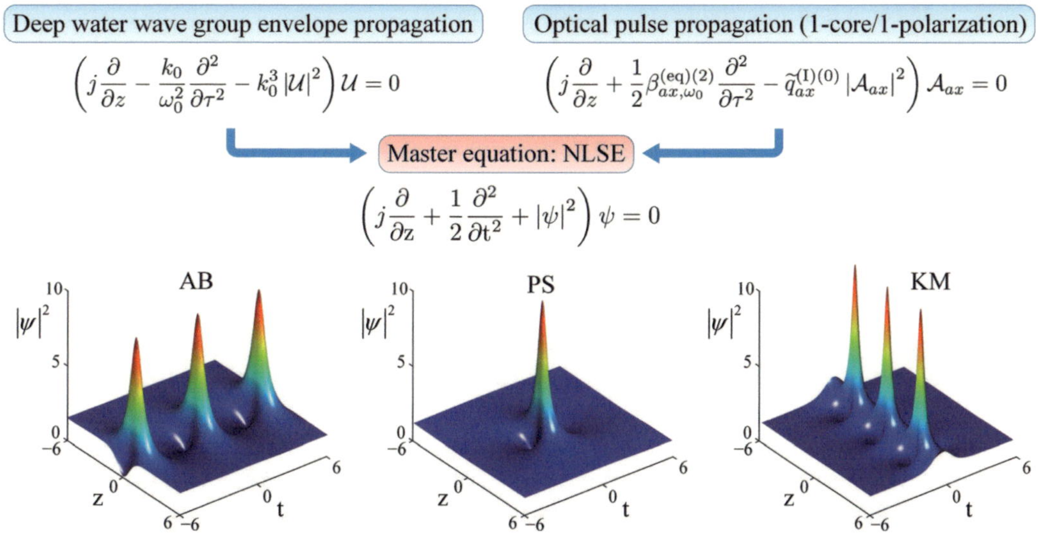

Figure 12. Analogy between fluid mechanics and optics. The NLSE describes the linear and nonlinear wave propagation in different physical systems. Analytical SFB solutions of the NLSE: Akhmediev breathers (ABs), the Peregrine soliton (PS) and the Kuznetsov-Ma (KM) solitons.

mechanics, general relativity or condensed matter physics, among others [16–23]. In fact, a specific example of solid-state physics has already been discussed in the previous subsection, the Anderson effect, relying on the immobility of an electron in a disordered lattice [21, 22]. As Anderson localization involves an interfering phenomenon, this effect has been extended to optics. In [21], Anderson localization has been discussed in two-dimensional photonic lattices, and in [22] it has been discussed its potential applications for medical imaging using disordered MCFs, as pointed out before. More broadly, additional strong disorder phenomena in optics such as the self-organized instability in MM-SCFs [123] can be generalized to MM-MCFs.

Another interesting example can be found in fluid dynamics in the studio of rogue waves on deep water. The giant oceanic rogue waves emerge from the sea induced by many different linear and nonlinear wave propagating effects [124]. Indeed, these nonlinear phenomena can be investigated from a fiber-optical analogy [125]. The nonlinear wave propagation on deep water and in a SM-SCF is described in both cases by a master equation: the nonlinear Schrödinger equation (NLSE), as shown in **Figure 12**.

It can be seen that both propagating equations present a similar form, and therefore, the theoretical results can be directly extrapolated from one field to another. Significantly, the emergence of rogue waves can be analytically studied from the solutions of the NLSE referred to as *solitons on finite background* (SFB) [126]. As a specific example, we include in **Figure 12** the Akhmediev breathers (ABs), the Peregrine soliton (PS) and the Kuznetsov-Ma (KM) solitons[21]. In a similar way, the coupled NLSEs (CNLSEs) have also been discussed in the literature to gain physical

[21]Many of these SFB solutions are termed in the literature as rogue waves. Nevertheless, the fundamental concept of rogue waves emerging unexpectedly from the sea requires additional statistical criteria only fulfilled by higher-order SFB solutions. In either case, the term rogue waves is commonly used for any analytical SFB solution of the NLSE.

insight between interacting rogue waves [127–130]. In this scenario, MCFs offer the possibility of investigating the collision of these nonlinear solutions by using the CLMT [Eq. (14)]. In fact, MCFs can be employed to elucidate the underlying wave propagation phenomena of any physical system with propagating equations of the form of the CNLSEs, for example, superposed nonlinear waves in coherently coupled Bose-Einstein condensates [130] or turbocharge applications in acoustics [131]. Remarkably, in acoustics, the CLMT reported in [60] can play an essential role. Time-varying multi-core cylindrical acoustic ducts can be engineered with the same modal properties as optical MCFs. Therefore, the presented theory can be employed to analyze the intermodal dispersion and the random medium perturbations in acoustic duct conductions.

On the other hand, additional exotic physical phenomena can also be explored in MCF media expanding the possibilities of the classical SCFs. For example, an optical pulse propagating through a SCF establishes a moving medium which corresponds to a space-time geometry. Specifically, this gravitational approach was employed in [20] to demonstrate a fiber-optical analogy of the event horizon in a black hole. Along this line, additional gravitational anomalies could be investigated in a MCF when adjacent cores perturb the space-time geometry created by an optical pulse propagating in a given core of the fiber.

Finally, it is worth mentioning that MCFs are being explored in other branches of experimental physics as in astronomy [132]. The main advantage of these new fibers is the reduced core-to-core distance which can be achieved in a single cladding. In particular, this property has revealed special interest because of the superior fill factor[22] to other approaches for creating spectroscopic maps of galaxies or detecting exoplanets. The Sydney-AAO Multi-object Integral field spectrograph (SAMI) project [133], responsible of performing a large spatial spectroscopy of galaxies, pioneered the introduction of MCFs in astronomical observatories.

5. Conclusions and outlook

Multi-core optical fibers have been developed during the last decade, remarkably within the context of SDM transmissions. In this chapter we have reviewed the main MCF types, the fundamental concepts of the linear and nonlinear propagation, and finally, their potential applications in diverse fields of science. In spite of the fact that the fundamentals of the MCF technology have been well elucidated in recent years, the main challenges in this topic involve the following points:

- The analysis of the longitudinal and temporal fluctuations of the crosstalk should be further investigated in the multi-mode regime. To this end, the CLMT of [61] could be extended to MM-MCFs. In addition, other theoretical models based on the Manakov equations [56, 57] can also be employed and extended to the femtosecond regime.
- Existing and additional MCF fabrication methods should be explored and optimized not only in the S + C + L optical bands, but also in the first and second transmission window. In general, the manufacturing cost of a MCF and the peripheral devices (fan-in/fan-out

[22] Fill factor: ratio between the cores and the total transversal area of the waveguide. Typically, in MM-SCFs (105/125) the fill factor is of the order of 0.7. Using a MM-SCF or a fiber bundle of SM-SCFs a lot of dead space cannot be observed.

connectors, lasers, amplifiers, photonic lanterns, power combiners, couplers, multiplexers, etc) should be reduced.

- The efforts in future MCF designs must be focused on the increment of the core density minimizing the IC-XT, the intermodal dispersion and the random linear birefringence induced by the microbends. The impact of external perturbations such as the macrobends and the fiber twisting should also be reduced in real-deployed MCF systems. Furthermore, new MCF designs should also be investigated for lensless endoscope integrating a high number of cores with a reduced evanescent field in the cladding. In this scenario, it has been proposed HB-MCFs with a random orientation of the principal axes in each core.
- Fronthaul connectivity performed by MCF-ROF transmissions should be spurred for the next-generation wireless systems, e.g. 5G cellular technology and Beyond-5G. In this line, selective-inscribed FBGs [31] and SUSY MCFs [96] will allow us to process the propagated optical signals between the OLT and the microcell.
- On-line MIMO processing of MDM transmissions using MM-MCFs should be developed to support real-time applications in backbone and access networks [134].
- Quantum communications are emerging as a fundamental key in network security [135]. Nowadays, quantum key distribution (QKD) is making the transition from the laboratory to field trials [136]. In this scenario, the QKD through MCF media should be further investigated for the next-generation optical SDM networks [137].

Acknowledgements

This work was supported by Spanish National Plan projects MINECO/FEDER UE XCORE TEC2015-70858-C2-1-R and HIDRASENSE RTC-2014-2232-3. A. Macho work was supported by BES-2013-062952 F.P.I. Grant.

Author details

Andrés Macho Ortiz* and Roberto Llorente Sáez

*Address all correspondence to: amachor@ntc.upv.es

Nanophotonics Technology Center, Universitat Politècnica de València, Valencia, Spain

References

[1] Ayhan Y, Cavdar IH. Optimum link distance determination for a constant signal to noise ratio in M-ary PSK modulated coherent optical OFDM systems. Telecommunication Systems. 2013;**4**:461-470

[2] Yu J, Dong Z, Chien H-C, Jia Z, Li X, Huo D, Gunkel M, Wagner P, Mayer H, Schippel A. Transmission of 200 G PDM-CSRZ-QPSK and PDM-16QAM with a SE of 4 b/s/Hz. Journal of Lightwave Technology. 2013;**31**(4):515-522

[3] Li L, Jijun Z, Degong D, Aihan Y. Analysis modulation formats of DQPSK in WDM-PON system. Optik. 2012;**123**:2050-2055

[4] Essiambre R-J, Kramer G, Winzer PJ, Foschini GJ, Goebel B. Capacity limits of optical fiber networks. Journal of Lightwave Technology. 2010;**28**(4):662-701

[5] Richardson D, Fini J, Nelson LE. Space-division multiplexing in optical fibres. Nature Photonics. 2013;**7**:354-362

[6] Bozinovic N et al. Terabit-scale orbital angular momentum mode division multiplexing in Fibers. Science. 2013;**340**(6140):1545-1548

[7] Van Uden RGH et al. Ultra-high-density spatial division multiplexing with a few-mode multi-core fibre. Nature Photonics. 2014;**8**:865-870

[8] Inao S et al. High density multi-core-fibre cable. In: The International Cable Connectivity Symposium (IWCS). 1979. pp. 370-384

[9] Inao S, Sato T, Sentsui S, Kuroha T, Nishimura Y. Multi-core optical fiber. In: The Optical Fiber Communication Conference and Exhibition (OFC). Washington: Optical Society of America; 1979. WB1

[10] Berdagué S, Facq P. Mode-division multiplexing in optical fibers. Applied Optics. 1982;**21**(11):1950-1055

[11] Saitoh K, Matsuo M. Multicore Fiber technology. Journal of Lightwave Technology. 2016;**34**(1):55-66

[12] Mizuno T, Takara H, Sano A, Miyamoto Y. Dense space-division multiplexed transmission systems using multi-core and multi-mode fiber. Journal of Lightwave Technology. 2016;**34**(2):582-591

[13] Temprana E et al. Overcoming Kerr-induced capacity limit in optical fiber transmission. Science. 2015;**348**(6242):1445-1447

[14] Li G, Bai N, Zhao N, Xia C. Space-division multiplexing: The next frontier in optical communication. Advanced in Optics and Photonics. 2014;**6**:413-487

[15] Puttnam BJ et al. Modulation formats for multi-core fiber transmission. Optics Express. 2014;**22**(26):32457-32469

[16] Wan W, Jia S, Fleischer JW. Dispersive superfluid-like shock waves in nonlinear optics. Nature Physics. 2007;**3**:46-51

[17] Fatome J, Finot C, Millot G, Armaroli A, Trillo S. Observation of optical undular bores in multiple four-wave mixing. Physical Review X. 2014;**4**:021022

[18] Dreisow F et al. Classical simulation of relativistic Zitterbewegung in photonic lattices. Physical Review Letters. 2010;**105**:143902

[19] Miri M-A, Heinrich M, Ganainy RE, Christodoulides DN. Supersymmetric optical structures. Physical Review Letters. 2013;**110**:233902

[20] Philbin TG et al. Fiber-optical analog of the event horizon. Science. 2008;**319**(5868):1367-1370

[21] Scwartz T, Bartal G, Fishman S, Segev M. Transport and Anderson localization in disordered two-dimensional photonic lattices. Nature. 2007;**446**:52-55

[22] Karbasi S et al. Image transport through a disordered optical fibre mediated by transverse Anderson localization. Nature Communications. 2014;**5**(3362)

[23] Saleh MF et al. Raman induced temporal condensed matter physics in a gas-filled photonic crystal fibers. Optics Express. 2015;**23**(9):11879-11886

[24] Sumimoto Electric. Sumitomo electric has developed new-type multi-Core optical Fiber for optical interconnects and realized highest-density multi-Core Fiber optic cable [internet]. March 26, 2015. Available from: http://global-sei.com/company/press/2015/03/prs022_s.html

[25] Xia C, Bai N, Ozdur I, Zhou X, Li G. Supermodes for optical transmissions. Optics Express. 2011;**19**(17):16653-16664

[26] Ryf R et al. Space-division multiplexed transmission over 3 × 3 coupled-core multicore fiber. In: Optical Fiber Communication Conference (OFC). San Francisco: Optical Society of America; 2014. Tu2J.4

[27] Hayashi T et al. Coupled-core multi-core fibers: High-spatial-density optical transmission fibers with low differential modal properties. In: European Conference On Optical Communications (ECOC); Valencia. IEEE. 2015

[28] Stone JM, Yu F, Knight JC. Highly birefringent 98-core fiber. Optics Letters 2014; **39**(15): 4568-4570

[29] Kim Y et al. Adaptive multiphoton Endomicroscope incorporating a polarization-maintaining multicore optical fibre. IEEE Journal of Selected Topics in Quantum Electronics. 2016;**22**(3):171-178

[30] García S, Gasulla I. Dispersion-engineered multicore fibers for distributed radiofrequency signal processing. Optics Express. 2016;**24**(18):20641-20654

[31] Gasulla I, Barrera D, Hervás J, Sales S. Spatial division multiplexed microwave signal processing by selective grating inscription in homogeneous multicore fibers. Scientific Reports. 2017;**7**(41727). https://www.nature.com/articles/srep41727

[32] López JEA, Eznaveh ZS, LiKamWa PL, Schülzgen A, Correa RA. Multicore fiber sensor for high-temperature applications up to 1000°C. Optics Letters. 2014;**39**(15):4309-4312

[33] Hayashi T, Taru T, Shimakawa O, Sasaki T, Sasaoka E. Design and fabrication of ultra-low crosstalk and low-loss multi-core fiber. Optics Express. 2011;**19**(17):16576-16592

[34] Ziolowicz A et al. Hole-assisted multicore optical fiber for next generation telecom transmission systems. Applied Physics Letters. 2014; **105**(081106)

[35] Mahdiraji GA, Amirkhan F, Chow DM, Kakaie Z, Yong PS, Dambul KD, Adikan FRM. Multicore flat Fiber: A new fabrication technique. IEEE Photonics Technology Letters. 2014;**26**(19):1972-1974

[36] Ishida I et al. Possibility of stack and draw process as fabrication technology for multi-core fiber. In: Optical Fiber Communication Conference (OFC). Anaheim: Optical Society of America; 2013. OTu2G.1

[37] Yamamoto J et al. Fabrication of multi-core fiber by using slurry casting method. In: Optical Fiber Communication Conference (OFC). Los Angeles: Optical Society of America; 2017. Th1H.5

[38] Feuer M et al. Joint digital signal processing receivers for spatial superchannels. IEEE Photonics Technology Letters. 2012;**24**(21):1957-1960

[39] Eriksson TA et al. Single parity check-coded 16QAM over spatial superchannels in multicore fiber transmission. Optics Express. 2015;**23**(11):14569-14582

[40] Rademacher G, Puttnam BJ, Luis RS, Awaji Y, Wada N. Experimental investigation of a 16-dimensional modulation format for long-haul multi-core fiber transmission. In: European Conference on Optical Communications (ECOC). Valencia: IEEE; 2015. P.5.10

[41] Puttnam BJ, Luis RS, Mendinueta JMD, Awaji Y, Wada N. Linear block-coding across >5 Tb/s PDM-64QAM spatial super-channels in a 19-core fiber. In: European Conference on Optical Communications (ECOC). Valencia: IEEE; 2015. P.5.6

[42] Marcuse D. Theory of Dielectric Optical Waveguides. Elsevier; 1974

[43] Winzer PJ, Essiambre R-J. Advanced modulation formats for high-capacity optical transport networks. Journal of Lightwave Technology. 2006;**24**(12):4711-4728

[44] Gloge D. Weakly guiding fibers. Applied Optics. 1971;**10**(10):2252-2258

[45] Fini JM, Zhu B, Taunay TF, Yan MF. Statistics of crosstalk in bent multicore fibers. Optics Express. 2010;**18**(14):15122-15129

[46] Koshiba M, Saitoh K, Takenaga K, Matsuo S. Multi-core fiber design and analysis: Coupled-mode theory and coupled-power theory. Optic Express. 2011;**19**(26):B102-B111

[47] Koshiba M, Saitoh K, Takenaga K, Matsuo S. Analytical expression of average power-coupling coefficients for estimating intercore crosstalk in multicore fibers. IEEE Photonics Journal. 2012;**4**(5):1987-1995

[48] Hayashi T, Sasaki T, Sasaoka E, Saitoh K, Koshiba M. Physical interpretation of intercore crosstalk in multicore fiber: Effects of macrobend, structure fluctuation, and microbend. Optics Express. 2013;**21**(5):5401-5412

[49] Macho A, Morant M, Llorente R. Experimental evaluation of nonlinear crosstalk in multi-core fiber. Optics Express. 2015;**23**(14):18712-18720

[50] Macho A, Morant M, Llorente R. Unified model of linear and nonlinear crosstalk in multi-core fiber. Journal of Lightwave Technology. 2016;**34**(13):3035-3046

[51] Ye F, Tu J, Saitoh K, Morioka T. Simple analytical expression for crosstalk estimation in homogeneous trench-assisted multi-core fibers. Optics Express. 2014;**22**(19):23007-23018

[52] Morant M, Macho A, Llorente R. On the suitability of multicore fiber for LTE-advanced MIMO optical fronthaul systems. Journal of Lightwave Technology. 2016;**34**(2):676-682

[53] Macho A, Meca CG, Peláez FJF, Morant M, Llorente R. Birefringence effects in multi-core fiber: Coupled local-mode theory. Optics Express. 2016;**24**(19):21415-21434

[54] Iizuka K. Elements of photonics volume I. Wiley; 2002

[55] Luis RS et al. Time and modulation frequency dependence of crosstalk in homogeneous multi-core fibers. Journal of Lightwave Technology. 2016;**34**(2):441-447

[56] Mecozzi A, Antonelli C, Shtaif M. Nonlinear propagation in multi-mode fiber in the strong coupling regime. Optics Express. 2012;**20**(11):11673-11678

[57] Mumtaz S, Essiambre RJ, Agrawal GP. Nonlinear propagation in multimode and multicore fibers: Generalization of the Manakov equations. Journal of Lightwave Technology. 2013;**31**(3):398-406

[58] Chiang KS. Coupled-mode equations for pulse switching in parallel waveguides. IEEE Xplore: IEEE Journal of Quantum Electronics. 1997;**33**(6):950-954

[59] Liu M, Chiang KS. Effects of intrapulse stimulated Raman scattering on short pulse propagation in a nonlinear two-core fiber. Applied Physics B. 2007;**87**:45-52

[60] Liu M, Chiang KS. Pulse propagation in a decoupled two-core fiber. Optics Express. 2010;**18**(20):21261-21268

[61] Macho A, Meca CG, Peláez FJF, Cortés-Juan F, Llorente R. Ultra-short pulse propagation model for multi-core fibers based on local modes. Scientific Reports. 2017;**7**(16457)

[62] Weiner AM. Ultrafast optics. Wiley; 2009

[63] Agrawal GP. Nonlinear fiber optics. Elsevier; 2013

[64] Zhu B, Taunay T, Yan M, Fini J, Fishteyn M, Monberg EM. Seven-core multi-core fibre transmissions for passive optical network. Optics Express. 2010;**18**(11):11117-11122

[65] Puttnam BJ, Delgado J-M, Sakaguchi J, Luis RS, Klaus W, Awaji Y, Wada N, Kanno A, Kawanishi T. 210Tb/s Self-Homodyne PDM-WDM-SDM Transmission with DFB Lasers in a 19-Core Fibre. In: European Conference on Optical Communications (ECOC). London: IEEE; 2013. TuC1.2

[66] Sakaguchi J et al. 305 Tb/s space division multiplexed transmission using homogeneous 19-Core fibre. Journal of Lightwave Technology. 2013;**31**(4):554-562

[67] Takara H et al. 1.01-Pb/s (12 SDM/222 WDM/456 Gb/s) crosstalk-managed transmission with 91.4-b/s/Hz aggregate spectral efficiency. In: European Conference on Optical Communications (ECOC). Amsterdam, Netherlands: IEEE; 2012. Th3.C.1

[68] Chandrasekhar S et al. WDM/SDM transmission of 10 x x128-Gb/s PDM-QPSK over 2688-km 7-core fiber with a per-fiber net aggregate spectral-efficiency distance product of 40,320 km·b/s/Hz. In: European Conference on Optical Communications (ECOC). Geneva, Switzerland: IEEE; 2011. Th13.C.4

[69] Takeshima K et al. 51.1-Tbit/s MCF transmission over 2520 km using cladding pumped 7-core EDFAs. In: Optical Fiber Communication Conference (OFC). Los Angeles, USA: Optical Society of America; 2015. W3G.1

[70] Sakaguchi J et al. Realizing a 36-core, 3-mode Fibre with 108 Spatial Channels. In: Optical Fiber Communication Conference (OFC). Los Angeles, USA: Optical Society of America; 2015. Th5C.2

[71] Shibahara K, et al. Dense SDM (12-core x 3-mode) Transmission over 527 km with 33.2 ns Mode-Dispersion Employing Low-Complexity Parallel MIMO Frequency-Domain Equalization. In: Optical Fiber Communication Conference (OFC). Los Angeles, USA: Optical Society of America; 2015. Th5C.3

[72] Igarashi K, et al. 114 Space-Division-Multiplexed Transmission over 9.8-km Weakly-Coupled-6-Mode Uncoupled-19-Core Fibers. In: Optical Fiber Communication Conference (OFC). Los Angeles, USA: Optical Society of America; 2015. Th5C.4

[73] Igarashi K et al. Ultra-dense spatial-division-multiplexed optical fiber transmission over 6-mode 19-core fibers. Optics Express. 2016;**24**(10):10213-10231

[74] Kobayashi T et al. 1-Pb/s (32 SDM/46 WDM/768 Gb/s) C-band dense SDM transmission over 205.6-km of single-mode heterogeneous multi-core fiber using 96-Gbaud PDM-16QAM channels. In: Optical Fiber Communication Conference (OFC). Los Angeles, USA: Optical Society of America; 2017. Th5B.1

[75] Tanaka K, Agata A. Next-generation Optical Access Networks for C-RAN. In: Optical Fiber Communication Conference (OFC). Los Angeles, USA: Optical Society of America; 2015. Tu2E.1

[76] Cvijetic N. Optical network evolution for 5G mobile applications and SDN-based control. In: Proceedings of 16th International Telecomm. Network Strategy and Planning Symposium. IEEE; 2014

[77] Öhlen P et al. Flexibility in 5G Transport Network: The Key to Meeting the Demand for Connectivity [Internet]. 2015. Available from: https://www.ericsson.com/

[78] Rappaport TS. Millimeter wave wireless communications for 5G cellular: It will work! In: Personal, indoor, and mobile radio communications conference. Washington, USA: IEEE. 2014

[79] Ericsson. 5G Energy Performance [Internet]. 2015. Available from: http://www.5gamericas.org/files

[80] Kohn U. Fronthaul Networks–a Key Enabler for LTE-Advanced [Internet]. 2014. Available from: https://oristel.com.sg/wp-content/uploads/2015/03/Fronthaul-Networks-A-Key-Enabler-for-LTE-Advanced.pdf

[81] Zhu M, Liu X, Chand N, Effenberger F, Chang G-K. High-capacity mobile fronthaul supporting LTE-Advanced carrier aggregation and 8×8 MIMO. In: Optical Fiber Communication Conference (OFC). Optical Society of America; 2015. M2J.3

[82] Cheng L, Liu X, Chand N, Effenberger F, Chang G-K. Experimental demonstration of sub-Nyquist sampling for bandwidth- and hardware-efficient mobile fronthaul supporting 128×128 MIMO with 100-MHz OFDM signals. In: Optical Fiber Communication Conference (OFC). Anaheim, USA: Optical Society of America; 2016. W3C.3

[83] Wang N, Hossain E, Bhargava VK. Backhauling 5G small cells: A radio resource management perspective. IEEE Wireless Communications. 2015;**22**(5):41-49

[84] Morant, M., Macho, A., and Llorente, R. Optical fronthaul of LTE-Advanced MIMO by spatial multiplexing in multicore fiber. In: Optical Fiber Communication Conference (OFC). Optical Society of America; 2015. W1F.6

[85] Karadimitrakis A, Moustakas AL, Vivo P. Outage capacity for the optical MIMO channel, IEEE transactions on information theory. IEEE Transactions on Information Theory. 2014;**60**(7):4370-4382

[86] Sano A, Takara H, Moyamoto Y. Large capacity transmission systems using multi-core fibers. In: OptoElectronics and communication Conf. Melbourne, Australia: IEEE. 2014

[87] Macho A, Morant M, Llorente R. Next-generation optical fronthaul systems using multicore fiber media. Journal of Lightwave Technology. 2016;**34**(20):4819-4827

[88] Gasulla I, Capmany J. Microwave photonics applications of multicore Fibers. IEEE Photonics Journal. 2012;**4**(3):877-888

[89] Yu S, Jiang T, Li J, Zhang R, Wu G, Gu W. Linearized frequency doubling for microwave photonics links using integrated parallel Mach–Zehnder modulator. IEEE Photonics Journal. 2013;**5**(4):5501108

[90] Gasulla I, Barrera D, Sales S. Microwave photonic devices based on multicore fibers. In: Transparent Optical Networks (ICTON). Graz, Austria: IEEE; 2014. Th.B6.4

[91] Garcia S, Gasulla I. Design of heterogeneous multicore fibers as sampled true-time delay lines. Optics Letters. 2015;**40**(4):621-624

[92] Lawrence RC. Silicon Photonics for Microwave Photonics Applications. In: Optical Fiber Communication Conference (OFC). Anaheim, USA: Optical Society of America; 2016. M2B.4

[93] Wang J et al. Subwavelength grating enabled on-chip ultra-compact optical true time delay line. Scientific Reports. 2016;**6**:30235

[94] Zainullin A, Vidal B, Macho A, Llorente R. Multicore Fiber Beamforming Network for Broadband Satellite Communications. In: Proceedings of SPIE Terahertz, RF, Millimeter,

and Submillimeter-Wave Technology and Applications X. San Francisco, USA: SPIE; 2017. 1010310

[95] Chumakov SM, Wolf KB. Supersymmetry in Helmholtz optics. Physics Letters A. 1994;**193**:51-53

[96] Macho A, Llorente R, Meca CG. Supersymmetric transformations in optical fibers. Forthcoming

[97] Song KY, Hwang IK, Yun SH, Kim BY. High performance fused-type mode-selective coupler using elliptical core two-mode fiber at 1550 nm. IEEE Xplore: IEEE Photonics Technology. 2002;**14**(4):501-503

[98] Hanzawa N. Mode multi/demultiplexing with parallel waveguide for mode division multiplexed transmission. Optics Express. 2014;**22**(24):29321-29330

[99] Chang SH et al. Mode- and wavelength-division multiplexed transmission using all-fiber mode multiplexer based on mode selective couplers. Optics Express. 2015;**23**(6): 7164-7172

[100] Corral JL, Rodríguez DG, Llorente R. Mode-selective couplers for two-mode transmission at 850 nm in standard SMF. IEEE Photonics Technology Letters. 2016;**28**(4):425-428

[101] Michaille L, Bennet CR, Taylor DM, Shepherd TJ. Multicore photonic crystal Fiber lasers for high power/energy applications. IEEE journal of selected topics in. Quantum Electronics. 2009;**15**(2):328-336

[102] Anderson J, Jollivet C, Van Newkirk A, Schuster K, Grimm S, Schülzgen A. Multi-Core Fiber Lasers. In: Frontiers in Optics/Laser Science. Los Angeles, USA: Optical Society of America; 2015. LTu2H.2

[103] Chuncan W, Fan Z, Chu L, Shuiseng J. Microstructured optical fiber for in-phase mode selection in multicore fiber lasers. Optics Express. 2008;**16**(8):5505-5515

[104] Jollivet C et al. Mode-resolved gain analysis and lasing in multisupermode multi-core fiber laser. Optics Express. 2014;**22**(24):30377-30386

[105] Li L, Schülzgen A, Chen S, Témyanko VI. Phase locking and in-phase supermode selection in monolithic multicore fiber lasers. Optics Letters. 2006;**31**(17):2577-2579

[106] Krummrich PM, Akhtari S. Selection of energy optimized pump concepts for multi-core and multi-mode erbium doped fiber amplifiers. Optics Express. 2014;**22**(24):30267-30280

[107] Abedin KS et al. Multicore erbium doped Fiber amplifiers for space division multiplexing systems. Journal of Lightwave Technology. 2014;**32**(16):2800-2808

[108] Mizuno Y, Hayashi N, Tanaka H, Wada Y, Nahamura K. Brillouin scattering in multicore optical fibers for sensing applications. Scientific Reports. 2015;**5**:11388

[109] Newkirk AV, López JEA, Delgado GS, Piracha MU, Correa RA, Schülzgen A. Multicore Fiber sensors for simultaneous measurement of force and temperature. IEEE Photonics Technology Letters. 2015;**27**(14):1523-1526

[110] Newkirk AV, Eznaveh ZS, López J.EA, Delgado GS, Schülzgen A, Correa RA High Temperature Sensor based on Supermode Interference in Multicore Fiber. In: Conference on Lasers and Electro-Optics (CLEO). San Jose, USA: IEEE/OSA; 2014. SM2N.7

[111] Villatoro J, Arrizabalaga O, López JEA, Zubia J, de Ocáriz IS, Multicore Fiber Sensors. In: Optical Fiber Communication Conference (OFC). Los Angeles, USA: Optical Society of America; 2017. Th3H.1

[112] Thompson AJ et al. Adaptive phase compensation for ultracompact laser scanning endomicroscopy. Optics Letters. 2011;**36**(9):1707-1709

[113] Andresen ER et al. Toward endoscopes with no distal optics: Videorate scanning microscopy through a fiber bundle. Optics Letters. 2013;**38**(5):609-611

[114] Andresen ER et al. Two-photon lensless endoscope. Optics Express. 2013;**21**(18):20713-20721

[115] Andresen ER et al. Measurement and compensation of residual group delay in a multi-core fiber for lensless endoscopy. Journal of the Optical Society of America B: Optical Physics. 2015;**32**(6):1221-1228

[116] Roper JC et al. Minimizing group index variations in a multicore endoscope fiber. IEEE Photonics Technology Letters. 2015;**27**(22):2359-2362

[117] Kim Y et al. Adaptive multiphoton endomicroscope incorporating a polarization-maintaining multicore optical fibre. IEEE Journal of Selected Topics in Quantum Electronics. 2015;**22**(3):6800708

[118] Sivankutty S et al. Extended field-of-view in a lensless endoscope using an aperiodic multicore fiber. Optics Letters. 2016;**41**(15):3531-3534

[119] Conkey DB et al. Lensless two-photon imaging through a multicore fiber with coherence-gated digital phase conjugation. Journal of Biomedical Optics. 2016;**21**(4):045002

[120] Stasio N, Moser C, Psaltis D. Calibration-free imaging through a multicore fiber using speckle scanning microscopy. Optics Letters. 2016;**41**(13):3078-3081

[121] Andresen ER et al. Ultrathin endoscopes based on multicore fibers and adaptive optics: A status review and perspectives. Journal of Biomedical Optics. 2016;**21**(12):121506

[122] Helmchen F, Denk F. Deep tissue two-photon microscopy. Nature Methods. 2005;**2**(12): 932-940

[123] Wright LG, Liu Z, Nolan DA, Li M-J, Christodoulides DN, Wise FW. Self-organized instability in graded-index multimode fibres. Nature Photonics. 2016;**10**:771-776

[124] Kharif C, Pelinovsky E. Physical mechanisms of the rogue wave phenomenon. European Journal of Mechanics. 2003;**22**:603-634

[125] Dudley JM, Dias F, Erkintalo M, Genty G. Instabilities, breathers and rogue waves in optics. Nature Photonics. 2014;**8**:755-764

[126] Akhmediev N, Korneev VI. Modulation instability and periodic solutions of the nonlinear Schrödinger equation. Theoretical and Mathematical Physics. 1986;**69**:1089-1093

[127] Wu CF, Grimshaw RH, Chow KW, Chan HNA. Coupled "AB" system: Rogue waves and modulation instabilities. Chaos. 2015;**25**:103113

[128] Zhong W-P, Belić M, Malomed BA. Rogue waves in a two-component Manakov system with variable coefficients and an external potential. Physical Review E. 2015;**92**:053201

[129] Manikandan K, Senthilvelan M, Kraenkel RA. On the characterization of vector rogue waves in two-dimensional two coupled nonlinear Schrödinger equations with distributed coefficients. The European Physical Journal B. 2016;**89**:218

[130] Mareeswaran RB, Kanna T. Superposed nonlinear waves in coherently coupled Bose–Einstein condensates. Physics Letters A. 2016;**380**:3244-3252

[131] Rämmal H, Lavrentjev J. Sound reflection at an open end of a circular duct exhausting hot gas. Noise control Engineering Journal. 2008;**56**(2):107-114

[132] Jovanovic N, Guyon O, Kawahara H, Kotani T. Application of Multicore Optical Fibers in Astronomy. In: Optical Fiber Communication Conference (OFC). Los Angeles, USA: Optical Society of America; 2017. W3H.3

[133] Croom S et al. The Sydney-AAO multi-object integral field spectrograph. MNRAS. 2012;**421**:872C

[134] Randel, S. et al. First Real-Time Coherent MIMO-DSP for Six Coupled Mode Transmission. In: IEEE Photonics Conference (IPC). Virginia, USA: IEEE; 2015. 15600579

[135] Scarani V, Bechmann-Pasquinucci H, Cerf NJ, Dušek M, Lütkenhaus N, Peev M. The security of practical quantum key distribution. Reviews of Modern Physics. 2009;**81**(3): 1301-1350

[136] Peev M et al. The SECOQC quantum key distribution network in Vienna. New Journal of Physics. 2009;**11**(7):075001

[137] Dynes JF et al. Quantum key distribution over multicore fiber. Optics Express. 2016;**24**(8): 8081-8087

Chapter 5

Fabrication of Polymer Optical Fiber Splitter Using Lapping Technique

Latifah S. Supian, Mohd Syuhaimi Ab-Rahman,
Norhana Arsad, Chew Sue Ping, Nani Fadzlina Naim,
Nurdiani Zamhari and
Syed Mohd Fairuz Syed Mohd Dardin

Additional information is available at the end of the chapter

http://dx.doi.org/10.5772/intechopen.71868

Abstract

This work involves in designing and developing a POF-based directional coupler/splitter using lapping technique and geometrical blocks. Two fiber strands were first tapered at the middle and they were attached to the geometrical blocks and lapped together. Design parameters that are used to develop this coupler/splitter are core diameter, D_c, etching length, L_e, bending radius, R_c, coupling length, L_c and pressure, F_c. All the parameters were taken into account during characterization and analysis of the designed coupler in order to find the most optimum prototype coupler/splitter. Characterizations are done by experimental set-up to test the efficiency, splitting ratio, coupling ratio, excess loss and insertion loss for all the couplers/splitters. Through the characterization process and analysis, the optimized coupler with high splitting ratio and low excess loss were identified. Throughout the experimental process, some of the fibers were improved and renewed in order to realize the design and development of the coupler using this technique. The device can also be utilized as an optical tap and the applications of the device are not only limited in in-house network but also in automotive applications. By using a platform, several splitting ratio can be obtained by integrating different core-cladding thickness and bending radius in order to get the desired splitting ratio and excess loss.

Keywords: polymer optical fiber, splitter, low-cost, lapping technique, green technology, short-haul communication system, geometrical blocks

1. Introduction

1.1. POFs for main medium in short-distance transmission

Polymer optical fibers (POFs) have many advantages compared to other communication medium such as glass fibers, copper cables and wireless communication system. Although glass fibers show high performance in terms of speed, higher bandwidth and minimal loss, POFs in the other hand, offer easy and cost-efficient processing and flexibility, whilst glass optical fibers are very brittle, expensive and the cost of installing overall system is very expensive [1]. Therefore due to the complication of installation using glass optical fibers, where the fibers is easily broken and very sensitive, POFs are more handy and easy to install.

For short-haul distance application, POFs show lower attenuation and data transmission works perfectly within distance less than 1 km. Thus, POFs are more suitable as for medium transmission for home-networking, optical data buses for automotive applications or industrial automation sector. Transmission method of using POFs have emerged as a highly-potential candidate for cost-effective and future-proof solution [2]. Glass optical fibers have replaced copper for telephone networks and backbone wiring for buildings, however, still unable to be used in short distance applications due to its cost.

Polymer optical fibers are widely known around 1960s after glass optical fiber was introduced as an effective transmission medium for optical communication. Over the decades, the performance of POFs have shown improvement in terms of transmission capability from having large attenuation as large as 300–20 dB/km and 3 dB/km for passive device such as optical splitter at visible wavelength [3]. Characteristics that give advantages for POFs include low insertion loss, low production cost, having thermal and mechanical stability and highly potential for mass production reliability. Although the loss of POFs is generally higher than silica or glass fiber, POFs are mostly used in intra office communication system where the distance requirement is only up to a few hundred meters where the losses are low and cost-effective. It provides cost-effective solutions to short distance applications such as local area networks (LAN) and high speed internet access and in vehicles [4].

Multimode POFs particularly is chosen as the fiber technology that is largely employed in short-distance communication applications such as in LANs and interconnects. It is also driven by the needs of higher bit rates and lower cost as cost is one of the important drives in short distance communications.

1.2. DIY kit

Although for short-haul communication system, POFs show lower attenuation and cost-effective compared to glass optical fibers, however, there are lack of industry attention received due to less of industry marketing and education of the end user on how to utilize POFs in the system. The lack of information for the end-users on how to install POF-based devices and components lead to the limitation of POF utilization among the customers or end-users. Although huge companies of automobile and medical equipment companies have already utilizes POFs, however, the lack of education of the end-users on the technique to install and implement the POFs in home-networking causes the industry to pay less attention on building POFs "do-it-yourself" components and devices [5, 6].

1.3. Green technology

About 3% of the world-wide energy is consumed by the information and communications technology (ICT) which contributes to the carbon dioxide (CO_2) emissions [7]. Telecommunication applications can have a direct impact on lowering greenhouse gas emissions and power consumption. Technical approaches of achieving green communication includes energy-efficient network architecture and protocol and energy-efficient wireless transmission techniques [7]. A green technology coupler/splitter is presented based on polymer optical fiber. The splitter has been fabricated by using harmless chemical solvent to etch the fiber and by using various radii of geometrical blocks as the platforms that are made of acrylic and aluminum.

Wavelength from eco-friendly light emitting diode (LED) is utilized to transmit signal. The red LED (650 nm) is capable to download and upload data through Ethernet cable or video signal. LED source used in the system is a solar powered product of semiconductor diode. Compared to incandescent light, light produced by LED is a cool light. The lifetime of LED surpass incandescent which has at most 50,000 hours. Compared to laser, LED is much safer source and little effort is needed to maintain and conduct the source.

The technique used to fabricate this device includes etching using harmless chemical solvent, acetone and also side-polishing. Both the techniques used are environmentally safe and easy to conduct. Acetone is a safe chemical solvent where studies suggest that acetone has low acute and low toxicity if being ingested or inhaled. It is not regarded as carcinogen, a mutagenic chemical or cause chronic neurotoxicity effects. It is mostly used in cosmetic products, processed food and other household products. Acetone has been rated as "generally recognized as safe" (GRAS) substance [8].

1.4. POF couplers

Polymer optical fiber coupler or splitter is a passive device that is built to perform functions required by optical communications such as isolator, circulator and attenuator [9]. Coupler or splitter is important device in the development of optical networks such as in transportations, local area network and for short-distance or in-house applications, in industrial automation or for sensor applications.

Optical couplers are used to combine two or more optical signal inputs from different paths into one output while splitters act oppositely. A particular message encoded as optical signal that needs to be delivered to several outputs at the same time can utilize 1 × N splitter so that the optical signal can be sent to different routes of optical fibers connected to intended end-users or destinations [10]. The requirements for POF couplers are having small excess loss and insertion loss, various power splitting ratios, easy to develop, can be mass produced and having low cost.

Common types of optical splitter are directional, distributive and wavelength-dependent. The mechanism involves can be characterized as diffusion type, area-splitting type and beam-splitting type. Evanescent wave coupling or radiative coupling is part of diffusion couplers. In evanescent wave coupling, two or more fibers are placed sufficiently closed to one another [11]. It is crucial to place the fibers in parallel over a finite distance known as coupling length so that the evanescent field from the primary fiber builds up a propagation field in the secondary fiber to provide two outputs.

A traditional silica optical coupler in one work is formed by placing the two polished-cladding fibers closely together making the light couple from the direct branch to the coupling branch by evanescent field as shown in **Figure 1**. By changing the polishing depth or tapering depth, the contact area and angle, certain modes can be selected and bandwidth can be enhanced in network systems [12]. When the light is transmitted to the coupler, some light will leak out to the branch and be transmitted as stable transmission mode due to the changed of waveguide structure by polishing. Some of the modes were selected by changing the angle of the contact areas of both fiber [12].

1.5. Existing technique of POF splitters/couplers

One of the advantages of polymer optical fibers as compared to other cable types is the simple connector fittings. Copper cables for high data transfer rates mostly require the connection of twisted pairs that must be individually shielded. At frequencies of several 100 MHz, however, cutting open the shielding over a distance of 1 cm results in a noticeable drop in quality of the connection. Glass fibers in the other hand, have a core diameter between 10 and 200 μm. Precise guides are required and glass fibers cannot simply be cut. The face must either be precisely cut by craving with diamond blade or the face must be polished after the cutting. Other advantages for POF is due to its material where the surface of plastics can be smoothed by both cutting and simple polishing and thermal smoothing of the surface is also possible for PMMA.

One of the existing couplers that have been fabricated is done by [7] where the team had demonstrated for the first time that the POF devices can be fabricated by hand using fused

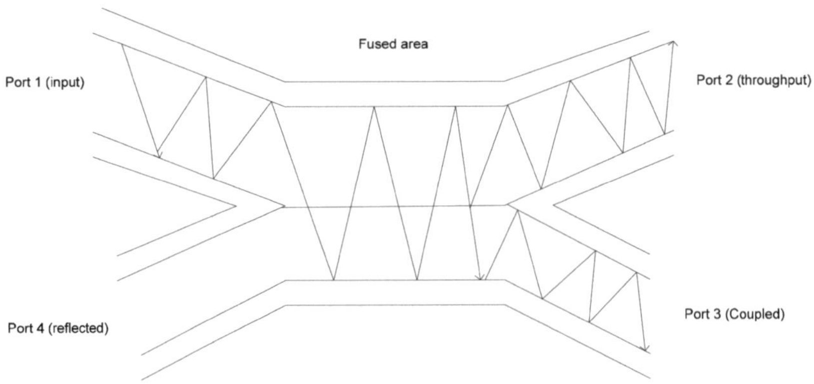

Figure 1. Lapping fibers.

technique. The temperature, stress and splitting technique are the most important parameters to fabricate low loss device. With some modification the device can be used for the extended function such as demultiplexer which is fabricated from uniformity optical splitter [7].

Other techniques include cutting and gluing where in this technique, POF is cut at certain angle using hot knife and glue is applied to attach the two segments of POF. Thermal deformation is a technique where two POF ends are shaped into semicircular form by thermoplastic deformation using hot plate flattening technique. Molding in the other hand is a technique where POF is assembled by a polymer waveguide. Lithography method is used to fabricate the mold [3].

1.6. Etching technique

There are several ways of technique to develop a coupler/splitter. In some cases, tapering is done unto the fibers in the process of developing the coupler. One of the tapering techniques is using chemical etching [4, 5]. Tapering offers unique optical properties that have application to couplers and sensors. The change of diameter can redistribute the modes within the core or remove selected modes. The penetration depth of the evanescent field and proportion of power within this field increases in the tapered section. The change in diameter is used for coupling fibers or interface fibers to devices [13]. This technique has been used by [13] where the cladding of the fiber is thinned using hydrofluoric acid where cladding layer of four micron thick is removed. A technique for removing the cladding of polymethylmethacrylate (PMMA) polymer optical fiber using chemical solvent is used to create etched tapers of a certain length in the middle part of the fiber [14]. PMMA POF is a thermoplastic with softening temperature of 75–80°C. Heat drawing sometimes done unto POF for tapering, however, polymer is not so compatible with drawing due to the latent stress within the structure from production and different physical properties of core and cladding [13]. Thus, etching is by far one of the best methods. The process is simple, low cost and requires no sophisticated equipment and is a safe process since it involves a harmless chemical solvent which is acetone.

PMMA is dissolved using organic solvents such as acetone and methyl isobutyl ketone (MIBK) in order to remove the polymer in concentric layers as required. The method requires no tension to be applied on fiber under etching process so as to prevent brittle stress fracture from occurring and break the fiber. Isopropyl alcohol is used to neutralize the solvent and leave the exposed core clean and grease-free. Once the region has been washed, it will return to PMMA physical and chemical properties.

By tapering the multimode optical fiber cladding, higher modes of the fiber are removed while some other modes are redistributed. As the tapered section is developed, the evanescent field and proportion of total power within this field increases in the affected region. This technique is conventionally applied to glass fibers, however, POF has becoming widely used in optical communication systems and this method can be as a potential towards fabricating a practical coupler using this technique.

The etching process of POF gives an optically smooth surface of similar quality to the original POF due to the polymer quality of POF. If the tapered region is cleanse and washed appropriately the core material section will remain its original properties of PMMA.

1.7. Polishing technique

Side-polishing is one of tapering technique that can be done unto fibers other than etching and fusing. It has been used by other researchers [12, 15, 16] in order to develop couplers/splitter. In this process, single strand multimode fiber is polished at the outer cladding layer for several micrometers. Then the polished fiber is lapped to the other fiber in a bent surface and aided by a thin film of UV curing adhesive is used between them as mode stripping. When the launched port is injected with light source, the mode coupling will occur between the lapped region and split the light accordingly. The strength of the evanescent light coupling is tuned in the range of 0–50% by translating the fibers in or out of alignment of each other [17]. The insertion loss of using this technique is below 5 dB. Polishing technique is considered as one of the simple method to fabricate a directional coupler over the years.

1.8. Macro-bending loss by radiation

Attenuation is a loss of optical power as light travels along the fiber as a result of loss mechanisms such as absorption, scattering and bending. One of the important concepts applied in this research is loss due to macro-bending. Macro-bends are bending that has relatively large radius of curvature compared to the fiber diameter. The loss is high when the bending is smaller. Any dielectric waveguide will radiate if it is bent [18]. Radiation loss occurs in the cladding even when the propagating ray is greater than critical angle. If radius of bending is smaller and wavelength of transmitted light is longer, the macrobending loss or power loss is due to the radiation [19]. Below a critical radius optical fiber bending, macrobending loss is significant. When light facing total internal reflection an electromagnetic disturbance which is known as evanescent wave penetrate the reflecting interface. The amplitudes of evanescent wave decay exponentially when it gets further than the reflecting interface because it cannot propagate in medium of lower refractive index. Non-uniformity in the reflecting interface may cause the evanescent wave to convert into propagating wave. In bending loss, evanescent fields extend to the cladding but decay exponentially with radial distance.

All rays in bent fiber are leaky where at some reflections they lose power by either refraction or tunneling. By stripping the cladding layers, bent fiber will escape from the fiber and the cladding will behave as if it were infinite. The power losses in bent step-index multimode fibers can depend on the cladding thickness. Decreasing the bend radius and increasing fiber core will increase the bending losses. As bend radius begins to decrease, more refracting rays are produced and more power is transferred to the cladding.

Low losses can be obtained in bends that have small radii of the where width of the channel is decreased and the refractive index contrast is increased. Bend losses of multimode channels are independent of wavelength. At some point from the center of the bend, the portion of the field in the cladding would have to exceed the speed of light and be radiated, thus reduce the power in the guided mode [20].

In this research, light that escapes from the bending is utilize to develop an optical coupler or splitter by varying the bending radius in order to obtain certain splitting ratio using mechanical platform of circular blocks and elliptical blocks [21].

1.9. New approach of using lapping technique

Directional coupler is a passive device where power exchanges between two waveguides that is placed in proximity to each other. When a certain power is launched into a waveguide, some of the power is transferred to an adjacent guide due to coupling. The power exchange depends on the interaction length and coupling strength along with other parameters. When two guides are parallel to each other, coupling coefficient is constant and the power launched into one guide will alternate back and forth between the two guides as long as they are close [22].

Complete power transfer occurs when phase velocities are perfectly synchronized and the interaction length is half the coupling length [16]. In [16], they developed a coupler with variable spacing where by gradually increasing the separation between the two guides where all or some of desired power may be transferred from one guide to the other in strong interaction region. By controlling the separation between the channels, the coupler can be realized as switch, splitter and power divider [23].

This research focuses on developing user-friendly and inexpensive splitters with various splitting ratio and low excess loss using POF splitter kit consists of circular blocks and elliptical blocks of varied radii and several pairs of splitters. This technique of splitter development has the advantage of low-cost installation, environmental-friendly with considerable low losses. The proposed fabrication of the coupler/splitter consists of two parallel fibers in contact with each other along a certain coupling length or know as lapping technique [24].

The technique used in this experiment is lapping technique where two similar length fibers are tapered in the middle region using harmless chemical solvent for particular time and the tapered regions of certain diameter, D_c and etched length, L_e are lapped to each other. The taper introduces variations in the effective refractive index of a waveguide and a change in coupling coefficient [16]. Chemical solvent, acetone is used to taper the fiber and stripped off the cladding layer. The thickness of cladding layer that is being stripped off depends on the duration of the etching process. The time duration for etching process in this research range from 30 to 120 minutes. The longer the duration process, the smaller the fiber cores obtained. At this diameter, the cladding layers have been fully etched and leave only the bare core. The aim of this process is to strip off the cladding layers so that when the bare cores are lapped to each other and bent, the rays that propagate in the first fiber can transfer to the second fiber. However, for some fibers the etching process strips off the whole cladding layer including the region that will not be lapped to the other fiber core, this situation will lead the unnecessary losses during splitting/coupling. Thus, the geometrical blocks that the fibers are attached to are customized using acrylic material that has similar refractive index so that it replaces the cladding layers that has been etched.

To stimulate the energy transfers between the primary fiber that has been injected with light source to the secondary fiber, macro-bending effect is a parameter used where the fibers are bent accordingly to the customized geometrical blocks of several bending radii, R_c. It is known that when an optical fiber is bent it radiates power to the surrounding medium. The radiation power in the fiber depends on radius of curvature and the difference between refractive indices of core and cladding. The size of bending radius of the blocks either circular or elliptical gives the coupling length, L_c, of the contact region of the lapping cores. Coupling length is important in order to obtain high splitting ratio and coupling efficiency.

Two forces, F_c are exerted upon blocks that hold the lapped splitter together, i.e., normal force and given force. The aim is to characterize the splitter when the fiber cores touches each other without external force and recorded as normal force and with external force or namely given force. Force is exerted upon the blocks and fibers in order to minimize the gap that exists between the two lapped fibers in order to observe and characterize the energy transfer of the splitter when different bending radii are formed along with particular core-cladding thickness at certain coupling length. The force exerted upon the fiber also has small impact on the coupling length between the two fibers.

The pair of fibers will be attached to different circular and ellipse-shaped blocks that consist of different bending radii, R_c. The tapering and bending contributes to the effective index of higher-order modes that approach cladding and when the tapering section is constant, the modes become cut-off and radiate which contributes to the loss.

The aim of varying the bending radii, R_c is to characterize and analyze the macro-bending effect on the splitter having varied diameter, D_c, varied etching length, L_e, coupling length, L_c with different load force, F_c. The aim of the design is to obtain optimum bending radius that gives the optimum splitting ratio and low excess loss.

Analytical work is considered where the parameters of core thickness, D_c, coupling length, L_c and distance between the two fibers, d, are taken into account in order to obtain the coupling efficiency between the two fibers. The bent fiber using circular and elliptical blocks encourages the modes to radiate out of the tapered section. The amount of power transfer is calculated using Coupled Mode Theory. The coupling length between the two fibers also affects the coupling efficiency [24].

1.9.1. Taper

Taper changes the fiber diameter or thickness which allows redistribution of the modes in the core or eliminates the modes. The penetration depth of the evanescent field and proportion of power within this field increases in the tapered section. The modified core-cladding thickness is used for coupling fibers or interface fibers to devices [13]. In [17] in their research stated that they polished the fiber for several micrometers and lapped the fibers together and when source is inserted, mode will couple between the lapping regions and split the light accordingly. By adjusting the alignment between the two lapping fibers, the strength of the evanescent light coupling can be tuned. However, [17] do not apply mechanical technique of circular and elliptical blocks to vary the splitting ratios, rather the alignment is varied. In [13] in the other hand manipulate the thickness of cladding in order to obtain desired results. In [25] states that at the end of the tapering section of the first fiber, the rays propagate in the cladding will be recaptured by the core. If the tapering section of parallel fibers is small, the coupling ratio is high. However, if the radius of the tapered section is too small, the coupling efficiency will decrease dramatically due to the lost rays while passing through the down taper.

For some tapered fibers in this research, whole surfaces were etched around the fiber creating extra taper regions that allows unnecessary losses of the splitter. Thus, circular and elliptical blocks were made of acrylic and when the fibers are placed in the groove, the non-lapping surfaces are surrounded by acrylic material. This material is aimed to reduce the losses of the splitter due to tapering of the fiber cladding.

1.9.2. Bending and radiation

Generally, loss is high when the bending is smaller. Radiation loss occurs in the cladding even when the propagating ray is greater than critical angle. If light facing total internal reflection, an electromagnetic disturbance occurs which is known as evanescent wave where it penetrate the reflecting interface. Non-uniformity in the reflecting interface may cause the evanescent wave to convert into propagating wave where in bending loss, evanescent fields extend to the cladding but decay exponentially with radial distance [26].

The loss in bent step-index multimode fibers not only depends on bending but also on the cladding thickness [11]. When a refracting ray hits a core-cladding interface, two rays are created where one ray is refracting at the cladding interface while the other ray is tunneling at inner core interface. At cladding interface, some rays will be reflected back and some will be refracted with considerable power content [27]. By decreasing the bend radius and increasing fiber core, losses will increase. As bend radius begins to decrease, more refracting rays are produced and more power is transferred to the cladding [28].

Some rays in the incident radiation are not bounded by core of the fiber rather the rays that propagate through the core-cladding interface and get into the cladding region. Due to the finite radius of curvature at the cladding surface, some of the rays will be reflected back into the cladding and propagates while some will radiate. The rays that propagate in the cladding are known as the cladding modes and coupling can occur with the higher-order modes of the core resulting in loss of the core power. According to [27], the refracting rays lost most part of their energy at the beginning of the bent section and then the rate of loss will be slower whilst the remaining losses afterwards are due to weakly leaky rays which are known as whispery gallery rays of tunneling rays.

Therefore, this research integrate the concept of taper, lapping and bending the splitter in order to increase or limiting the splitting or coupling of rays between the two lapping fibers in order to gain certain splitting ratios.

1.9.3. Coupling length

The coupling efficiency describes the total power of coupling between the two fibers depending on the distance, fiber core thickness and length of the contact region. In [14] stated that cross type coupler is not able to achieve high coupling efficiency due to its short coupling length. Coupling length must be long in order to achieve high coupling efficiency such as parallel type coupler. For short coupling length, coupling is dominant only for higher order modes. In cross type coupler, the radiation loss is increased by increasing the pressure since the coupling length is quite short in order to gain optimum result. In [14] shows that as the coupling length is longer, the coupling efficiency will be stabilized. However, high coupling efficiency is achieved at coupling length between 6 to 10 mm and from 18 to 22 mm.

1.9.4. Distance of cores

In [29] agrees that the distance between the two fibers affects the coupling efficiency among other considered parameters. Although coupling length is important, however, the optimum efficiency of coupling not only depends on the optimum length but also on the distance between the two cores. In [29] shows that when the gap is zero between the lapping cores, high coupling efficiency is achieved, however, when gap exists or distance over the two cores is 1.01, the coupling efficiency drops to less than 30%. When the gap or distance over the two cores is further increased to 1.05, the efficiency drops to less than 5%. The distance between the two fibers affects the coupling efficiency strongly.

According to [25] states that when the ray of light propagates along the tapered section of the lapping fiber, the angle of incidence on the core surface decreases with each reflection and the high-order modes may have cutoff points in the down-taper section and therefore will leak out the core. If the cladding modes encounter the air-cladding interface at incident angles smaller that the critical angle, the modes will radiate away from the fiber at the air-cladding interface. However, if the claddings of the two lapping fibers are closed enough over appropriate length, the light in the cladding of the fiber will be transferred to the second fiber.

Therefore, load is accounted in this research in order to minimize if not eliminate the gap between the two lapping cores. The difference of characterization is compared when normal force is exerted and external force is exerted.

1.10. Fiber preparation development

Apart from the platform design, the development of the coupler/splitter also comprises of three sets of fibers tapered by etching method and side polishing. The first set of fibers were prepared by fully etching using harmless chemical solvent, acetone. During the etching process, the tapered length of each fiber was set to be 25 mm long and the duration of each pair of the fiber strands was varied from 30 to 120 minutes. For second set of fibers using fully etching method, the etching length of tapered length of the fibers were varied for each pair between 4 and 25 mm long and the duration of etching also was set to 60 minutes. The third set of fiber pairs were tapered using side polishing method and side etching where only one side of the fiber strand was polished and etched to clean the rugged surfaces. The tapered length of the fiber pair of each coupler/splitter is also varied from 4 to 25 mm long and the duration of etching is about 60 minutes. The effect of etching duration leads to the different diameter of core-cladding of the fiber pairs. The longer the etching process, the smaller the core thickness diameter. **Figure 2(a)–(c)** show the surface of the fibers before etching process, **Figure 2(a)** and post etching fiber, **Figure 2(b)**.

This shows that the physical property of the fiber strand is sustained when the fiber is tapered using chemical solvent. However, when the fiber is etched while it is in bending state, brittleness occurs and leads the fiber to break apart. **Figure 2(c)** shows the physical state of the fiber

http://dx.doi.org/10.5772/intechopen.71868

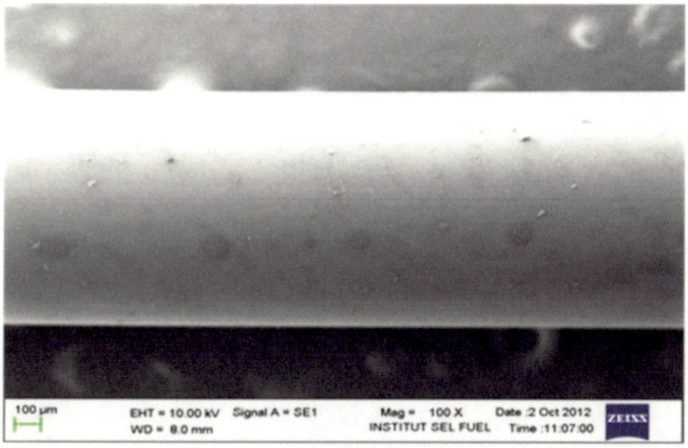

(a)

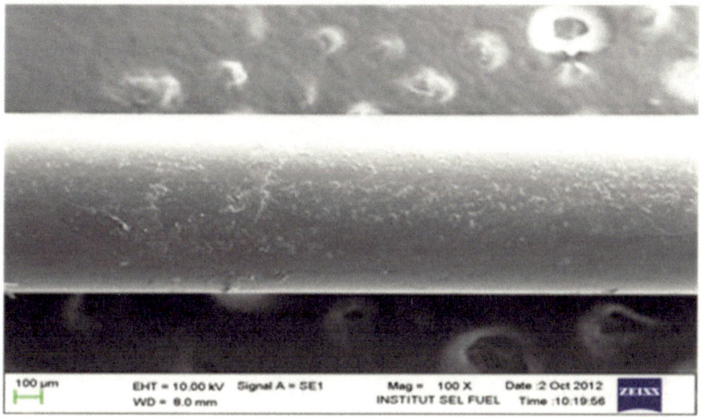

(b)

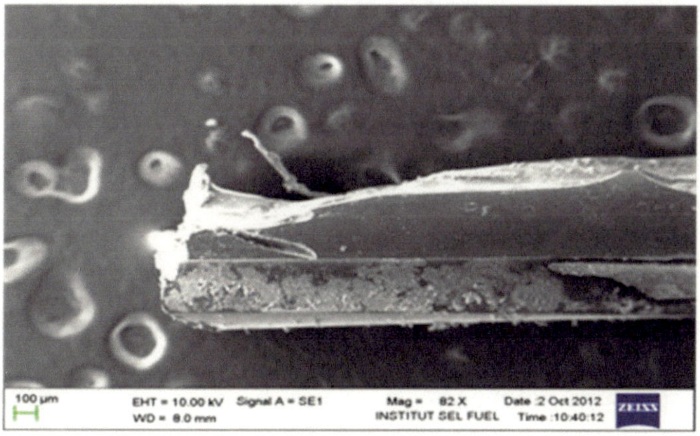

(c)

Figure 2. Physical surface of the (a) pre-etch (b) post-etch fiber and (c) fiber breaks due to brittleness.

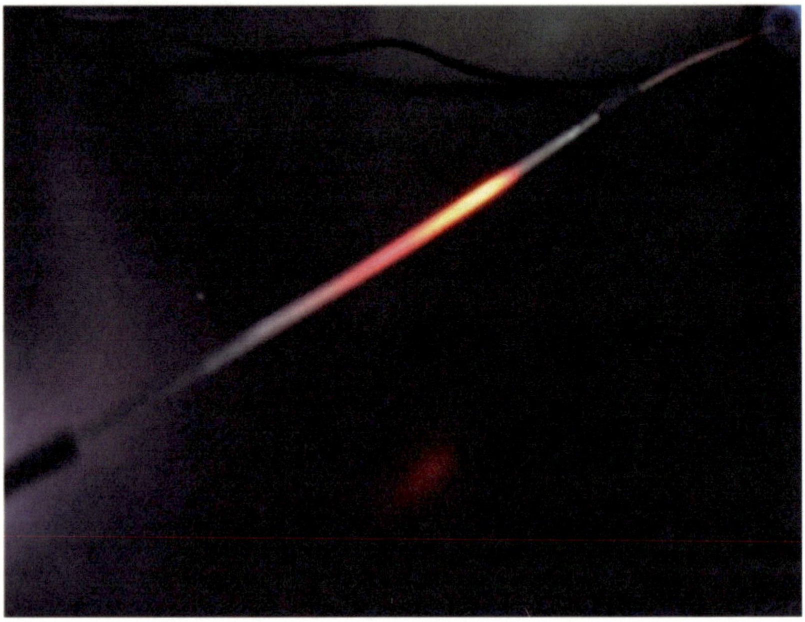

Figure 3. The light transmitted over the etched fiber.

when brittleness causes the fiber to break. **Figure 3** shows the light signal transmitted over the etched fiber and it shows that the light spread out at the etched region only and due to this behavior, insertion loss increases and output power decreases.

1.11. Platform development

The coupler/splitter platform is one of the important parts of developing this directional coupler using lapping technique and geometrical blocks. The platform of the coupler is customized using acrylic material. The platform of the fiber coupler/splitter is consisted of mainly circular/elliptical blocks of various radii, made of acrylic material that has similar refractive index as the cladding of the fiber.

1.11.1. Circular

The first phase of platform development involves the development of circular blocks made of acrylic material as shown in **Figure 4**. The circular block functions to hold the etched fibers while the fibers will be bent accordingly to the bending radius of the circular blocks. A groove of 1 mm is carved along the edge of the blocks in order to place and hold the fibers as shown in **Figure 5**. The tapered area of the fiber is facing outward of the groove and will be lapped to the other tapered surface so that coupling or splitting between the two fibers can occur. The etched area or surface that is not facing or lapping to the other fiber is covered by the groove of the blocks that are made from acrylic. Acrylic material has similar refractive index as the cladding layer, $n = 1.402$. Therefore, the unlapping surface of the etched area is covered by similar cladding layer refractive index. Thus, when the propagating modes are traveling along the etched fiber, some of the modes will be transferred to the other lapped fiber while some the modes that radiated out of the unlapping area is bounded by similarly refractive index of cladding, that is the groove of the block. Therefore, this will decrease the loss. The pivot is designed to hold the etched fibers that are placed at the groove so that they do not move and the tapered surfaces are lapped to each other. The screws of the pivots are used to loosen and tighten the pivot accordingly so when other block of bending radius is placed, the fiber can be bent according to the bending radius of the circular blocks.

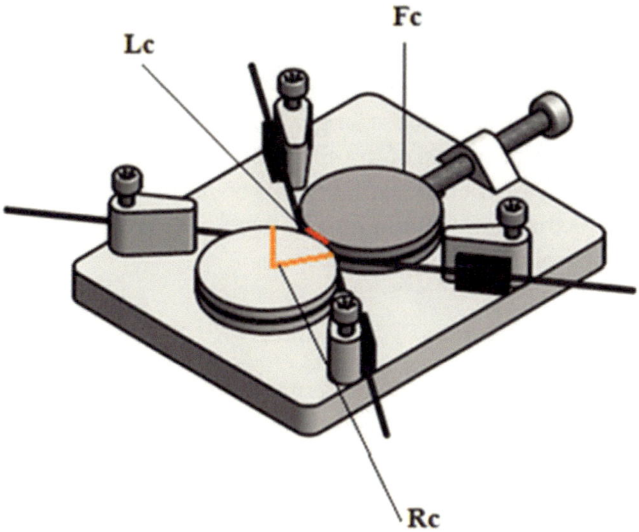

Figure 4. Circular blocks platform and fibers lapped.

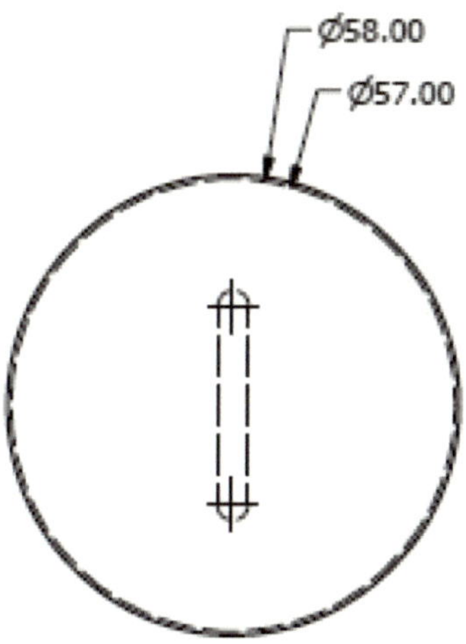

Figure 5. The bottom view of a circular block with groove showing at the round edge of the block.

1.11.2. Elliptical

Elliptical blocks shape are designed apart from circular blocks as another geometrical shape in order to study the effect of different bending radius, R_c, and coupling length, L_c, between the two lapped fibers. Apart from circular blocks having bending radius or curve that is more critical than elliptical shape, circular blocks are used to mainly study the effect of bending radius when the fibers of the coupler are bent at certain bending radii. Elliptical shapes in the other hand, when fibers are attached to the elliptical blocks and bent, the bending radii are less curvier than that of circular blocks, however, the curves of ellipse shape blocks are more flatter thus the coupling length between the two lapped fibers are longer than that of circular blocks. Bending does play part in order to stimulate the transfer of modes from the first fiber to the second one, however, another parameter that also play an important part is coupling length, L_c. The bending radius of elliptical shapes range from 10 to 29 mm. The bigger the bending radius of the elliptical shapes, the longer the lapping region between the two lapping cores. **Figure 6** shows the dimensions of the elliptical platform together with the force gauge embedded in the design. The experiment platform is big as to characterize and analyze which bending radius range is the most optimum to be used and developed as an efficient optical coupler/splitter using lapping technique.

1.11.3. Semi-elliptical

The third phase of the development is using semi-elliptical shaped blocks and platform with spring embedded as shown in **Figure 7**. The semi-elliptical shaped blocks have bending radius of 30, 40 and 50 mm. No force gauge is used on this platform because the force is exerted unto the blocks and fibers by the spring embedded in the platform.

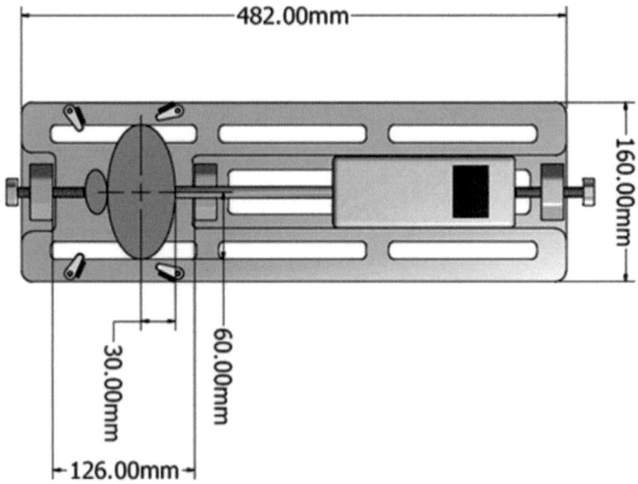

Figure 6. Elliptical blocks platform design measurement with ellipse-shaped blocks.

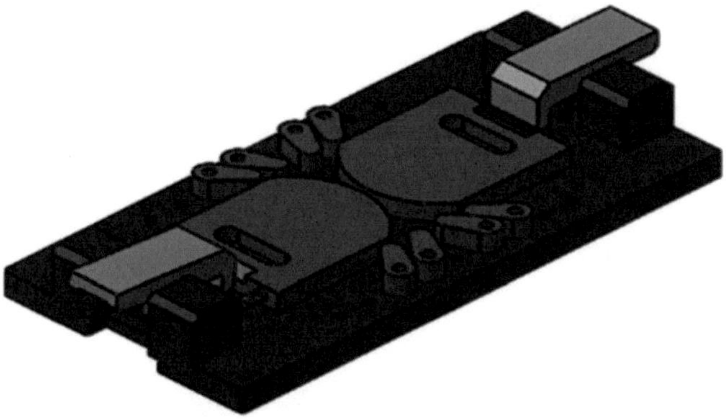

Figure 7. Coupler/splitter experimental platform using semi-elliptical blocks and embedded spring.

1.12. Coupling efficiency by integration of CMT and Hertz's law

The amount of power transfer is relatively in accordance to the coupling length. Thus, in this study, two very important theories are applied where the force exerted on the fiber through geometrical blocks relates to elliptical point contacts of Hertz's Law [30] and the amount of force put upon the fibers determines the radius of contact area or coupling length which brings to Couple Mode Theory. The propagation of modes between the two fibers is studied analytically and coupling efficiencies are obtained by varying the load force, coupling lengths and the distances between the two fibers.

To obtain an efficient coupling ratio or splitting ratio, the coupling length between the two lapped fibers must be long in order to obtain an adequate level of coupling efficiency. In this research, two similarly fibers were tapered at the middle region with particular diameter of core-cladding. They are attached to the circular blocks/elliptical blocks of certain bending radius that determines the bending angle of the lapped fibers that helps the transfer of energy from primary fiber to the secondary fiber. The performance of the splitter/coupler is analyzed through the relationship of the distance between the two fiber waveguides and the load put upon the blocks and fibers which gives effect in the coupling length. For the multimode step index fiber, a group of modes exist as according to parameters assigned by the optical waveguides. Between reflections of each of the propagation rays, each ray travels in straight line and Snell's Law determines the reflection on the interface [31].

Coupling efficiency is calculated by first specifying the coupling length, L_c, which in this design is assumed to be directly relative to radius of contact area, *c*, of Hertz's ellipsoid. Coupling efficiency is done by integrating the coupling coefficient and coupling length.

$$\eta = \frac{P_{A\rightarrow B}}{P_{in}} \leqq \frac{1}{N}\int_0^1 \sin\left(\frac{(2^{1/4}\,\Delta^{1/4})\,L_c}{\sqrt{\pi k}\,n_0\,D_c^{\,3/2}}\,t\,(1-t)^{1/4}\right)^2 dt \tag{1}$$

$$\eta = \left(\int_0^1 \sin\,(C_{coef}\cdot L_c)^2\,dt\right) \tag{2}$$

Here the relationship between CMT and Hertz's can be focused on coupling length or twice the radius of contact area of the elliptical point contacts of two spheres.

$$\eta = \int_0^1 \sin\left[\left(\frac{1}{\sqrt{\pi}}\frac{\sqrt{NA(n0, n1)}}{\sqrt{k(\lambda)\cdot(D_c)\cdot(n0)}}\right)\left(\frac{(t)\cdot(1-t)^{1/4}}{D_c}\right)(2)\left(\sqrt[3]{\frac{3\cdot F\cdot(Re)}{4\cdot E}}\right)\left[(F1)\left(\frac{R1}{R2}\right)\right]\right]^2 dt \quad (3)$$

This expression will be used to vary the distance between the two fibers having different load F_c and the coupling efficiency of the splitter can be determined accordingly.

1.13. Performance

During characterization of the couplers/splitters, the experimental set-up is first prepared as shown in **Figure 8**. Geometrical blocks of different radii are placed on the platform with pair of tapered cladding secured in the groove of the circular/elliptical blocks. At each of the ports of splitter/coupler except at the input port, power meter is set to take readings of the output power. Red LED of $\lambda = 650$ nm is injected through input port and normal force of $F_n = 0.3$ lbF is exerted upon the middle region of the coupler/splitter through the geometrical blocks as shown in **Figure 8**. Normal force is the reading at which the two cores of the fibers touch each other without any external force. Output power is taken accordingly at each port. Then, given force or external pressure is exerted upon the blocks and fibers, namely $F_c = 3.0$ lbF. This external pressure is presumed to minimize the air gap that existed between the two lapped cores.

The experiment was repeated twice as in the first test, normal force is exerted upon the blocks and the fiber cores and in the second test, external force is exerted upon the blocks and the fiber cores. Based on the data collected, efficiency of each coupler/splitter can be measured. The efficiency of a coupler/splitter, σ can be defined as power ratio of overall output, ΣP_o against the power input, P_1. The mathematical equation that refers to the coupler/splitter efficiency is:

$$\Sigma P_o = P_2 + P_3 + P_4 \quad (4)$$

$$\sigma\,(\%) = \frac{\Sigma P_o}{P_1} \times 100\% \quad (5)$$

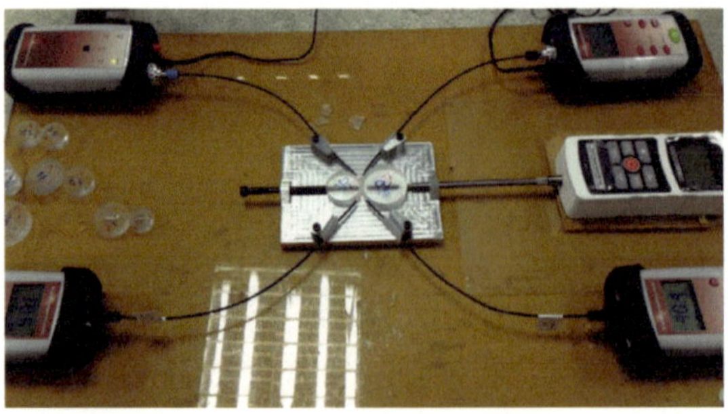

Figure 8. Experimental set-up platform with force gauge.

P_2, P_3 and P_4 are the output power of the light signal that propagates out of the throughput port, P_2, coupled port, P_3 and reflected port, P_4, while P_1 is the input port of the injected light signal.

The objective of this work which is to apply theories of CMT and Hertz's Law in studying the effect on coupling efficiency by manipulating the cores and coupling lengths between the fibers are analyzed and the optimum efficiency obtained shows that when the two fiber cores are closed to each other, the efficiency lies between 40% and 70% depending on the coupling length and distance is zero. When force exertion is small, the coupling length decreases thus coupling efficiency decreases to less than 50%. It decreases when the force, F_c is less and the coupling length, L_c is shorter. The efficiency decreases as the distance, d, between the two fiber cores is bigger. The optimum range of efficiency achieved based on coupling length, L_c, depends on the fiber core size, D_c. The diameter of the cores affects the efficiency where optimum efficiency is achieved at shorter coupling length range when the core diameter is smaller. This study gives an insight of the optimum distance and fiber core size of the lapping fibers used in the experiment.

Experimental results show the splitters having different splitting ratios as high as 80% to as low as 1%. Each of the splitter has different bending radius, R_c and different tapered length, L_e. Macro-bending effect shows that different bending radius of circular and elliptical blocks allow different bending angle for each splitter. Small angle of bending leads to radiation of rays that propagate along the bent section. The radiation is enhanced by the tapered regions at the bent section where some of the cladding layers are etched to allow coupling between the lapping sections. Different core-cladding thickness may influence the amount of rays coupling into second fiber. Large bending radius in the other hand slows the radiation rate, however, helps the splitting and coupling by lengthen the lapping region or coupling length. Diameters of 0.88 mm of splitters show optimum splitting ratios such as 80%, 70% and 60%. However the losses are high due to the surfaces of the tapered fibers that were etched wholly around the surfaces. The non-lapping section may contribute to the losses. Diameters of 0.77 mm in the other hand give splitting ratios between 20% and 50% with considerable losses. All bending platform of circular, elliptical and semi-elliptical blocks contribute to these range of splitting ratios. Side-polished and side-etched splitters and fully etched splitters with varied etched length are mostly used with the platforms to obtain optimum results. Long tapered region as given by fully etched fibers with constant etched length shows very low splitting ratios with considerable high losses. However, shortest tapered region does not give the most optimum results. Therefore, depending on each parameter, particular tapered length, diameter and bending have to be considered into the design to obtain desired optimum results.

1.14. Maintenance and reproducibility

Another important attributes that represents a coupler include its flexibility, maintenance and reproducibility. The device is flexible since the fiber pair and blocks can be exchanged in order to get desired splitting ratio. Even though the blocks needs to be exchanged for obtaining desired splitting ratio, the blocks and fibers are ready to be fitted into platform and they are highly durable. Since the splitter is custom-made, the splitter can be reproduced by handling each process of fabrication meticulously to the etching rate, polishing rate and bending size in order to achieve uniform and persistent results.

1.15. Installations and performance

Installations and performances are very important attribute to any coupler or splitter developed. Although a coupler is a small component in a system, it does play an important role throughout the whole performance. Although the coupler/splitter developed has considerable higher loss than the ones in the market line, the ability of the coupler to achieve several splitting ratio has high potential where improvements and expandability of the device can be done in order for the device to compete in the market line.

The device developed is easy to install using prepared platform and mix matched fiber pair and blocks where varied splitting ratio can be obtained using one platform. Although targeted loss is considerably higher than marketed splitter, however, the device is applicable for signal transmittance and applications. The device developed is green technology-based since the production process is using eco-friendly material, harmless solvent, moreover, LED source is used which is very safe for consumers.

1.16. Research future Prospect

The limitation of "DIY" kit can be overcome by this design. Since the design of this splitter gives several solution of splitting ratios, users that demand different value of splitting ratios can utilize the splitters. Although the prototype design for users are yet to be finalized, however, based on the results shown, the platform shows good performance and can be realized as DIY kit. Apart from that, the since the cost of POF is low and materials and tools needed to build the platform are inexpensive and does not involve high-end expensive machine, this device has low-cost production and very economic. The material used to develop this device such as POF, harmless chemical solvent, acrylic and aluminum are safe and green technology based.

1.17. Summary

New technique of developing an optical coupler using POF and mechanical platform using lapping technique is discussed and analyzed in this chapter. Three different categories of splitters are prepared where the first category of splitters has constant etching length of 25 mm and the diameter of the tapered section is varied. The second category is splitters that have been etched with different length between 4 and 25 mm but having constant diameters of 0.88 mm. The third category is splitters that have been polished and etched at one side of the fibers only having different etching length between 4 and 25 mm and constant diameter of 0.77 mm. Three different platforms are also built. The first platform having small angle of circular blocks, the second platform having larger angle of elliptical blocks and the third platform having intermediate angle between small and large angle of semi-elliptical blocks with spring embedded. Varied angles represented by the bending radii of the blocks are chosen and designed in order to study the bending effect of the splitter. Different bending radius with combination of fiber diameter and coupling length leads different coupling behavior between the two lapping regions and therefore provides different splitting ratios.

http://dx.doi.org/10.5772/intechopen.71868

Author details

Latifah S. Supian[1]*, Mohd Syuhaimi Ab-Rahman[2], Norhana Arsad[2], Chew Sue Ping[1], Nani Fadzlina Naim[3], Nurdiani Zamhari[4] and Syed Mohd Fairuz Syed Mohd Dardin[1]

*Address all correspondence to: sarah@upnm.edu.my or cawa711@gmail.com

1 Department of Electrical and Electronics Engineering, Faculty of Engineering, Universiti Pertahanan Nasional Malaysia, Kuala Lumpur, Malaysia

2 Department of Electric, Electronics and Systems Engineering, Faculty of Engineering and Built Environment, Universiti Kebangsaan Malaysia, Bangi, Selangor, Malaysia

3 Faculty of Electrical Engineering, Engineering Complex, Universiti Teknologi MARA, Shah Alam, Selangor, Malaysia

4 Department of Electrical and Electronics Engineering, Faculty of Engineering, Universiti Malaysia Sarawak, Kota Samarahan, Sarawak, Malaysia

References

[1] IGI Consulting. Plastic Optical Fiber Market & Technology Assessment Study. Boston, MA: IGI Group, Information Gatekeepers Inc.; 2011

[2] Khoe GD, Boom H, Monroy IT. High Capacity Transmission Systems. Polymer Optical Fiber. Stevenson Ranch: American Scientific Publisher; 2004

[3] Nalwa HS. Polymer Optical Fiber. Stevenson Ranch: American Scientific Publisher; 2004

[4] Supian LS, Ab-Rahman MS, Arsad N. Etching technique study for POF coupler fabrication using circular blocks. Optik Journal. 2014;**125**:893-896. Elsevier

[5] Supian LS, Ab-Rahman MS. Polymer optical fiber coupler fabrication using chemical etching and lapping technique. In: 4th International Conference on Photonics (ICP 2013); Malacca, Malaysia; IEEE; 2013. pp. 184-186. DOI: 10.1109/ICP.2013.6687108. ISBN: 978-1-4673-6073-9

[6] Ghatak A. Optics. New Delhi: McGraw Hill Education; 2013

[7] Ab-Rahman MS, Guna H, Harun MH, Supian L, Jumari K. Integration of ecofriendly splitter and optical filter for low-cost WDM network solution. In: Optical Fiber Communication and Devices. InTech; 2012. ISBN: 978-953-307-954-7

[8] Strategic Services Division. Acetone. [online]. Available from: http://hazard.com/msds/mf/baker/baker/files/a0446.html [Accessed: Jan 15, 2015]

[9] Lin CF. Optical Components for Communications: Principles and Applications. Boston: Kluwer; 2010

[10] Chen CL. Optical Directional Couplers and their Applications. Hoboken, USA: John Wiley & Sons, Inc.; 2007

[11] Love JD, Durniak C. Bend loss, tapering and cladding-mode coupling in single-mode fibers. IEEE Photonics Technology Letters. 2007;**19**(16):1257-1259

[12] Ji F, Xu L, Li F, Gu C, Gao K, Ming H. Simulation and experimental research on polymer fiber mode selection polished coupler. Chinese Optics Letter. 2008;**6**(1):16

[13] Merchant DF, Scully PJ, Schmitt NF. Chemical tapering of polymer optical fiber. Sensors and Actuators A: Physical Elsevier. 2000;**76**(1-3):365-371

[14] Ogawa K, McCormick AR. Multimode fiber coupler. Applied Optics. 1978;**17**(13):2077-2079

[15] Tanaka T, Serizawa H, Tsujimoto Y. Characteristics of directional couplers with lapped multimode fibers. Applied Optics. 1980;**19**(20):2019-2024

[16] Findakly T, Chen CL. Optical directional couplers with variable spacing. Applied Optics. 1978;**17**(5):769-773

[17] Kawase LR, Santos JC, Silva LPC, Ribeiro RM, Canedo J, Werneck MM. Comparison of different fabrication techniques for POF couplers. In: The International POF Technical Conference, Cambridge, Massachusetts. Boston, MA: Information Gatekeepers, Inc.; 2000. pp. 68-71

[18] Gloge D. Bending loss in multimode fibers with graded and ungraded core index. Applied Optics. 1972;**11**(11):2506-2513

[19] Barnoski MK, Friedrich HR. Fabrication of an access coupler with single-strand multimode fiber waveguides. Applied Optics. 1976;**15**(11):2629-2630

[20] Musa S, Borreman A, Kok AM, Diemeer MBJ, Driessen A. Experimental study of bent multimode optical waveguides. Applied Optics. 2004;**43**:5705-5707

[21] Supian LS, Ab-Rahman MS, Arsad N, Ramza H. Study of macro-bending of polymer fiber in multimode POF couplers development by lapping technique. International Journal of New Computer Architectures and their Applications (IJNCAA). 2014;**4**(1):39-47. The Society of Digital Information and Wireless Communications. (ISSN: 2220-9085)

[22] Badar AH, Maclean TSM, Gazey BK, Miller JF, Shiraz HG. Radiation from circular bends in multimode and single-mode optical fibres. IEEE Proceedings. 1989;**136**(3):147-151

[23] Supian LS, Ab-Rahman MS, Arsad N. The study on pressure effect on characterization of directional coupler for different thickness of fiber cores in POF splitter development. In: The 2nd International Symposium on Telecommunication Technologies (ISTT 2014). Langkawi, Malaysia: IEEE; 2014. pp. 298-302. ISBN: 978-1-4799-5981-5/14

[24] Supian LS, Ab-Rahman MS, Ramza H, Arsad N. Characteristics study of multimode directional coupler by elliptical point contacts and CMT. In: 2nd International Conference on Applications of Optics and Photonics Proceedings of SPIE. San Francisco, USA. Vol. 9286 92863K-1. 2014. DOI: 10.1117/12.2063541

[25] Li YF, Lit WY. Coupling efficiency of a multimode biconical taper coupler. Journal of Optical Society of America A. 1985;**2**(8):1301-1306

[26] Marcuse D. Curvature loss formula for optical fibers. Journal of Optical Society of America. 1975;**66**(3):216-220

[27] Durana G, Zubia J, Arrue J, Aldabaldetreku G, Mateo J. Dependence of bending losses on cladding thickness in plastic optical fibers. Applied Optics. 2003;**42**:997-1002

[28] Winkler C, Love JD, Ghatak AK. Loss calculations in bent multimode optical waveguides. Optical and Quantum Electronics. 1979;**11**:173-183

[29] Ogawa K. Simplified theory of the multimode fiber coupler. The Bell System Technical Journal. 1977;**56**(5):729-745

[30] Johnson KL. Contact Mechanics. Cambridge: Cambridge University Press; 1985

[31] Snyder AW, Love JD. Optical Waveguide Theory. New York: Chapman and Hall; 1983

Chapter 6

Plastic Optical Fibre Sensor System Design Using the Field Programmable Gate Array

Yong Sheng Ong, Ian Grout, Elfed Lewis and
Waleed Mohammed

Additional information is available at the end of the chapter

http://dx.doi.org/10.5772/intechopen.71451

Abstract

Extrinsic optical fibre sensor (OFS) systems use a fibre optic cable as the medium for signal propagation between the sensor and the sensor electronics using light rather than electrical signals. A range of different optical fibre sensors have been developed and electronic hardware system designs interfacing the sensor with external electronic systems devised. In this chapter, the use of the field programmable gate array (FPGA) is considered to implement the circuit functions that are required within a portable optical fibre sensor system that uses a light emitting diode (LED) as the light source, a photodiode as the light receiver and the FPGA to implement the system control, digital signal processing (DSP) and communications operations. The capabilities of the FPGA will be investigated and a case study sensor design introduced and elaborated. The OFS system will be based on the FPGA and will provide wireless communications to an external supervisory system. The chapter will commence with an overview of OFS systems and the typical architecture of the system. Then the FPGA will be introduced and discussed as a hardware alternative to a software programmed processor that is currently widely used. A case study will then be presented with a discussion into design considerations.

Keywords: optical fibre sensor, field programmable gate array, electronic system design, SPR sensor, Python

1. Introduction

Recent advances in optical fibre sensing technology have been rapid and widespread stimulated through achievements in developing sensors for applications such as environmental monitoring, mechanical structure stress and strain monitoring, and biomedical [1–4]. Advances in sensor fabrication, sensitivity enhancement and the availability of new materials have demonstrated the potential of using optical sensors in new applications. The aim is to harness the advantages that optical components can provide over electrical component alternatives in a wide range of sensor applications. However, identifying and using suitable enabling technologies, supported through experience in achieving repeatable results from suitable experiments for new OFS industrial applications, should be recognised as key in moving forward from a research idea into application [5].

It is widely accepted that initial OFS designs were fragile, reliable only under laboratory conditions and with material costs higher than those used in other sensor technologies [6, 7]. They were also difficult to produce and this limited their ability for practical use. However, with an increase in the availability of new materials, the proliferation of inexpensive 3D printing and other additive manufacturing technologies that enable rapid prototyping and analysis, sensor design and packaging performance have been improved dramatically. These OFS packaging improvements have enabled optical fibre sensor (OFS) deployment [8] in a wide range of different applications as referenced above. An OFS system is however not complete without supporting instrumentation [9]. It requires a suitable electronic system to implement system functions that include system control, sensor stimulus generation, data acquisition (DAQ) and storage, data analysis (high-speed digital signal processing (DSP) operations) and serial communications interfacing. To date, OFS system deployment is still largely dependent on laboratory-based sub-systems, particularly the light source and optical receiver that interface to a computer through a data acquisition (DAQ) module or proprietary software via universal serial bus (USB). A typical OFS system needs to use high-cost and complex instruments to provide the necessary control and measurement capabilities, and usually depends on a human operator to set-up the light source and access the sensor results. To make optical sensors more accessible, the use of high quality but lower cost equipment is required. Some improvements have been noted in recent system designs where operations were performed using a software programmed microcontroller based arrangement [10–12]. The potential also exists to create supporting instrumentation that can harness inexpensive and physically smaller components to make an OFS system portable (i.e. a battery operated system capable of operating independently for extended periods of time, small physical size, programmable with wireless communications and ability to be carried by an individual, e.g. a first responder). The low cost aspect however must not compromise the quality of the system operation and it must also have reliable operation in the intended application. The inert nature of optical fibre also allows it to perform in-situ measurements in harsh environments whilst the sensitive electronic circuits can be shielded at an appropriate location and remotely from the environment being sensed. In addition, for complex photonic system design, photonic integrated circuit (IC) technology is a promising solution to achieve lower cost and miniaturisation, but is still not yet as mature as electronic technology [13].

This chapter will consider the electronic system design requirements for a portable OFS system based on the use of a microcontroller alternative. Specifically, the field programmable gate array (FPGA) is used to implement the necessary digital logic functions in hardware including light emitting diode (LED) (light source) switching and current control and photodiode (light receiver) signal sampling and signal conditioning. A plastic optical fibre (POF) arrangement is considered, although the focus of the chapter will be on the FPGA based electronic system design approaches and potential solutions. The chapter will consider the system design, both analogue and digital parts, and will develop a discussion into the use of the FPGA as a hardware based alternative to a software based processor design. The available hardware resources within the FPGA will be considered and approaches to using these resources will be discussed. The FPGA, as an alternative, provides potential benefits to system design and operation by using configured hardware that provides the potential for high speed operation, parallel processing (concurrent operation), high-speed embedded DSP operations and low-voltage/-low-power capability. The available hardware resources within the FPGA, along with the capability for low-voltage and low-power operation, mean that a powerful whilst portable sensor system can be developed and deployed. This chapter will consider the potential benefits of using the FPGA in such sensor systems and the design choices required. The chapter will commence with an introduction to OFS systems and electronic system design using the FPGA. It will then discuss the use of the FPGA within an existing OFS system and the benefits the FPGA can provide before providing a conclusion.

In this chapter, the discussion will be based on a single sensor arrangement to demonstrate the use of the FPGA. However, the FPGA would be easily capable of interfacing to multiple sensors due to the availability of configurable hardware within the FPGA and the availability of a large number of programmable digital pins which can be configured to act as inputs, outputs or bi-directional pins. It would therefore be practical for the FPGA to be used in multiple (multipoint) and quasi-distributed OFS systems of the form shown in **Figure 1**. Additionally this concept may ultimately be extended to use in fully distributed systems, e.g. Rayleigh and Brillouin scattering based systems, utilising the considerable on-board capabilities of the FPGA in fulfilling the complex signal process required from these systems. However, this article is constrained to assessing the feasibility of a potential distributed OFS application where multiple 'single point' sensors are distributed in location but connected to

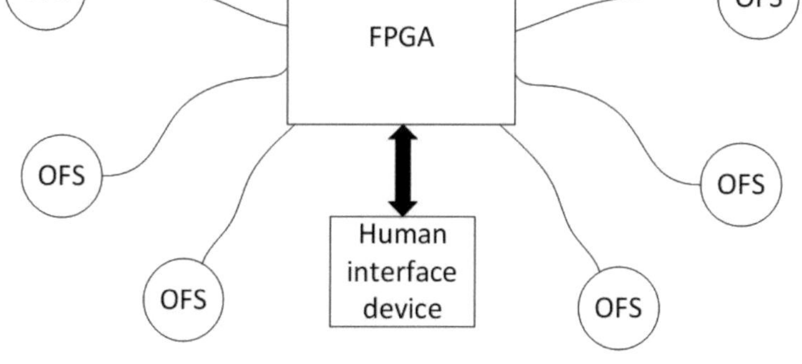

Figure 1. Distributed sensing with multiple 'single-point' sensors.

a single FPGA. The high number of available FPGA I/O pins are especially useful in allowing independent driving of multiple light sources and taking multiple light intensity measurements. Those measurements can be taken in a parallel manner and could be useful in phenomena detection [14]. Whilst the focus of this chapter is on a single 'single point' sensor, the basic design could be duplicated with other similar sensors and using the same FPGA to realise distributed sensing.

The chapter is structured as follows. Section 2 describes the OFS system operation and the design choices in the use of suitable components. Section 3 introduces the FPGA and how digital and mixed-signal electronic circuits and systems can be designed using the FPGA. This section also highlights a number of the advantages that the FPGA has over software programmed processor alternatives. Section 4 discusses the feasibility of use of the FPGA in an OFS system and Section 5 provides a real system based case study design using POF to provide an optical interface to the surface plasmon resonance (SPR) sensor. Section 6 concludes the paper.

2. Optical fibre sensor systems

2.1. Introduction

In this section, the structure and operation of a typical optical fibre sensor system is discussed. A typical OFS system essentially comprises of a light source, sensing element, optical detector and sensor data readout parts (external communications). A simplified architecture of such a system is shown in **Figure 2**. Light is guided from the light source to the optical receiver using an optical fibre. The sensing element is the part that transforms a change in the monitored parameter into a corresponding change in one or more of the physical properties of the guided light. Those properties include purely intensity, wavelength, polarisation orientation and phase. A change in the properties of the transmitted light through the sensor can be created by different physical mechanisms. For example, the chemical properties of a material being sensed. To capture the change in the propagated light properties, an appropriate light source

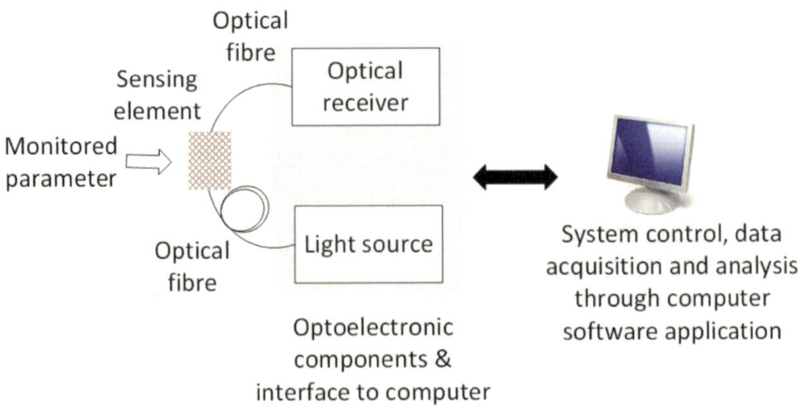

Figure 2. Simplified architecture of an OFS system (either intrinsic or extrinsic).

and matching optical receiver are required. The light source provides the sensor light stimulus of a particular wavelength or band of wavelengths to propagate to the sensing element. This light is modulated by the sensor and the resulting signal is captured by optical receiver. The optical parts of the system are linked together using an optical fibre cable. POF is becoming an increasingly popular choice in many sensors when compared to glass optical fibre (GOF) due to its low-cost, ease of connection and flexibility (allowing it to be mechanically bent and shaped to the requirements of the sensor housing and location). The source and receiver are electrically connected to the computing device that provides an optical-electrical interface, system control, communications and data analysis. The complexity of an OFS system is usually dependent on light modulation method used, the type of sensor and whether a single or multiple sensors are required. For example, distributed sensing requires complex light sources, coupling optics and receivers to obtain a sufficiently high-resolution result [14]. Distributed sensing ideally enables the measurement of any point along the optical fibre and is used in applications such as temperature, strain and acoustic sensing. Other OFS types include multi-point sensing or single point sensing [15, 16]. A multipoint sensor is similar to distributed sensing and therefore often also requires complex instrumentation although the cost of the interrogator unit is often much less than the fully distributed sensor. On the other hand, a single point sensor usually is the simplest form that requires less complex instrumentation. In all cases, the sensors may be classified as being either intrinsic or extrinsic. An intrinsic sensor uses the fibre itself as the sensor whilst in an extrinsic sensor, the fibre is used only used for light propagation to and from the sensor element which is therefore an external device. Clearly in the case of distributed sensors, e.g. Rayleigh or Brillouin, these are examples of intrinsic sensors.

The individual components of the OFS system are considered in the following sections.

2.2. Optical fibre

An optical fibre is a flexible optically transmitting waveguide that is drawn from glass or plastic [7]. In its simplest form, it is composed of three parts. Referring to **Figure 3**, these are the core, cladding, and coating. Both the core and cladding are made of a suitable dielectric

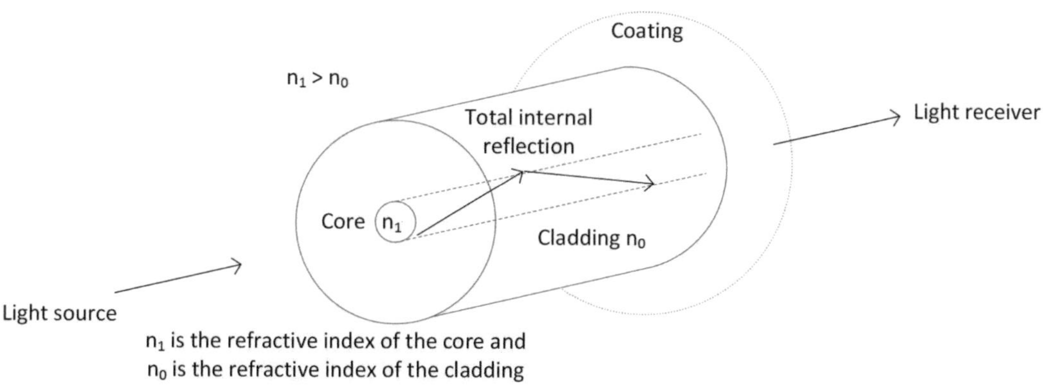

Figure 3. Optical fibre structure.

material, usually doped silica glass or a polymer. The core usually has a cylindrical cross-section and has higher refractive index (n) than the cladding that surrounds it. The coating is an additional layer of material, often a toughened polymer that provides protection from physical damage whilst maintaining flexibility. The interface of the core and cladding can be considered as a perfect abrupt but completely smooth boundary and the refractive index difference between the materials allows for total internal reflection to occur whilst minimising scattering losses [17]. In the case of total internal reflection, all of the light energy is ideally preserved within the core and allows the light to transmit through the fibre not only when the fibre is straight, but also when bent (up to a limit known as the minimum bend radius) as would be typically seen with a flexible fibre [7].

A number of fibre types are commercially available for different purposes and can be broadly categorised into two basic types: silica (glass) fibre or polymer (plastic) fibre. Glass fibre has widespread use in communications applications [18] due to its low-cost and material properties. It allows for extremely low loss long distance transmission of data with the need for fewer repeaters to boost signal strength due to losses in the transmission medium than would be required with electrical signal transmission. POF has found use in relatively short distance light transmission, e.g. in local area networks and, due to its flexibility, in situations that involve bending of the fibre, e.g. as in the case of automotive based networks. POF is generally large core fibre (i.e. a large fibre diameter, being typically 0.25 to 2 mm in diameter) that can bend into a smaller radius than can a silica fibre. There are some exceptions with much recent research being focused on single mode POF, e.g. for strain sensing [19]. Despite the material differences, both types of fibre operate under same principle of allowing the transmission of light. A major advantage of all optical fibres is that due to their material properties, an optical fibre is immune to external electrical interference. For example, unlike copper wiring, there is no cross-talk between signals in different close-proximity cables and no pickup of environmental noise, e.g. from proximal electrical machines or faulty electrical equipment. Recently, the application of the optical fibre has been extended from primarily a communications medium to the use as a sensor. An OFS introduces the idea of the modulation of a transmitted light signal as it propagates through a fibre and through different physical phenomena that are aimed to be measured [20]. Other specialist fibre types are available which result from novel structures or alternative materials. Photonic Crystal or Photonic Bandgap Fibres (PCF or PBF) rely on hollow or honeycomb (or many other sophisticated designs) patterned cores but currently their use is confined to some specialist applications, e.g. Supercontinuum Light sources [21]. Other specialist fibre materials are also available, e.g. Chalcogenide and ZBLAN for mid-infrared (MIR) operation, but again their use has been confined to some specialist sensing activities e.g. gas sensing [22].

2.3. Light source and optical detector

The light source and optical detector are fundamental parts of an OFS system. Both are required to have the necessary operating characteristics to be useful, are required to operate reliably over a long periods of time and to provide trustworthy data to the user. Light sources used to support low-cost optical fibre sensors are LED based and in some cases low-cost semiconductor laser diodes. Other systems use a broadband white light source and spectrometer arrangement. The choice of which light source to use depends on the required application. For example, the laser provides a very narrow wavelength range, whereas the white light source provides a wide band of wavelengths. LEDs also provide a relatively small band of wavelengths (broader than laser diodes) that is governed by the semiconductor material band gap energy. The light source can also be categorised as either a directional or diffused light source. A directional light source emits the majority of the light in one direction and with a narrow angle of spread. A diffused light source emits light in a wide range of directions equally and requires optical components to efficiently couple it into waveguide. Some OFS applications may be able to utilise an unpolarised light source, others may need a polarised light source. In addition, there may be a requirement to the use a coherent rather than an incoherent light source. The most commonly used optical detectors for low-cost optical fibre sensors are the photodiode, the avalanche photodiode and the phototransistor. These optical detectors translate optical intensity into an electrical signal (a current proportional to the intensity of the light that the device is sensitive to). In addition, the light dependent resistor (LDR) can sometimes be used which converts input light intensity to generate a change in electrical resistance. However the LDRs tend to be inefficient and slow compared to the photodiode/phototransistor type detectors. A single current output is produced which corresponds to an integral result of the intensities at all the detected light wavelengths. If a system is to be able to discriminate signals at discrete spectral values, an optical detector needs to be used in conjunction with an optical grating or filter [23]. The spectral bandwidth of the detector must also be aligned with the light source spectrum and the spectral transmission of the sensor. Other parameters to consider in choice of detector include response time and the noise figure.

In considering the use of the LED and photodiode based arrangement, the digital sub-system must provide the necessary digital-analogue-digital signal interface. A typical system arrangement is shown in **Figure 4**. On the transmission side, the LED requires a drive current to output a specific light intensity. This is produced by a voltage-to-current (VI) converter that receives a voltage and outputs a current proportional to the input voltage. As the signal originates from a digital sub-system, the digital output is required to be converted to its analogue equivalent

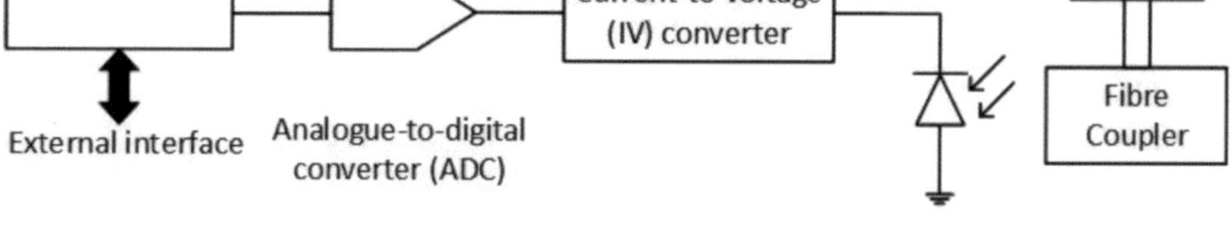

Figure 4. LED based light source current control and photodiode receiver circuit.

through a digital-to-analogue converter (DAC) with suitable speed and resolution. **Figure 4** assumes that the DAC output is a voltage rather than a current. The light is coupled to the POF and propagates to the sensor. Following transmission through and modulation by the sensor, the POF propagates the returning light signal to be received by the distal photodiode. The output of the photodiode is a current proportional to the intensity of the input light received. This is converted to a voltage though a current-to-voltage (IV) converter and applied to the input of an analogue-to-digital converter (ADC) with suitable speed and resolution. This set-up assumes that the ADC is a voltage input rather than a current input device. The digital subsystem captures and processes the digitised value. Results can then be processed and transmitted to a user or an external electronic system. The photodiode characteristics (light intensity range and range of wavelengths that can be detected) must be matched to the LED light source, otherwise the measurements do not accurately reflect the behaviour of the measured quantity.

2.4. Referencing

The accuracy of an OFS measurement depends on how reliably the value of the modulated signal reflects the behaviour of physical quantity being measured. However, other effects may cause the signal to change and therefore, their influence must ideally be isolated and removed in order to focus solely on the physical quantity being measured. In practice, this condition is not easy to achieve due to various factors. For example, the propagating light suffers from multiple attenuation factors in its propagation path which include losses from optical components and uncertainty introduced by optoelectronic component variability. To compensate for known effects that the system can be calibrated to account for, potential deviations from the ideal should be recognised and recorded. A common method is to apply an additional light signal that is used as a reference and is subjected to the same environmental influence as for measurement signal. Thus, the reference signal can calibrated to the measurement signal. There are different techniques that can be used to introduce a reference signal into system including spatial, spectral, temporal, frequency separation or through a combination of these methods [24]. Such an approach also requires some form of signal processing, analogue or digital, and hence this must be factored into the design. Increasingly, digital signal processing (DSP) techniques are used and in the case of software based processors or hardware configured programmable logic devices (PLDs), can readily be included in order to provide for programmable and local sensor signal conditioning.

2.5. Wavelength and intensity interrogation

Some sensing elements react only to a narrow band of wavelengths. These sensor systems use a simple intensity interrogation technique where light is collected as a single value regardless of wavelength in order to provide a single light intensity reading. This technique is normally used with narrow bandwidth or monochromatic light sources. A low-cost choice for the light source in intensity interrogation based systems includes the LED and semiconductor laser diode. However, the LED is less complex to use from the viewpoint of driver circuitry and does not have the safety concerns that exist with some laser diodes. The optical receiver can be a photodiode or phototransistor. Referring to **Figure 5**, this represents the main choices to be made. The light source (**Figure 5**) is either an LED, low-cost laser diode (a) or a white light

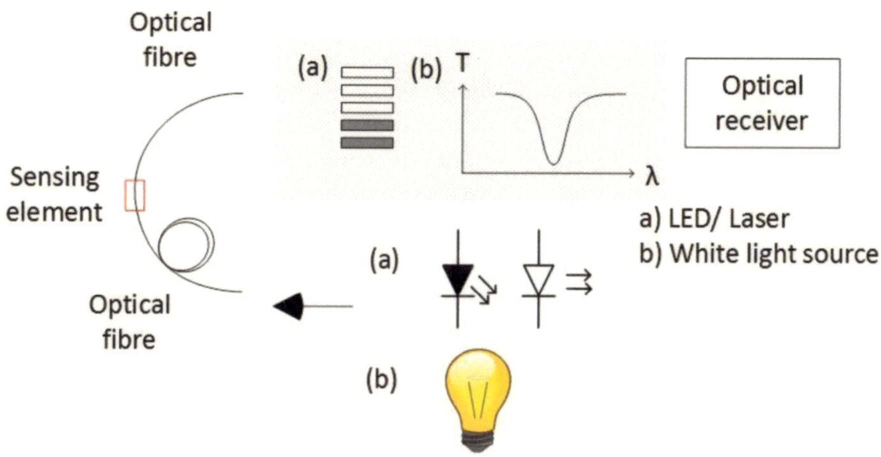

Figure 5. OFS system using a white light source and spectrometer or laser/LED and photodiode/phototransistor.

source (b). The choice of receiver is based on the light source. For the LED or laser diode, the receiver (**Figure 5**) provides only intensity interrogation (a) and for the white light source, either intensity or wavelength resolved interrogation (b).

It should also be noted and considered that some information can only be captured by observing at the signal spectrum instead of the cumulative intensity integrated across the whole spectrum. To extract such information, the signal needs to be spectrally resolved with adequate (spectral) resolution that will enable the desired measurand to be accurately determined and recorded. This technique, also known as wavelength interrogation, usually requires the use of a white light source and spectrometer as a detector.

Figure 6 shows the principle of operation of the spectrometer.

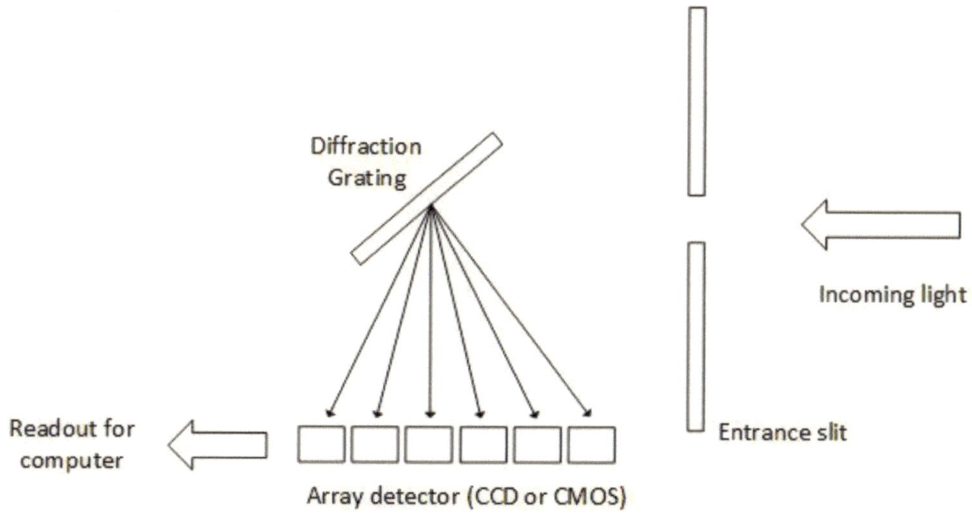

Figure 6. Principle of operation of the spectrometer.

The spectrometer is becoming increasingly widely used as an optical receiver. It receives a collimated input and disperses it into spectral components by an optical component such as a diffraction grating. The different spectral components are then registered by an array detector such as a charge-coupled-device (CCD) or complementary metal oxide semiconductor (CMOS) camera. To read out the spectral data as a digital representation of the captured optical signal, the spectrometer normally incorporates a suitable computer interface. The whole series of tasks require high precision optical elements and the spectrometer often needs to be recalibrated at regular intervals by a qualified technician. The cost of a spectrometer system is a factor that normally prohibits its use in situations with limited (financial) resources [25], although their cost and robustness is rapidly improving. Different parameters are best measured using the optimum combinations of suitable light sources and optical receivers. For example, surface plasmon resonance (SPR) sensing typically usually uses a white light source and spectrometer to measure a shift of wavelength. For purely light intensity interrogation sensing, it is possible to use a variety of light sources including the LED and laser diode, and optical receivers including the photodiode, phototransistor or the LDR. To access the information gathered by an OFS, the sensor information needs to be read out and displayed. A common acquisition module for light intensity interrogation is the DAQ module with microcontroller control. For other more complex instruments, the DAQ module is usually built into the instrument. A computer is universally used for data read-out using a common serial interface such as a wired USB or general purpose interface bus (GPIB), or a wireless interface such as Wi-Fi, Bluetooth or ZigBee. Software such as *LabVIEW*™ from National Instruments or *MATLAB*™ from The Mathworks company are commonly used to implement system control and data analysis. To identify examples of developed systems, **Table 1** provides a summary of the attributes of five reported systems. These arrangements use a variety of approaches, but are currently only suitable for use in a laboratory environment.

System configuration is generally similar for all sensor system implementations and this structure is shown in **Figure 7**. Each part can be implemented using different technologies and the choice of implementation is dependent on the requirements of the application. The required

Authors	Jiang et al. [26]	Zhao et al. [27]	Di et al. [28]	Mahanta et al. [29]	Stupar et al. [30]
Sensing mechanism	Surface plasmon resonance	Surface plasmon resonance	Intensity interrogation	Intensity interrogation	Intensity interrogation
Light source	White light source	White light source	LED	He-Ne laser source	LED
Optical receiver	Spectrometer	Spectrometer	Phototransistor	Light dependent resistor	Photo Darlington transistor
Data acquisition	Data acquisition card	Spectrometer	Data acquisition card	Data acquisition card	Microcontroller
Computer interface	USB	USB	DB37 cable	USB	ZigBee
Data readout	Computer	Computer	Computer	Computer	Computer
Data analysis	LabVIEW	–	WaveScan 2.0	LabVIEW	LabVIEW

Table 1. Examples of reported OFS system operation.

http://dx.doi.org/10.5772/intechopen.71451

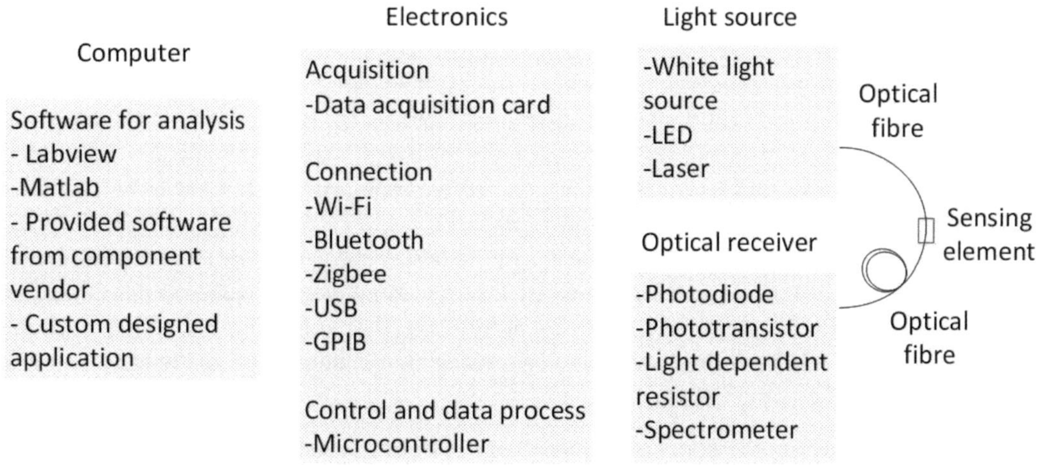

Figure 7. Typical sensor system – choice of sub-system components.

resources are identified in terms of the computing (PC), the electronics (electronic-optical-electronic signal interfacing), the light source and the optical receiver.

3. The field programmable gate array

3.1. Introduction

A programmable logic device (PLD) is a device with configurable logic and flip-flops linked together with programmable interconnect. Unlike a processor [microprocessor (μP), microcontroller (μC) or digital signal processor (DSP)] which is programmed to implement a software program, the PLD is configurable (programmable) in hardware. It is configured to implement different digital logic hardware circuits. The architecture provides flexibility in use and re-programmability to quickly re-target the same device to a different application. Early PLDs were based on the architecture of the simple PLD (SPLD) which could implement basic combinational logic and synchronous sequential logic (counter and state machine) circuits. Development of the SPLD led to the complex PLD (CPLD) which essentially is an array of SPLDs with programmable interconnect within a single device and hence, can implement more sophisticated functions. The CPLD has found uses in applications such as instrumentation and control. An alternative to the CPLD is the FPGA. This has a different architecture which allows for high-speed digital signal processing (DSP) operations which would not be possible in a CPLD. An advantage of the PLD is that it the same device can be configured and reconfigured in a short time, thus saving resources in terms of time and money to the need to develop new digital hardware. Any new update or correction can be simply done by downloading a new configuration bit stream into it.

3.2. The FPGA architecture

The FPGA is an integrated circuit (IC) that can be electrically configured to implement a digital hardware design. It consists of a matrix of configurable logic blocks (CLBs) connected via programmable interconnects and through input/output blocks to interface with other devices. The logic blocks consist of look-up tables (LUTs) constructed using simple memories that store Boolean functions to perform logic operations. The LUT can handle any kind of logic function, but the complexity of the LUT depends on the manufacturer. In addition to the LUT, and with the growing demands to create more complex digital circuits and systems using the FPGA, special purpose blocks such as the DSP slice, embedded memory (Block RAM in FPGAs from Xilinx Inc.) and other functions have been introduced into FPGA. Today, a range of FPGAs from different vendors are available. **Table 2** provides a summary of the devices available from the key programmable logic vendors. The key differences between CPLD and the FPGA are in the device architectures and the complexity of functions that its basic unit is able to perform. The basic unit of CPLD is known as macro cell, using simple sum-of-products combinatorial logic functions and an optional flip-flop.

For a particular design, the configuration pattern is stored in memory within the device or external to the device. The configuration would initially be downloaded into the FPGA memory from an external circuit in the form of a bit stream file. Depending on the particular FPGA, the internal memory may non-volatile memory [electrically erasable programmable read only memory (EEPROM) or Flash] or volatile memory [static read only memory (SRAM)]. When using volatile memory, the configuration will need to be loaded from external memory into FPGA every time it powered on in order to set-up the FPGA. Currently, SRAM based FPGA dominate the market, however the availability of Flash memory based FPGAs is increasing and these provide lower power consumption and no boot time (i.e. the time to start operating once the device power supply is provided).

3.3. FPGA design flow

In many sensor designs, a software programmed processor, typically a μC, is programmed to provide the necessary digital sub-system operations (referring to **Figure 4**). The design flow for

Vendor	FPGA families	SPLD/CPLD	Other relevant devices	Notes
Xilinx Inc.	Virtex, Kintex, Artix and Spartan	CoolRunner-II, XA CoolRunner-II and XC9500XL	3D ICs, Zynq PSoC	PSoC – Programmable system on a chip
Intel Corporation	Stratix, Arria, MAX, Cyclone and Enpirion	–	Stratix SoC FPGA	FPGAs were formerly from Altera. Altera MAX family also included CPLDs.
Atmel Corporation	AT40Kxx family FPGA	ATF15xx and ATF75xx CPLD families, ATF16xx and ATF22xx SPLD families	–	–
Lattice Semiconductor	ECP, MachX and iCE FPGA families	ispMACH CPLD family	–	–
Microsemi	PolarFire, IGLOO2 and RTG4 FPGA families	–	SmartFusion2 SoC FPGA	FPGAs were formerly from Actel

Table 2. Summary of key FPGAs and vendors (including SPLD and CPLD devices).

http://dx.doi.org/10.5772/intechopen.71451

processor based design is well-known to many people and would be chosen based on existing knowledge and experience. However, hardware configured devices such as the FPGA can provide advantages to the system capabilities and operation if the capabilities of the FPGA were known and the ability to design with the FPGA existed. There are a number of advantages of using an FPGA within the digital sub-system and these include:

- High speed DSP operation capabilities such as digital filtering and FFT (fast Fourier transform) operations.
- Concurrent operation which means that operations in hardware can be run in parallel.
- A high number of digital input and output pins for connecting to peripheral devices.
- Programmable I/O standards such as LVTTL, LVCMOS, HSTL and SSTL.
- In-built memories for temporary data storage.
- In-built hardware multipliers for DSP operations.
- In some FPGAs, in-built ADCs are available for analogue input sampling.
- Availability of macros such as FFT blocks for high-speed DSP operations.
- Ability to develop custom architectures suited to the particular application which does not restrict the designer to using the capabilities and limitations of existing processor architectures.

'But how is a design created using programmable logic, specifically with reference to the FPGA?' is an important question to ask. Details of the design flow would be specific to a particular device vendor and each vendor would provide an integrated development environment (IDE) for their devices. The terminology used also differs between vendors. However, **Figure 8** shows a

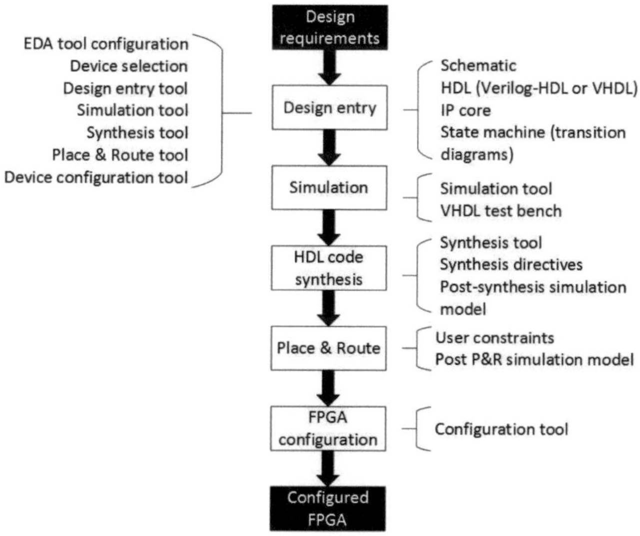

Figure 8. Simplified FPGA design flow.

simplified view of a typical FPGA design flow. Starting with the design requirements and understanding of the capabilities of the target FPGA, the FPGA design tool is used and the design description created. This can be in the form of a circuit schematic, hardware description language (HDL) description, a predesigned IP core or a FSM state transition diagram. An advantage of using an HDL is that the design description is in the form of an ASCII text file and if suitably written, is independent of any particular target device. This means that the HDL code can be synthesised to a different device and this allows for flexibility in design choice and the ability to update designs in the future if so required. The design is initially simulated using a suitably defined test bench (also referred to as a test fixture) before synthesis into logic and implementation within the FPGA (design place and route (P&R) into the FPGA hardware). After synthesis and P&R, timing details in terms of additional signal delays due to logic and interconnect propagation delays can be identified and included in the simulation studies. This is particularly important for high-speed operation as the timing of the signals propagating through the hardware need to be taken into consideration otherwise incorrect circuit operation may occur. Finally, the design can be configured into the FPGA and physical testing of the hardware can then be undertaken. At each step, depending on the simulation and test results, it may be necessary to modify aspects of the design in order to achieve correct circuit operation.

4. FPGA based OFS system

4.1. Introduction

Within an OFS system, the operation of the system in the majority of solutions is controlled by the deployment of one or more software programmed processors. There would be many reasons for adopting such an approach, particularly designer understanding and experience. Using programmable logic as an alternative requires a different design approach and a different set of designer skills. The design philosophy must be switched from '*how can I achieve the required functionality using the available sequential software functions?*' to '*how can I achieve the required functionality in concurrent hardware?*'. When considering the use of the FPGA, there must be benefits for adopting such an approach. The FPGA can:

- Provide the ability to handle real-time system operation with a high sampling rate of sensor data and other DSP operations that other, traditional microcontroller only based approaches would have problems in undertaking. For example, [31] discusses the use of the FPGA within wireless sensor networks (WSNs). In Ref. [32], a phase-sensitive optical time-domain reflectometer using the FPGA is discussed.
- Undertake high-speed DSP operations in hardware using embedded macros.
- Be readily reconfigured in the field for system upgrades.
- Allow the user to rapidly prototype the design in the same way as software programmed microcontroller while provide concurrent processing capability that overcome processor performance and precision limitations.

- Adapt its hardware configuration to a target application needs. As the FPGA does not have a fixed architecture, it is not limited to a predefined set of possible operations and this allows for a custom design which matches the operating requirements of the target application (speed, power consumption, circuit size) to be developed.

In this section, the use of the FPGA is considered as a digital core (sub-Section 4.2) to integrate functions (sub-Section 4.3) for optically based sensor systems.

4.2. The FPGA as a digital core

With reference to the digital sub-system in **Figure 4**, this can be considered in terms of the functions it needs to implement. Specifically the required functions can be considered as:

- Sensor interfacing.
- Control.
- Storage (memory).
- DSP.
- Communications to an external electronic system.
- User interfacing.

Figure 9 shows a simplified block diagram for an FPGA based design that implements the above functions. The FPGA operation control signal timing is determined by the external clock it uses (e.g. a 100 MHz clock) and control signals from external peripheral devices (e.g. an external power-on reset circuit) as well as user control which has been configured into the

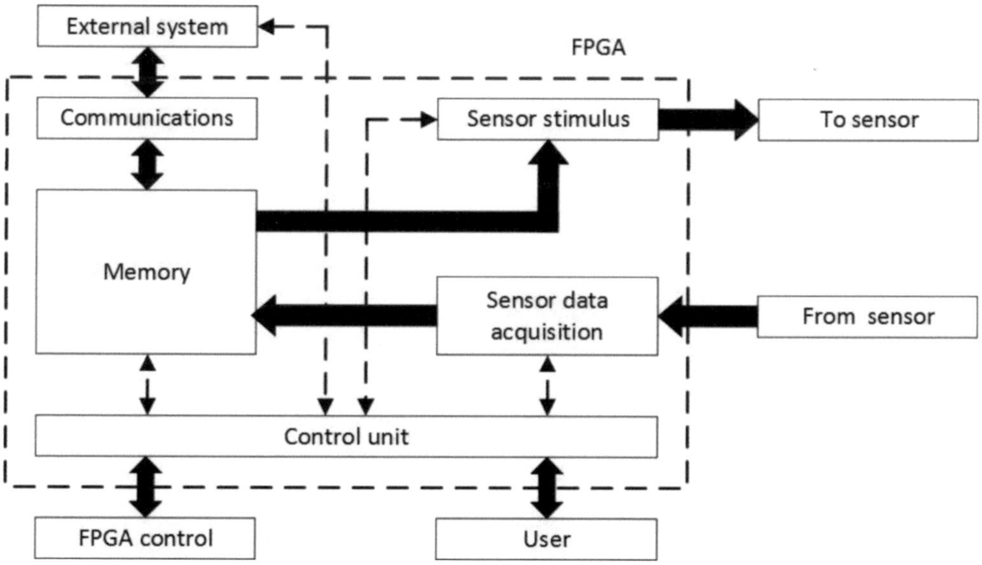

Figure 9. FPGA core functions example.

device. User control can come from an external electronic system using wired or wireless communications and from human interaction devices such as a keypad or touch-screen device. This requires a suitable *Control unit* to be designed for controlling the other circuits configured into the FPGA. Signals to/from an external system would be via a suitable *Communications* circuit that allows both control signal and data transfer. Local memory, bistables (latches and flip-flops) and SRAM are used for temporary storage. *Memory* may be internal to the FPGA only, or external. Output to the sensor in the form of *Sensor stimulus* would be in the form of a voltage that may need translating to a current. For example, an LED may be driven by a DC current level or by a modulated signal using a signal encoding scheme such as pulse-width modulation (PWM). Sensor response would be sampled using a *Sensor data acquisition* circuit. Building on this architecture, additional operations, such as DSP, can be included.

4.3. Integrating functions into the FPGA

An OFS system based on a digital core within an FPGA will need to ensure that the circuit configured into the FPGA meets the system requirements. A simple implementation would consist of a basic communications protocol, along with a basic signal generator (LED current control) and a signal capture (photodiode current sampling) along with memory for temporary storage of values (where the memory requirements are small then latches and flip-flops would be sufficient, whereas where the memory requirements are larger, static random access memory (SRAM) and Flash memory would be required). Such a basic system would not utilise many of the available hardware resources within the FPGA and in such a case, a smaller and simpler CPLD would probably be sufficient. The power of the FPGA however becomes apparent where higher operating frequencies and embedded DSP functions are introduced. For example, analogue signal measurements are influenced by external parameters such as ambient temperature and noise introduced into the signal from surrounding electronic circuitry and the ambient environment. This would be particularly noticeable where low-level signal measurements are required. To improve system performance, signal processing techniques such as oversampling, averaging and correlation can be used as well as digital filtering and the FFT. For these functions to operate in an application such as a real-time system, the calculations to be performed need to operate at high frequency and utilise high-speed memories for temporary storage. All these requirements map effectively to modern FPGAs and so can make the FPGA the natural choice for a high performance embedded sensor arrangement. With the increased interest and use of the FPGA in DSP operations, recent FPGA architectures have evolved with high-speed operation and complex DSP in mind. Generally, FPGAs today incorporate the following circuitry. With reference to the Xilinx FPGA families, these can include:

- CLB: The configurable logic block is a basic building block of the FPGA and incorporates circuitry such as the look-up table (LUT), a flip-flop and a multiplexer.
- Distributed RAM: Small RAM within the LUT.
- Slices: Slices are groups of LUTs and flip-flops.
- Block RAM (dedicated two-port memory), first-in first-out (FIFO) and error correction checking (ECC) cores.

- Singled-ended and differential input/output cells (I/O): For higher speed signals, differential signals are preferred and the programmable I/O cells can be configured for single-ended or differential signal operation and with different I/O standards.
- DSP slices: Essentially high-speed cells that can be used to create DSP operations.
- Analog(ue) mixed-signal (AMS) core: XADC, in-built ADC.
- Configuration bit stream encryption blocks incorporating methods such as AES, HMAC and DNA.

Although the discussion into the use of the FPGA has so far concentrated on the development of custom hardware designs, with hardware architectures suited to a target application needs, it is possible and in some cases preferable to utilise a software programmed processor. A processor itself is simply hardware with logic gates and memory connected in such a way to implement a processor architecture. The FPGA therefore can be used to implement a processor core and so be configured to act as a software programmed processor. Either a pre-defined processor architecture can be implemented within the FPGA as an intellectual property (IP) core with a description provided in Verilog-HDL or VHDL, or the designer can develop their own architecture processor. Therefore, the FPGA can implement a design which is based on hardware logic only, be based on a processor only or, and where the benefits of using the FPGA can be truly seen, can implement a hardware-software co-design within a single device. In addition, with the hardware resources available in many FPGAs, multiple processors can be implemented within a single FPGA and so architectures consisting of arrays of interconnected processors operating concurrently with hardware only based designs for high-speed computations can be developed.

5. Case study design

5.1. Introduction

Modern OFS systems are becoming ever more sophisticated as such systems today are targeting more complex problems than previously considered [33]. This includes the need for more sophisticated and computationally intensive data analysis methods. Today, an OFS system that uses a personal computer (PC) and software data analysis software such as *LabVIEW*™ or *MATLAB*™ has little difficulty to undertake such analysis tasks. However, the PC may not necessarily be an ideal data processing unit for in-field measurements or real time monitoring in remote areas due to its size and power consumption requirements. To make this system portable for personnel to carry, and for remote deployment over extended periods of time, it has to be physically small, lightweight, low-power and battery powered. The system should be as easy as possible for operator personnel to use as they should not be expected to understand the detailed electronic circuit operation or the arrangement that forms the system. It would also be better if it is capable of real-time data processing that provides timely and useful information via modern telemetry. Real-time data processing also reduces storage and transmission requirements that both consume precious resources in a portable system [34].

This section provides a case study into an OFS system using the surface plasmon resonance (SPR) principle. The objective of this case study was to demonstrate integration of the FPGA into an OFS system based on SPR. SPR is a popular sensing mechanism for biomedical optical sensors as it is sensitive to a change in the refractive index of a surrounding medium. It is a phenomenon that occurs when photon momentum is matched with plasmon momentum and thus photon energy efficiently transfers to the surface plasmon. This results in absorption of specific wavelengths of light that are then visible in the resulting light spectrum and the absorption wavelengths are then used to determine the properties of the measurand. The surface plasmon effect exists only in p-polarised (transverse magnetic) waves, so it is a requirement that a p-polarised wave is generated that oscillates in the same orientation in order to be excited. A propagating photon in vacuum or air does not have enough momentum to excite SPR [35]. The common technique to excite SPR optically is by using total internal reflection to create an evanescent wave that has a higher momentum. The matching condition depends heavily on the surrounding environment which makes it suitable for use as a sensor. **Figure 10** shows the principle of operation for the sensor used.

The electronic system design was also created to be modular in order to accommodate different sensing mechanisms. This is readily managed using POF as the interface medium since the sensor housing can be changed by simply removing and inserting the transmitting and receiving fibres. The system architecture developed is discussed and the strategy to implement refractive index measurement using a combination of the FPGA with a tri-colour (RGB: red, green, blue) LED and photodiode is proposed. The FPGA functionality is identified in a later section. System control and data acquisition with a host PC was created using the Python open

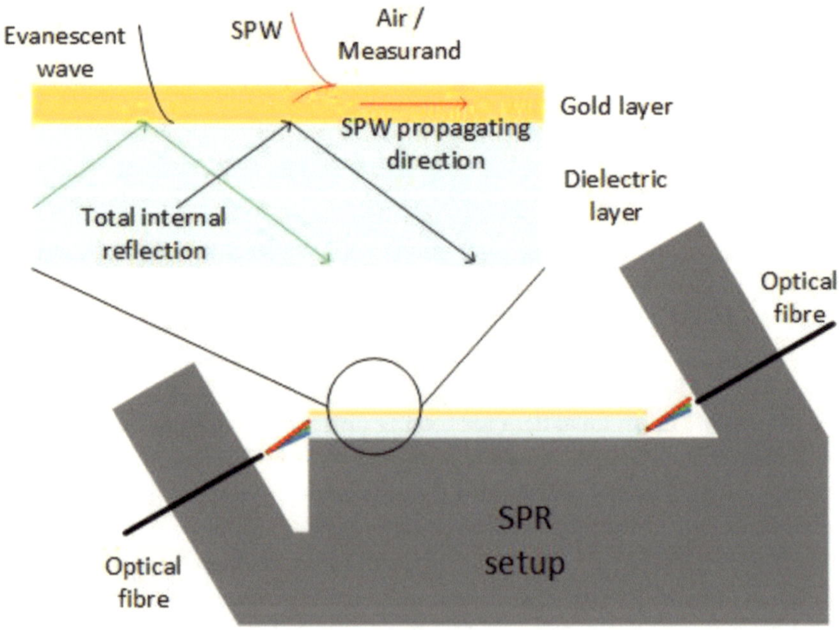

Figure 10. Surface plasmon sensor. Schematic diagram of total internal reflection and surface plasmon wave (SPW) generated during SPR.

source programming language. However, the FPGA operation does not rely on a PC connection to work. The FPGA used was the Xilinx Artix-7 [36] FPGA and this was embedded within the Arty development board from Digilent [37]. The FPGA operates on a 100 MHz clock frequency on this particular development board, so all digital signals were timed to this master clock. Connections to an external printed circuit board (PCB) with additional hardware was made using the available PCB header connections. This supports the capability to readily configure the FPGA as well as enabling the FPGA to be powered from the USB port +5 V power supply for development. In the final set-up, a battery pack provides the power supply for the FPGA. The digital logic operating within the FPGA was developed using VHDL design entry and synthesised to the Artix-7 architecture. Design entry and simulation using VHDL test benches were undertaken using the Xilinx Vivado toolset.

5.2. System architecture

The circuit hardware design, as shown in **Figure 11**, was separated into five sub-systems: OFS, power supply, digital core, light signal generation and sensing module. The power supply module comprises a low-dropout linear voltage regulator circuit for providing the necessary power supply to the light signal generation and sensing modules. A separate linear voltage regulator is used to isolate the photodiode circuit from digital noise which could be present on the power supply from the FPGA board. These regulators also provide the system flexibility to accommodate different battery voltage levels.

The light signal generation and sensing modules are additional circuits that support the opto-electronic components and their corresponding circuits. For light signal generation, a current source was designed to control the light through a digital signal (two level voltage) from the FPGA. As a tri-colour LED was used, the FPGA was required to provide a sequential series of

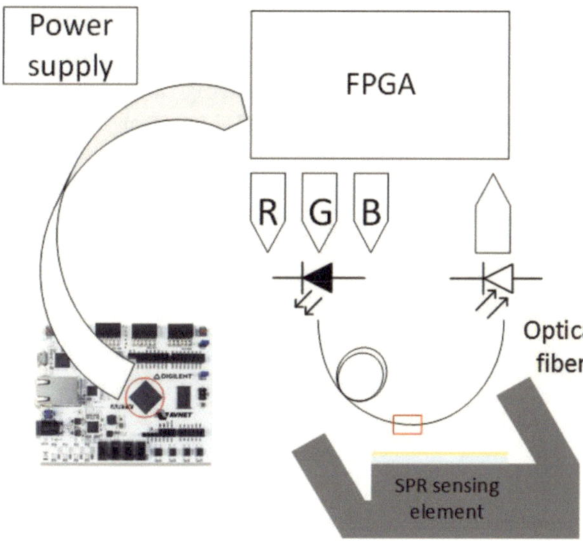

Figure 11. Case study system architecture.

red, green, and blue light pulses in a current range from 0 mA to 20 mA for each LED. With the receiver, the analogue response from photodiode circuit as a voltage was sampled using the in-built 12-bit 1 MSPS (mega sample per second) ADC (XADC) and the sampling of the responses were synchronised with the XADC conversion timing within the FPGA. Therefore, the intensity of the individual colour pulses could be recorded separately. The on-chip ADC is then used to translate the voltage reading into a 12-bit digital signal. The photodiode current is initially converted to a voltage using a high-gain transimpedance amplifier and then amplified. For a high attenuation sensing mechanism, the received light intensity is low and a high-gain transimpedance amplifier using an operational amplifier (op-amp) hence required. However, this leads to noise related issues and so the choice of a suitable op-amp along with the use of noise reduction techniques are required in the realised electronic circuit. To reduce noise in this design, PCB ground planes and an aluminium shield fixed on top of the receiver circuit was required. Although these additions reduced the resulting noise levels, additional noise reduction should also be considered. A 3D printed cap for the LED allowed a bare-end of the POF to collect light immediately from LED without any focus aid and coupled the light into the sensor housing without the need for a connector. A photodiode that requires no connector was also chosen to be the optical receiver for the modulated light. The OFS is an extrinsic sensor; gold film coating on top of a glass slide and mounted on top of a 3D printed housing to position the fibre at an angle facing the edge of the glass slide. It was designed to excite SPR for different measurands with minimum change in the blue light region.

The digital output from the FPGA is fed into an op-amp based current driver circuit. The digital logic 1 and 0 values from the FPGA are represented by two voltage levels of 3.3 V and 0 V. It is assumed in this discussion that the voltage levels produced are exactly 3.3 V and 0 V. The voltage-to-current conversion is achieved using an op-amp as a unity gain amplifier with a transistor in the feedback loop as shown in **Figure 12**. The transistor is included in this configuration as a current amplifier as it can achieve higher current output than the op-amp alone. A 20 mA current was allowed to flow when a logic 1 output is given and no current otherwise (logic 0). A current sensing resistor, R_E, is used to provide a voltage input for the op-amp. The average current then can be controlled using the pulse width modulation (PWM) control technique which is generated using a PWM generator circuit within the FPGA. The FPGA then receives an analogue output from a transimpedance circuit that translates the generated current into a voltage.

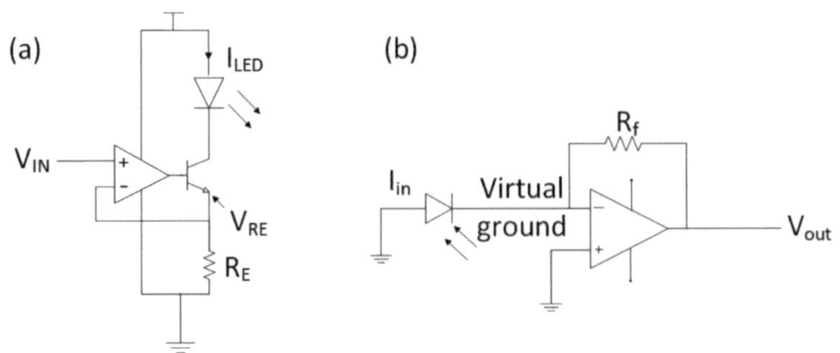

Figure 12. (a) Current source and (b) transimpedance amplifier circuit.

The Artix-7 FPGA board used also has some additional and useful features which include external SRAM, external Flash memory and the ability to be powered from the USB connector or any 7–15 V power supply voltage source. The Artix-7 FPGA operates with a core voltage of 1 V, with 3.3 V I/O and a 100 MHz master clock. The XADC has access to nine external analogue input channels and is capable of performing self-calibration in order to correct for offset and gain errors, provided that the ADC reference voltage is reliable (i.e. an accurate and stable value). On-board peripherals include a USB-UART (UART: universal asynchronous receiver transmitter) integrated circuit that allows a computer to communicate with FPGA for configuration and data transfer during normal operation. This wired communication is especially useful during development and testing. However, in the realised system and once the FPGA has been configured, a ZigBee wireless interface is used for normal data transmission and this allows the sensor hardware to be operated at a distance from the computer. This FPGA board undertakes multiple functions including system control, data processing, communication, LED control and analogue-to-digital conversion (ADC) for sensing module. Its functionality is discussed detail in the next section.

5.3. FPGA functions

To support the system operation as previously discussed, the FPGA function as a digital core (sub-system) that itself is composed of different sub-systems. The circuit configured within the FPGA is shown diagrammatically in **Figure 13**. This is a block diagram representation of the VHDL code. These sub-systems include data acquisition, control system, communication, LED pulse generation. All these sub-systems were initially described in VHDL and operate concurrently. The control system includes an XADC acquisition control unit that keeps the timing of the XADC and LED pulse generation units in synchronisation. It allows the user to set the

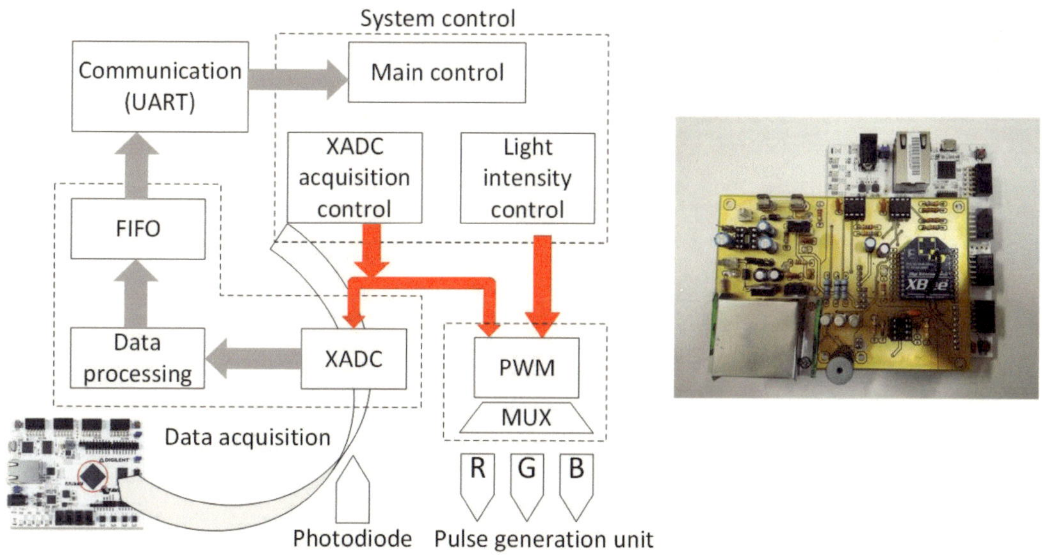

Figure 13. Schematic of the digital design implemented within the FPGA.

number of measurement to be taken. The light intensity control unit is tasked to regulate the duty cycle of the PWM output for each LED. The generated signal is then passed to the digital pin output set by multiplexer. The data acquisition system involves the use of the XADC that is capable to sample voltage reading from 0 to 3.3 V into 12-bit digital signal and can be controlled in order to accommodate different settling times for the signal conditioning circuit or different sampling rates. The acquired data bit size is bigger than the UART protocol (with eight bit data, one start bit and one stop bit), and so is required to be separated into two parts. A protocol was developed for data transmission which included a header to indicate each different light colour. Within the FPGA, sampled data is processed and the processed data stored in memory storage before transmitted to PC when requested to do so. Serial communication is implemented using UART protocol with a Baud rate of 115,200 baud. At this stage, the data is then processed with computer.

5.4. Control and data acquisition using Python

In order to interface the PC to the FPGA and then to the user, a USB or ZigBee serial interface is used. The FPGA and the computer then communicate using UART protocol so that the computer can send commands to the FPGA and the FPGA can respond with sensor data results. Therefore, a suitable programming language was required to achieve the computer-to-FPGA and computer-to-user communications. In this work, the Python open source language is used to implement system control, data acquisition and data presentation. A graphical user interface (GUI) was built using *Tkinter* and *Matplotlib* modules for Python which enabled live data presentation. The GUI is shown in **Figure 14** with left graph showing all red, green, and blue intensities plotted together and three smaller graph at the right side shown three

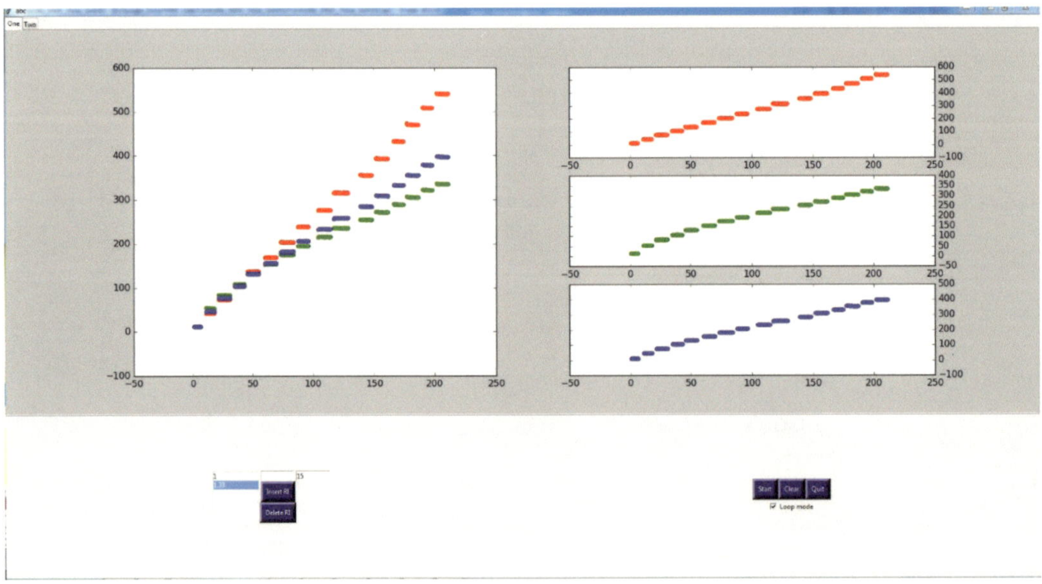

Figure 14. GUI created from Python.

individual colour intensity plots. Data labels for data frame are selected at lower left region of GUI. The lower right region are for system control: measurement initiation and process termination. The *Pyserial* module was used for communication with the FPGA by using the computer COM port in order to carry out FPGA control and sensor data acquisition. The acquired data stream is then formatted using the *NumPy* and *Pandas* modules. The data structure used was a data frame using the *Pandas* module. This data frame was considered as it allowed the sensor data to be serialised into a minimal and useful *JSON* format that could then be stored in a database or readily transferred elsewhere. In addition, a *MySQL* database was set up for data storage and management.

Test results are shown in **Figure 15**. The top plot shows the sensor results with air as the measurand and the bottom plot with water. A comparison is made between the spectrometer measurement using an Ocean Optic QE65000 and the Industrial Fiberoptics IF-D91 photodiode output. The spectrometer integration time was set to 500 ms. The line plot showing the spectrometer reading is shown on the left y-axis and the overlap bar plot showing the photodiode measurement for intensity reading is shown on the right y-axis. The wavelength x-values are only for the spectrometer reading while the photodiode just overlap with it for comparison. In this view, both the spectrometer and photodiode's y-values are represented in plot using arbitrary units. The photodiode measurement and spectrometer readings for different measurands has shown good agreement in terms of the trend of the signal during the measurand change. However, the magnitude of the changes are not exactly the same for the spectrometer and photodiode and this can be attributed to different spectral responsivity. The plot shows that the blue colour measurement is relatively stable for both the photodiode and spectrometer. The blue colour measurement shows a weak response to both measurands and it is therefore suitable for use as the reference signal. It provides the system with a self-calibration capability which can be built in hardware within the FPGA.

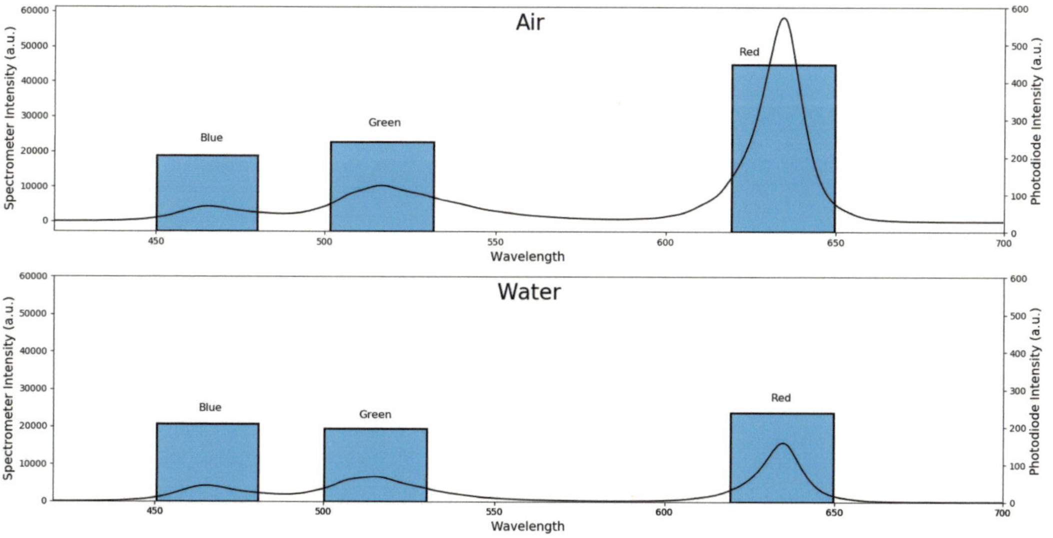

Figure 15. OFS test results visualisation.

6. Conclusions

In this chapter, the use of programmable logic has been discussed in relation to the design and implementation of OFS systems using POF technology. Specifically, the focus of the discussion was based on the use of the FPGA to provide hardware functionality as a complement to a software based processor approach. In this case, the necessary digital operations were performed in hardware using a synthesised VHDL description of the digital logic functions. The FPGA provided the necessary signals to control the output light intensity of a LED source using current control. An RGB LED was used, and hence three wavelengths of light could be created and combined. The light was propagated to a SPR sensor using standard 1 mm diameter POF and was received by a photodiode detector, also using POF. The chapter has included a case study of a SPR sensor based on POF, and the system has been designed, built and characterised when tested in air and submersed in water. With the functions created in this particular application, the building blocks of the FPGA based sensor systems capable of measurement of a wide range of parameters could be developed. FPGA control, data capture and visualisation was achieved on PC using the Python open source language.

Acknowledgements

The authors would like to thank the European Union Erasmus Mundus LEADERS (Leading mobility between Europe and Asia in Developing Engineering Education and Research) scholarship programme in its support of this work.

Author details

Yong Sheng Ong[1], Ian Grout[1]*, Elfed Lewis[1] and Waleed Mohammed[2]

*Address all correspondence to: ian.grout@ul.ie

1 Department of Electronic and Computer Engineering, University of Limerick, Limerick, Ireland

2 BU-CROCCS, School of Engineering, Bangkok University – Rangsit Campus, Bangkok, Thailand

References

[1] Grattan L, Meggitt B, editors. Optical Fiber Sensor Technology: Advanced Applications-Bragg Gratings and Distributed Sensors. New York, USA: Springer US; 2000. Vol. X. p. 385. DOI: 10.1007/978-1-4757-6079-8

[2] Alwis L, Sun T, Grattan KTV. Optical fibre-based sensor technology for humidity and moisture measurement: Review of recent progress. Measurement. 2013;**46**:4052-4074. DOI: 10.1016/j.measurement.2013.07.030

[3] O'Keeffe S, McCarthy D, Woulfe P, Grattan MW, Hounsell AR, Sporea D, Mihai L, Vata I, Leen G, Lewis E. A review of recent advances in optical fibre sensors for in vivo dosimetry during radiotherapy. The British Journal of Radiology. 2015;**88**. DOI: 10.1259/bjr.20140702

[4] Ashley J. Welch, Martin JC van Gemert, editors. Optical-Thermal Response of Laser-Irradiated Tissue. 2nd ed. The Netherlands: Springer Netherlands; 2011. XIV, 958 p. DOI: 10.1007/978-90-481-8831-4

[5] Yang M, Li S, Jiang D. Review on optical fiber sensing technologies for industrical applications at the NEL-FOST. In: EWSHM – 7th European Workshop on Structural Health Monitoring; July; Nantes, France. 2014

[6] Lecler S, Meyrueis P. Intrinsic optical fiber sensor. In: YasinM, Harun SW, Arof H, editors. Fiber Optic Sensors. InTech; 2012. DOI: 10.5772/27079. https://www.intechopen.com/books/fiber-optic-sensors/intrinsic-optical-fibersensor

[7] Grattan KTV, Sun T. Fiber optic sensor technology: An overview. Sensors and Actuators A: Physical. 2000;**82**(1–3):40-61. DOI: 10.1016/S0924-4247(99)00368-4

[8] Alwis LSM, Bustamante H, Roth B, Bremer K, Sun T, Grattan KTV. Evaluation of the durability and performance of FBG-based sensors for monitoring moisture in an aggressive gaseous waste sewer environment. Journal of Lightwave Technology. 2016;**35**(16): 3380-3386. DOI: 10.1109/JLT.2016.2593260

[9] Mescia L, Prudenzano F. Advances on optical Fiber sensors. Fibers. 2014;**2**:1-23. DOI: 10.3390/fib2010001

[10] Laskar S, Bordoloi S. Microcontroller-based instrumentation system for measurement of refractive index of liquid using bare, tapered and bent fibre as sensor. IET Optoelectronics. 2013;**7**(6):117-124. DOI: 10.1049/iet-opt.2013.0052

[11] Bram Van Hoe, Graham Lee, Erwin Bosman, Jeroen Missinne, Sandeep Kalathimekkad, Oliver Maskery, David J. Webb, Kate Sugden, Peter Van Daele, Geert Van Steenberge. Ultra small integrated optical Fiber sensing system. Sensors. 2012;**12**(9):12052-12069. DOI: 10.3390/s120912052

[12] Suryadi, Puranto P, Adinanta H, Waluyo TB, Priambodo PS. Development of microcontroller-based acquisition and processing unit for fiber optic vibration sensor. Journal of Physics: Conference Series. 2016;**817**(012042):1-6. DOI: 10.1088/1742-6596/817/1/012042

[13] Ibrahim SK, Farnan M, Karabacak DM, Singer JM. Enabling technologies for fiber optic sensing. In: Optical Sensing and Detection IV; April 29. Brussels, Belgium: SPIE; 2016. DOI: 10.1117/12.2234975

[14] Safia AA, Al Aghbari Z, Kamel I. Phenomena detection in mobile wireless sensor networks. Journal of Network and Systems Management. 2016;**24**(1):92-115. DOI: 10.1007/s10922-015-9342-z

[15] Guemes A, Fernandez-Lopez A, Soller B. Optical fiber distributed sensing – Physical principles and applications. Structural Health Monitoring. 2010;**9**(3):233-213. DOI: 10.1177/1475921710365263

[16] Pfrimer FWD, Koyama M, Dante A, Ferreira EC, Antonio Siqueira Dias J. A closed-loop interrogation technique for multi-point temperature measurement using fiber Bragg gratings. Journal of Lightwave Technology. 2014;**32**(5):971-977. DOI: 10.1109/JLT.2013.2295536

[17] Werneck MM, Allil RCSB. Optical fiber sensors. In: Krejcar O, editor. Modern Telemetry. 2011. DOI: 10.5772/25006. https://www.intechopen.com/books/modern-telemetry/optical-fiber-sensors

[18] Sun S-S, Dalton LR, editors. Introduction to Organic Electronic and Optoelectronic Materials and Devices. Boca Raton, Florida, USA: CRC Press; 2008. 936 p

[19] Lee GCB, Van Hoe B, Yan Z, Maskery O, Sugden K, Webb D, Van Steenberge G. A compact, portable and low cost generic interrogation strain sensor system using an embedded VCSEL, detector and fibre Bragg grating. In: SPIE 8276, Vertical-Cavity Surface-Emitting Lasers XVI, 82760E; January 21; San Francisco, California, USA. 2012. DOI: 10.1117/12.907810

[20] Fidanboylu K, Efendioglu HS. Fiber optic sensors and their applications. In: 5th International Advanced Technologies Symposium (IATS'09); May 13–15; Karabuk, Turkey. 2009. p. 1-6

[21] Woyessa G, Pedersen JKM, Fasano A, Nielsen K, Markos C, Rasmussen HK, Bang O. Simultaneous measurement of temperature and humidity with microstructured polymer optical fiber Bragg gratings. In: 25th Optical Fiber Sensors Conference (OFS); 24–28 April; Jeju, South Korea. 2017. DOI: 10.1117/12.2265884

[22] Muda R, Clifford J, Chambers P, Mulrooney J, Merlone-Borla E, Gili F, Dooly G, Fitzpatrick C, Lewis E. Simulation and measurement of carbon dioxide exhaust emissions using an optical fibre based mid infrared point sensor. Journal of Optics A: Pure and Applied Optics. 2009;**11**:1-7. DOI: 10.1088/1464-4258/11/5/054013

[23] Lee B. Review of the present status of optical fiber sensors. Optical Fiber Technology. 2002;**9**:57-79. DOI: 10.1016/S1068-5200(02)00527-8

[24] Udd E, Spillman WB Jr, editors. Fiber Optic Sensors: An Introduction for Engineers and Scientists. 2nd ed. John Wiley & Sons, Inc.; 2011. 512 p. DOI: 10.1002/9781118014103

[25] Santos JL, Farahi F, editors. Handbook of Optical Sensors. Boca Raton, Florida, USA: CRC Press; 2017. 718 p

[26] Jiang X, Meng Q. Design of optical fiber SPR sensing system for water quality monitoring. In: International Conference on Computational Science and Engineering (ICCSE 2015); July 20–21; Qingdao, Shandong, China. 2015. pp. 123-127

[27] Zhao Y, Deng Z-Q, Hu H-F. Fiber-optic SPR sensor for temperature measurement. IEEE Transactions on Instrumentation and Measurement. 2015;**64**(11):3099-3104. DOI: 10.1109/TIM.2015.2434094

[28] Di H, Xin Y, Sun S. Electric current measurement using fiber-optic curvature sensor. Optics and Lasers in Engineering. 2016;**77**:26-30. DOI: 10.1016/j.optlaseng.2015.07.009

[29] Mahanta DK, Laskar S. Power transformer oil-level measurement using multiple fiber optic sensors. In: 2015 International Conference on Smart Sensors and Application (ICSSA); IEEE; 2015. p. 102-105. DOI: 10.1109/ICSSA.2015.7322519

[30] Stupar DZ, Bajic JS, Manojlovic LM, Slankamenac MP, Joza AV, Zivanov MB. Wearable low-cost system for human joint movements monitoring based on Fiber-optic curvature sensor. IEEE Sensors Journal. 2012;**12**(12):3424-3431. DOI: 10.1109/JSEN.2012.2212883

[31] Antonio de la Piedra, An Braeken and Abdellah Touhafi. Sensor systems based on FPGAs and their applications: A survey. Sensors 2012;**12**(9):12235-12264. DOI: 10.3390/s120912235

[32] CK Madsen, T Snider, R Atkins, J Simcik. Real-time processing of a phase-sensitive distributed fiber optic perimeter sensor. In: SPIE Proceedings Vol. 6943: Sensors, and Command, Control, Communications, and Intelligence (C3I) Technologies for Homeland Security and Homeland Defense VII, 694310; April 16; SPIE; 2008. DOI: 10.1117/12.777987

[33] Wang Y, Negri LH, Kalinowski HJ, Mattos DS, Negri GH, Paterno AS. Hardware embedded fiber sensor interrogation system using intensive digital signal processing. Journal of Microwaves, Optoelectronics and Electromagnetic Applications. 2014;**13**(2). DOI: 10.1590/S2179-10742014000200003

[34] Plaza AJ, Chang C-I, editors. High Performance Computing in Remote Sensing. Boca Raton, Florida, USA: CRC Press; 2007. 496 p

[35] Pitarke JM, Silkin VM, Chulkov EV, Echenique PM. Theory of surface plasmons and surface-plasmon polaritons. Reports on Progress in Physics. 2007;**70**(1):1-87. DOI: 10.1088/0034-4885/70/1/R01

[36] Xilinx Inc. Artix-7 FPGA, 7 Series FPGAs Data Sheet: Overview, DS180 (v2.4) [Internet]. March 28, 2017. Available from: https://www.xilinx.com/support/documentation/data_sheets/ds180_7Series_Overview.pdf [Accessed: July 7, 2017]

[37] Xilinx Inc. Artix-7 35T Arty FPGA Evaluation Kit [Internet]. 2017. Available from: https://www.xilinx.com/products/boards-and-kits/arty.html [Accessed: July 7, 2017]

Chapter 7

Fiber Optical Tweezers for Applying and Measuring Forces in a 3D Solid Compartment

Chaoyang Ti, Minh-Tri Ho Thanh, Yao Shen,
Qi Wen and Yuxiang Liu

Additional information is available at the end of the chapter

http://dx.doi.org/10.5772/intechopen.71757

Abstract

We developed an inclined dual fiber optical tweezers (DFOTs) for simultaneous force application and measurements in a 3D hydrogel matrix. The inclined DFOTs provide a potential solution for cell mechanics study in a three-dimensional matrix.

Keywords: fiber optics, optical trapping and manipulation, optical tweezers, mechanical properties measurement, cell mechanics study

1. Introduction

Optical tweezers (OTs) have been widely used in manipulating micro- or nanoscale particles and measuring nanometer-scale displacements since Arthur Ashkin pioneered the field in the early 1970s [1]. OTs have enabled significant advances in a range of applications in in biological and physical researches, such as the study of the motion of individual motor proteins [2, 3], rheology measurements of cell cytoskeleton response to external stimuli [4, 5], and mechanical properties of polymers [6]. Conventional OTs rely on high-numerical-aperture objective lenses. The intrinsic limitations, such as being bulky, expensive, the lack of flexibility, and the requirement of substrate transparency, significantly hinder the application of conventional OTs in emerging biophysical topics such as cellular mechanics in a three-dimensional (3D) environment.

Both mechanical forces generated by cells [7] and external forces applied on cells [8] have been shown not only to determine cell behaviors, but also to regulate biological development such as proliferation [9] and differentiation [10]. Currently, most of cellular force characterization has been carried out on deformable and homogeneous cell substrates [11, 12]. Cellular traction force can be backed out by measuring the local deformation of substrate. These techniques are also called traction force microscopy (TFM). Polyacrylamide gel is one of the most commonly used substrates for TFM. Its linear elastic properties lie in the range of deformation induced by single cells. Therefore, Polyacrylamide gel has been wildly used to measure cellular forces based on the substrate deformation during the cell attachment and migration [7], to investigate cell behaviors under various conditions of extracellular matrix (ECM) (e.g., by varying substrate stiffness) [13, 14], and to study traction forces of cells fully encapsulated in an elastic gel matrix [15].

In TFM, the mechanical properties of the substrate have to be pre-characterized in order to obtain the cell traction forces from the measured deformation field. Generally, local mechanical properties of the substrates, such as polyacrylamide gel, is measured by using atomic force microscopy (AFM) [16, 17]. However, the AFM based microrheology measurements require physical contact, and the resolution of lateral force measurements is limited. Micropost arrays [18] are an alternative substrate structure that has been used to replace polyacrylamide gel. However, these techniques cannot measure cellular forces in 3D compartments. Measurements of cellular traction forces in 3D compartments are important, since the behaviors of cells are different in a native environment, which is always inhomogeneous and 3D, compared to a 2D substrate [19]. Legant et al. quantified traction forces of cells embedded in a 3D hydrogel matrices by tracking a large number of fluorescent beads around the cells [15]. However, it is challenging to provide real-time measurements, since the post-process is time-consuming. In addition, a homogeneous elastic medium is required in order to back out the cellular forces for all the techniques mentioned above.

In this chapter, we demonstrate a versatile and flexible fiber optical trapping system, namely the inclined dual fiber optical tweezers (DFOTs) [20, 21], for measuring forces on and applying forces to particles embedded in a 3D compartment, without requiring any physical contact with the particles. It can reach anywhere in a liquid medium and does not require the substrate transparency. The trap can work with particles high above the substrate or those encapsulated inside a solid 3D compartment but close to the surface. Since the inclined DFOTs create traps without relying on the bulky objective lens, they have great potential to be miniaturized and integrated.

We demonstrate that the maximum force provided by the inclined DFOTs is comparable with that of conventional optical tweezers. For example, in our experiment with the inclined DFOTs, we obtained an optical force of ~20 pN on a 4.63 μm silica bead when an optical power at each fiber tip was 100 mW. By comparison, the maximum optical trapping forces provided by the conventional OTs is around 10 pN per 100 mW for micrometer-sized beads [22]. Moreover, compared with conventional OTs, the inclined DFOTs could allow a higher optical trapping power before possible photodamage occurs to a trapped cell. This is due to the following two reasons. (1) The optical power is distributed over a large area of cell surfaces from both sides of the inclined DFOTs, and hence the intensity is lower on the cell. By comparison, conventional optical tweezers focusing all the power in a sub-micron spot, resulting in a higher intensity and likelihood of photodamage with the same power. (2) The optical beam size in the inclined DFOTs is around 30 μm at the trap position (beam intersection), so the effective trapping power illuminated on a 5 μm bead is much less than the power emitted from the fiber tip.

In this work, the calibration of the optical trapping spring constant was carried out on microscale silica beads in water as well as those embedded in 3D polyacrylamide gel compartments. In addition, the *in-situ* characterization of polyacrylamide gel stiffness was performed by optical trapping measurements, and the results are in agreement with AFM measurements. Since there is no requirement for the polyacrylamide gel to be homogeneous in the optical trapping measurements, our results imply that the inclined DFOTs can be potentially used to characterize local mechanical properties of a 3D inhomogeneous, nonlinear medium. In addition, by varying the optical power of the inclined DFOTs, we can change the effective spring constant on the particles encapsulated in a polyacrylamide gel compartment. These results indicate that the inclined DFOTs can be used as a powerful tool to apply forces to biologic samples, for example cells, and to measure their responses simultaneously in a native 3D inhomogeneous environment.

2. Methodology

2.1. System setup and working principles

In the optical trap, optical forces arise from the momentum transfer during the scattering or refraction of incident photons [23]. When a dielectric particle, with a refractive index higher than the surrounding medium, is illuminated by a light beam, it will change the direction of light beam and in turn experience a force that is described as the sum of two components: a scattering force directed along the light beam and a gradient force pointing to the region of maximum light intensity. Traditionally, an optical trap is created by focusing a single laser beam with a high numerical aperture objective, which is called objective based OTs. Unlike objective based OTs, the inclined DFOTs create traps based on two inclined optical beams that can apply two sets of gradient forces and scattering forces, as shown **Figure 1(a)**. Once the four optical forces balance with all other forces that the particle is subject to, such as viscous drag force and gravity, a 3D optical trap can be successfully created.

The setup of the inclined DFOTs was based on our previous work [20, 24, 25]. Briefly, we set up the inclined DFOTs on an inverted microscope platform. Each fiber was mounted on a common board and was aligned with respect to the other via a miniature 3D translational stage and a micro 1D rotational stage, as shown in **Figure 1(a)**. The fiber inclination angle and fiber separation was also controlled by adjusting these stages. The position of the optical trap was then adjusted by manipulating the common board via another 3D translational stage.

The measurement setup is shown in **Figure 1(b)**. Light from a 980 nm laser diode (AC 1405-0400-0974-SM-500, Eques) was split into two lensed fibers (Nanonics Imaging, Ltd) through

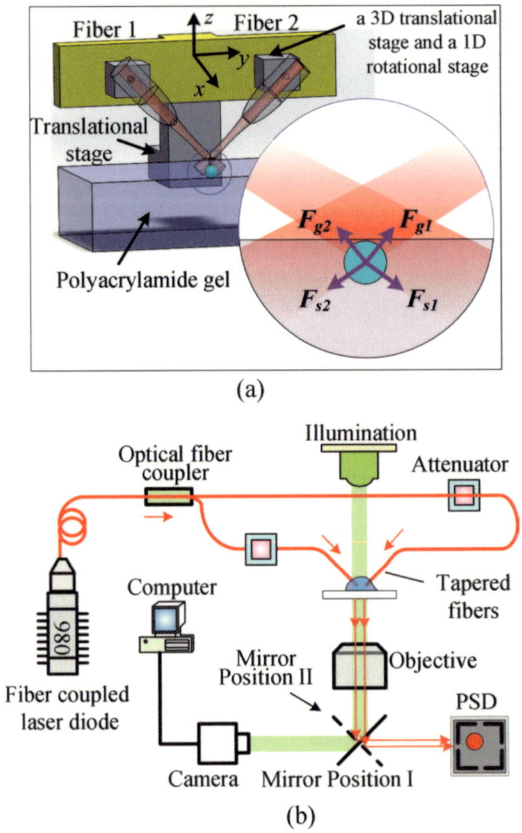

Figure 1. (a) Working principles and the setup schematic of the inclined DFOTs. The inclined DFOTs can be moved by a single translational stage that controls the position of the aluminum plate (top rectangular board in the figure). (b) Schematic of the measurement setup of the inclined DFOTs. F_s = scattering force; F_g = gradient force.

a 3 dB coupler. The equalized optical power outputs from the two lensed fiber ends were ensured by attenuators. When a micro silica bead in water or in polyacrylamide gel was trapped, the light scattered by the bead was collected by an objective lens and detected by a position-sensitive detector (PSD) (DL100-7-PCBA3, First Sensor), which enabled precision measurements of bead displacements in the x, y, and z directions with a nanometer resolution. All the fibers in the system are single mode at the wavelength of 980 nm.

2.2. Power spectrum calibration of optical spring constant

In this paper, calibration of the optical tweezers was accomplished by the power spectrum analysis method [26]. In the power spectrum analysis, the motion of a particle was confined by the restoring "spring" forces provided by the optical trap as well as the surrounding medium (such as the polyacrylamide gel) and can be expressed by [26]

$$m\ddot{x}(t) + \gamma\,\dot{x}(t) + k_x\,x(t) = \left(2\,k_B\,T\,\gamma_0\right)^{1/2}\eta(t), \tag{1}$$

and the one-sided power spectrum of the particle displacement can be expressed as [23, 25]

$$S_{xx}(f) = \frac{k_B T}{\pi^2 \gamma (f_0^2 + f^2)}, \tag{2}$$

where T is the absolute temperature, k_B is the Boltzmann's constant, γ represents the hydrodynamic drag coefficient ($\gamma = 6\pi\eta a$ for the Stokes drag on a spherical particle with a radius of a and a medium viscosity of η), and f_0 stands for the corner frequency that is related to the trapping spring constant k by

$$k = 2\pi\gamma f_0. \tag{3}$$

In our experiments, the motion of the trapped particle was detected by the PSD output signals that were recorded in a period of 5 seconds at a sampling rate of 50 kHz. Its power spectra were obtained by complex Fourier transform of the PSD signals [26]. The sampling rate in the experiment was much larger than the measured corner frequency, which in our case is tens of Hz. The exponential distributed power spectrum data were then blocked and transformed to Gaussian distribution, which was used to follow the least square curve fitting. Being fitted to a Lorentzian in Eq. (1) with a frequency range of 10–800 Hz, the blocked power spectra provided the calibration results of the spring constants following Eq. (2). The calibration of the spring constant of the inclined DFOTs does not require excitation of particle motion at any specific frequencies. The calibration is passive and relies only on monitoring the confined Brownian motion of trapped particle. Therefore, the inclined DFOTs can be used for *in-situ* and real-time measurements.

2.3. Experimental measurements in water

We first investigated the dependence of the spring constant of the inclined DFOTs on the optical power in water, which allowed us to better characterize the inclined DFOTs. The results obtained in water set the base for polyacrylamide gel characterization that was described in Section 2.4. Silica beads (Bangs Laboratories, Inc.) with a density of 2.0 g/cm^3 and a diameter of 4.63 μm were used in all experiments. A diluted silica bead solution, with a ratio of deionized (DI) water to original bead solution (10.2%, 0.5 g in weight) at 6000:1 was ultrasonicated in order to reverse bead aggregation. One drop of diluted bead solution (~0.2 ml) was added on a coverglass, where the trapping experiment was carried out. In order to reduce the sidewall effects, we accomplished the optical trap calibration with beads 3D trapped around 35 μm above the coverslip [22].

2.4. Experiment measurements in polyacrylamide gel

In this experiment, the inclined DFOTs were used to apply tunable forces on a bead embedded in the polyacrylamide gel without any physical contact. In the meantime, the resultant motion of the bead was monitored and measured by the PSD to enable spring constant calibration of both the optical trap and the polyacrylamide gel. By varying the optical power, the optical force provide by the inclined DFOTs was readily tuned, which enables the calibration of the effective spring constant of the hydrogel.

2.4.1. *Polyacrylamide gel sample preparation*

Polyacrylamide gels were fabricated on 25 × 25 mm^2 coverslips. In all the experiments, we prepared the polyacrylamide gel solution with the desired concentration of 5% acrylamide and 0.04% bisacrylamide in a pH 8.2 HEPES buffer. By adding 3 μl tetramethylethylenediamine (TEMED) and 10 μl of 10% ammonium persulfate (APS) solution for each 1 ml of polyacrylamide gel solution, we can initiate the polymerization process. The polyacrylamide gel solution was then quickly transferred to a glutaraldehyde-activated coverslip and covered by a second plasma-cleaned coverslip, on which dried silica beads with a diameter of 4.63 μm were attached previously. A polymerized polyacrylamide gel with a desired thickness of ~ 50 μm was achieved by controlling the volume of polyacrylamide gel solution on each coverslip to be 30 μl. The gelation was complete after curing the gel solution for 15 minutes at the room temperature. We obtained the polyacrylamide gel sample with silica beads embedded on the its top after peeling off the second coverslip. Due to the deformation of the polyacrylamide gel during the polymerization process, not all the beads are on the polyacrylamide gel surface. Beads close to the surface of the polyacrylamide gel were selected for the optical trapping experiments.

2.4.2. *Optical trapping experiment of polyacrylamide gel calibration*

The optical trapping experiment of polyacrylamide gel calibration was carried out by analyzing the confined motion of the bead, which resulted from the confinement effects of both the optical trap and the polyacrylamide gel. A bead close to the polyacrylamide gel top surface was first identified, and the inclined DFOTs were moved to this pre-identified location so that the bead was in the optical trap. The power spectrum measurements enabled the calibration of an effective spring constant. The tunable optical spring constant was achieved by varying the optical power, which allowed the polyacrylamide gel stiffness to be calibrated.

2.4.3. *AFM measurements of polyacrylamide gel moduli*

In this work, AFM [16, 17] was used to characterize local viscoelastic properties of polyacrylamide gel around the bead that was studied in the inclined DFOTs. An Asylum Research MFP3D-BIO AFM (Asylum Research, CA) was used for the measurements. Briefly, The AFM cantilever was first moved to the location above the polyacrylamide gel top surface around the selected bead that was studied by the inclined DFOTs. The cantilever was then lowered down to create an approximate 2 μm indentation in the gel. A small oscillating indentation at the tip was generated by driving the cantilever sinusoidally with an amplitude of 25 nm and a frequency $f = 10$ Hz. The gel elastic modulus (E') and the viscous modulus (E''), which are also referred to as the storage and loss moduli, respectively, were measured based on the cantilever force and indentation. The experimentally measured E' and E'' are 1469.9 ± 555.9 Pa and 533.2 ± 243.4 Pa, respectively (see Section 3.2.2). The viscosity of the polyacrylamide gel was calculated using $\eta = E''/2\pi f$.

3. Experimental results and discussion

3.1. Experimental results in DI water and discussion

In this section, we will demonstrate the capability of the inclined DFOTS to apply and measure 3D forces in aqueous environments.

3.1.1. 3D trapping of yeast cells in DI water

3D trapping of silica beads with the inclined DFOTs was demonstrated in our previous work [20, 24, 25]. Here, we show that the inclined DFOTs are also applicable to 3D trapping of biological samples. In this experiment, the inclined DFOTs were used to trap and manipulate a living yeast cell in all three dimensions in DI water, as shown in **Figure 2**. The trapped yeast cell was moved via controlling the aluminum board (see yellow block in **Figure 1(a)**) by a single 3D stage (not shown in **Figure 1(a)**). The maximum moving speed of the trapped cells was dependent on the optical power. For example, the maximum moving speed was measured to be around 20 μm/s before the yeast cells escaped from the trap, when the optical power is 6.8 mW from each fiber. Manipulating the 3D positions of living cells bestows on the inclined DFOTs the capability of relocation and assembly of living biological particles with micrometer size.

3.1.2. Optical trapping spring constant calibration in water

Silica beads with the size of 4.63 μm were used to carry out the calibration of the trapping spring constant. The uniformity in shape and material properties allows silica beads to serve as appropriate samples for evaluating the capability of the inclined DFOTs.

The optical trapping spring constant can be measured by the bead displacement power spectrum when a silica bead is trapped in three dimensions in water. **Figure 3(a)** and **(c)** show the typical power spectrum data of a bead trapped in water in the x and y axes, respectively. The corresponding corner frequencies obtained by fitting the blocked power spectra to Lorentzian are 86.0 ± 3.3 Hz in the x axis (**Figure 3(a)**) and 74.4 ± 3.1 Hz in the y axis (**Figure 3(c)**). We notice that the spring constants in the x and y directions are different, which are mainly due to different optical field distributions along the two directions, as shown in **Figure 1(a)**. The linear dependence of optical spring constant on optical power along the x and y directions is shown in **Figure 3(b)** and **(d)**, respectively. The maximum power used in the experiment is 120 mw and the corresponding spring constants of the inclined DFOTs are 22.1 ± 1.0 pN/μm and 21.5 ± 0.9 pN/μm in the x and y directions, respectively. The tunable optical forces can be determined by measuring the displacement of the trapped particle. According our previous results [20], the linear range of the force-displacement relationship is around −1 to 1 μm, and the maximum bead displacement in the inclined DFOTs is around 2 μm before it escapes from the trap. When the bead displacements is between 1 and 2 μm, the trap is not stable and the bead may easily escape.

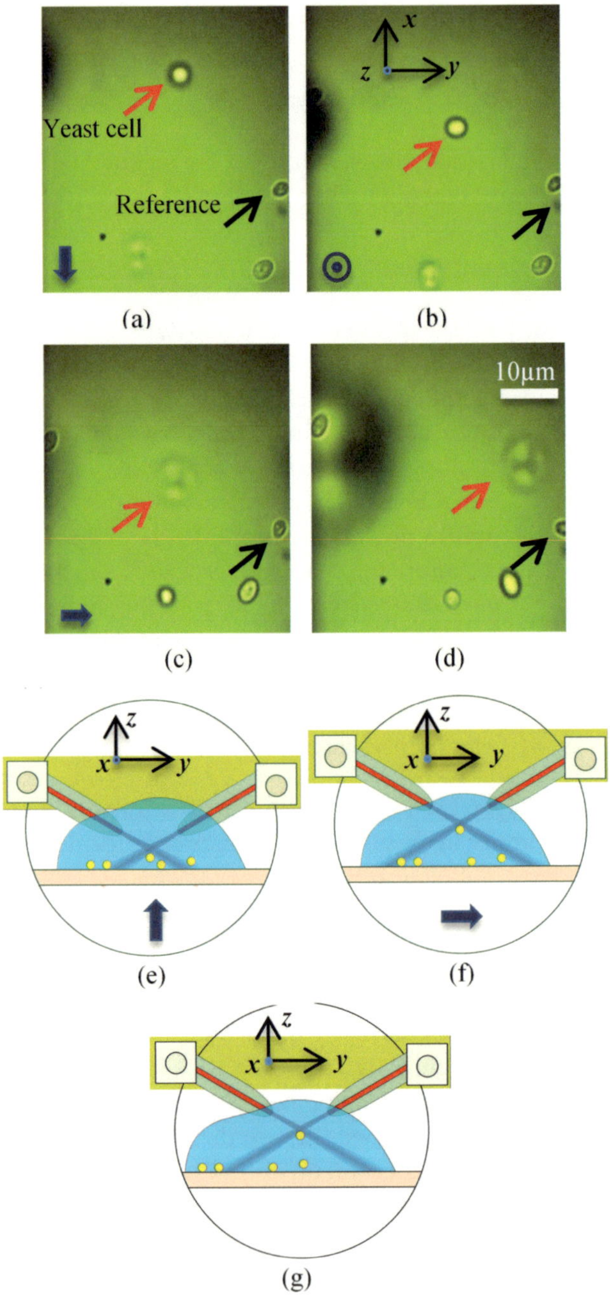

Figure 2. 3D trapping of living yeast cells with the inclined DFOTs in water [24, 25]. (a–d) Microscope images of trapping a yeast cell. The yeast cell was trapped (a) and manipulated in the (b) $-x$ direction, followed by (c) $+z$ and (d) $+y$ directions. The corresponding next movements of the optical trap are shown in the lower left corners of (a–c) and at the bottom of (e–f). The upper arrows represent the trapped yeast cell, and the lower arrows are the reference yeast cell. (e–g) Schematics showing the positions of the fibers and trapped beads in (b–d), respectively. The shadow shown on the left-hand side of (a–d) is the trapped fiber tip. The optical power from each fiber taper was 6.8 mW.

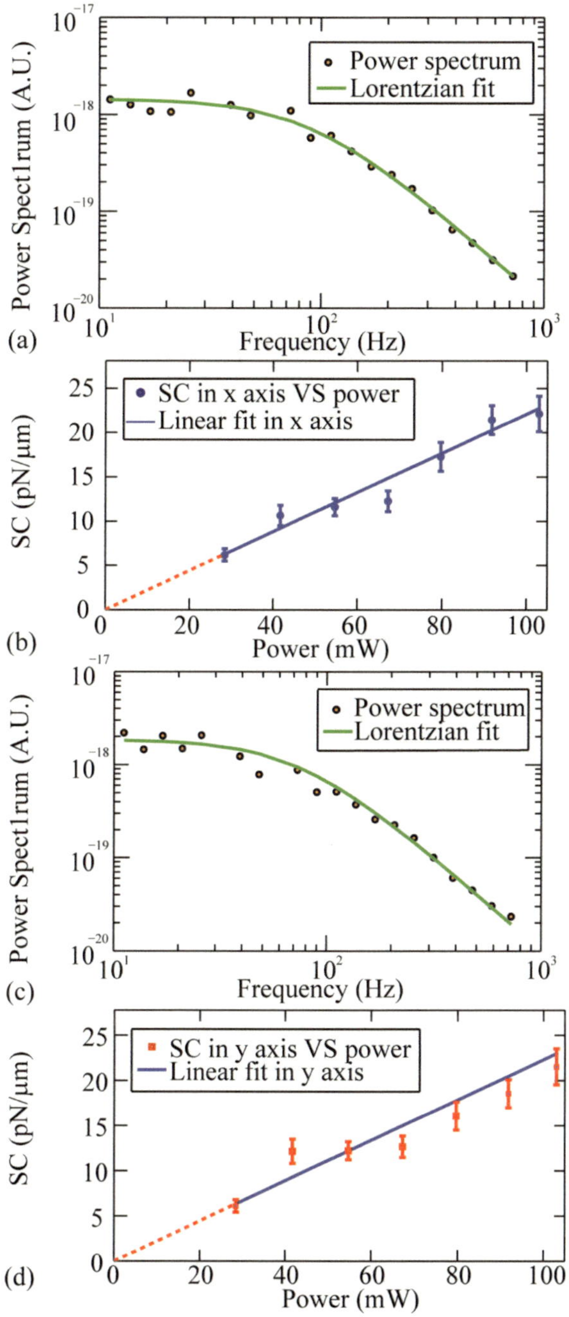

Figure 3. Experimentally measured power spectrum data of a 4.63 μm bead trapped in water (hollow circles) and Lorentzian fitting (solid curves) at 91.76 mW in the (a) *x* and (c) *y* directions. Optical trapping spring constant as a function of optical power in the (b) *x* and (d) *y* directions. The linear fitting (solid line) of spring constant on optical power passes point (0, 0). The extension is the dashed line. The optical power drawn in the figure is the power emitted by each fiber. SC stands for spring constant.

It is noted that the optical force can be calculated by multiplying optical trapping spring constant by the bead displacement. The results in **Figure 3(b)** and **(d)** indicate that the inclined DFOTs can provide forces ranging from sub-pN to tens of pN. It implies that the inclined DFOTs can be used to measure cellular forces, since various cellular forces lie in this range, such as those generated by neural growth cone [27] and by a single actin filament [28]. Compared with the force sensing range of AFM, which is in the range of 10 pN to 100 nN [29], the force range of inclined DFOTs may extend the range of the biological applications and measurable quantities that may be challenging for AFM.

The error bars in **Figure 3(b)** and **(d)** are the 95% confidence interval ranges of the Lorentzian curve fitting. It can be seen that there is a linear relationship between the spring constant and the optical power in both x and y directions. The results in water confirm that the optical force can be well characterized, which enables the inclined DFOTs to exert controlled optical forces as well as measurements of external forces on the trapped particles.

There are data points in **Figure 3(b)** and **(d)** scattering around the fitted linear curve. This may be that the z-dimensional equilibrium position (see **Figure 1(a)**) of the trapped bead is dependent on the optical power. It could introduce a nonlinearity between the spring constant and optical power. In addition to the errors of the Lorentzian fitting, other sources of the uncertainties include electronic noise, external noise, as well as errors of the bead diameter.

3.2. Experimental measurements of polyacrylamide gel and discussion

3.2.1. Experimental results measured by inclined DFOTs

In the experiments, the effective spring constant was changed by tuning the optical power. We obtained the dependence of effective spring constant on power in the y direction, as shown in **Figure 4(b)**. Each single data point in **Figure 4(b)** is acquired from the Lorentzian fitting of a set of power spectrum data at a fixed power as described in Section 2.2. A typical set of power spectrum data is shown in **Figure 4(a)**. It is seen that there exists a linear relationship between the effective spring constant and the optical power, as shown in **Figure 4(b)**. In the experiment, since the effective spring constant is attributed to both the optical trapping and the polyacrylamide gel, the intrinsic stiffness of the polyacrylamide gel with no laser illumination can be obtained to be 0.012 ± 0.005 N/m, which is based on the intersect of the fitted curve with the vertical axis. The error bar is determined by the standard deviation of 4 independent measurements. We show one of the spring constant-power curves in **Figure 4(b)**. This enables the inclined DFOTs to characterize the mechanical properties of solid media *in-situ*. Optical forces on embedded beads can be calculated by measuring the bead displacement once the equivalent spring constant of polyacrylamide gel is characterized.

Since the calibration of the material equivalent spring constant is accomplished by monitoring the bead confined Brownian motion, the inclined DFOTs can used to measure the stiffness of nonlinear materials. We estimate that the effective root-mean-square displacement of the Brownian motion is around 0.6 nm for a bead embedded in a material with an equivalent spring constant of 0.012 N/m at the room temperature. Most of the materials can be considered to be linear and uniform in such a small range as 0.6 nm, so the material equivalent

http://dx.doi.org/10.5772/intechopen.71757

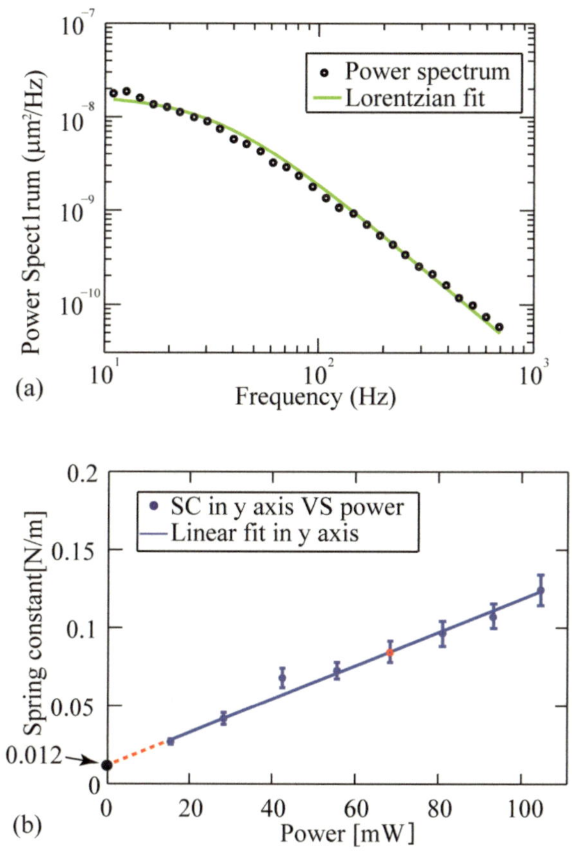

Figure 4. (a) A typical set of experimental power spectrum data fitted by a Lorentzian in the polyacrylamide gel at the power of 67.0 mW, which corresponds to the fifth data point in (b). (b) The effective y-axis (see **Figure 1(a)**) spring constant as a function of optical power for a bead embedded in the polyacrylamide gel. The intersection with the vertical axis of 0.012 N/m indicates the polyacrylamide gel stiffness. SC stands for spring constant.

spring constant can be calibrated *in situ*. As a result, unlike other measurements where material linearity is important, such as traction force microscopy, the optical trapping measurement of the material properties (and hence the forces) is not dependent on the linearity or homogeneity of the materials.

In addition, unlike traditional AFM measurements which require physical contact, no physical contact is needed for the inclined DFOTs. This enables the inclined DFOTs to work with particles embedded in 3D matrices. Although the traction force microscopy has been demonstrated in a homogenous 3D medium, it is still challenging to realize real-time measurements due to the requirement of tracking large numbers of fluorescent beads [15]. By comparison, the inclined DFOTs have a potential to provide a real-time, versatile, and non-invasive way to measure the material properties inside a 3D heterogeneous and nonlinear medium.

It is noted that laser illumination can affect the mechanical properties of polyacrylamide gel in the experiment. For example, under laser illumination, local temperature changes originating from optical absorption of the trapped particle can change the surrounding medium viscosity [30]. The polyacrylamide gel stiffness has also been observed to change under localized laser illumination [31, 32]. Moreover, polyacrylamide gel mechanical integrity can also be changed [31] due to the nature of polyacrylamide gel swelling. In addition, the effective polyacrylamide gel property may be influenced due to the imperfect contact condition at the interface between bead and surrounding polyacrylamide gel. In our experiments, we noticed differences in data obtained in water and polyacrylamide gel, mainly in the slope of spring constant-power curve, shown in **Figures 3(b)**, **(d)**, and **4(b)**. Although we do not fully understand how the polyacrylamide gel has been changed by laser illumination, we observed the linear dependence of corner frequencies on optical power, because the spring constant-power data in polyacrylamide gel can be well fitted by linear functions. According to the repeatability of our optical trapping measurements of the polyacrylamide gel stiffness, we believe the laser-induced polyacrylamide gel changes are reversible. As a result, the capability of the inclined DFOTs to measure the intrinsic polyacrylamide gel stiffness will not be influenced by the laser induced polyacrylamide gel changes, since the measurement of the intrinsic polyacrylamide gel stiffness with the inclined DFOTs is determined by the intersect of the linear fitting curve on the vertical axis.

3.2.2. *Experimental results measured by AFM*

AFM based microrheology measurements [16, 17] were also used to characterize the local viscoelasticity of polyacrylamide gel. In this work, AFM was used to serve as a reference to verify the inclined DFOTs measurements. To ensure the statistical measurements, we characterized the viscoelastic properties of polyacrylamide gel at 19 different locations around the bead of interest. The typical AFM force and indentation curves, shown in **Figure 5(a)**, can be fitted by sinusoidal function to remove noise. **Figure 5(b)** plots the force as a function of indentation. Its raw data and the corresponding fitted data is shown as the blue and red curves, respectively. The hysteresis in **Figure 5(b)** indicates the viscoelasticity of polyacrylamide gel, which can be used to back out the elastic and viscous moduli. Following the analysis method in Ref. [16], the elastic and viscous moduli of the polyacrylamide gel can be calculated and obtained to be 1469.9 ± 555.9 Pa and 533.2 ± 243.4 Pa, respectively. The mean values and standard deviations of the moduli were determined by 19 independent measurements. As a result, based on the viscous moduli of the polyacrylamide gel, we obtained its viscosity to be $8.5 \pm 3.9\ \mathrm{Pa} \cdot \mathrm{s}$.

3.2.3. *Comparison of the inclined DFOTs and AFM results*

In the experiments, the optical trapping experiments measure the effective spring constants (with a unit of N/m) on a silica bead, which is dependent on bead size, the bead location, and the elastic modulus of polyacrylamide gel. However, AFM measures the elastic modulus (with a unit of Pa). To compare them, we used the commercial finite element analysis software (COMSOL) to calculate the spring constant on a bead embedded in the top surface of polyacrylamide gel using the elastic modulus measured from the AFM experiments. We then compared the spring constant obtained by the inclined DFOTs measurements with that calculated from COMSOL.

http://dx.doi.org/10.5772/intechopen.71757

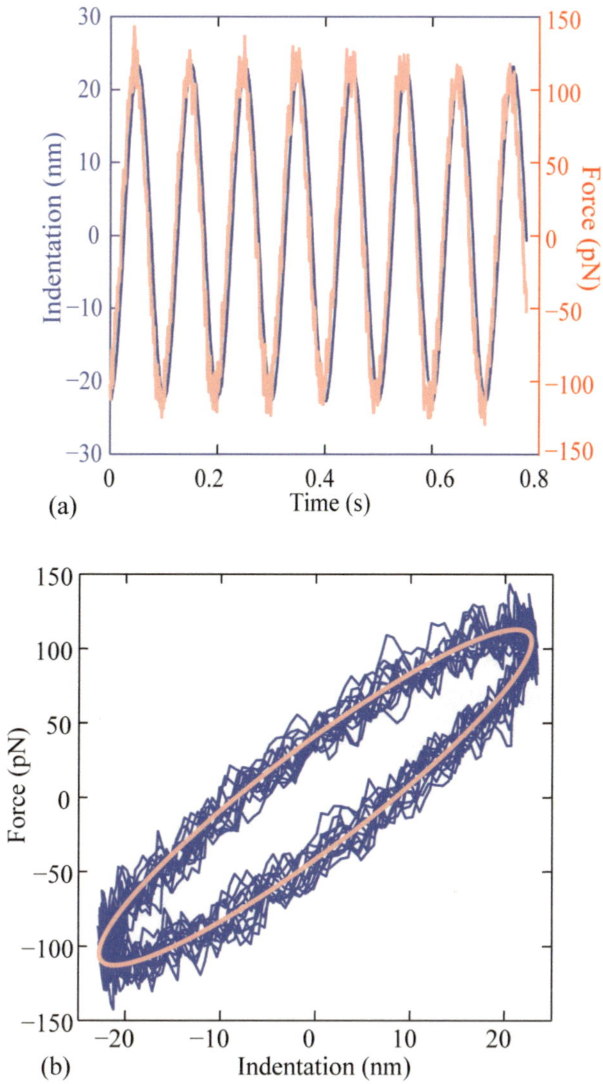

Figure 5. AFM microrheology measurements of the polyacrylamide gel. (a) A typical measurement of the sinusoidal force ("thick" curve) and indentation ("thin" curve) as functions of time. (b) Force as a function of indentation. The "thin" curve is obtained from the raw data in (a), and the smooth "thick" curve is from the sinusoidal fitting of the data in (a).

We use the Kelvin-Voigt model to describe the viscoelasticity of polyacrylamide gel in the simulation. By applying a 20 nN static body force applied parallel to the polyacrylamide gel surface, we obtain the displacement of a silica bead embedded in the polyacrylamide. The polyacrylamide gel stiffness is then calculated based on the resultant bead displacement and the applied force.

According to the procedures of polyacrylamide gel preparation, the beads selected for optical trapping measurements were always close to the top surface of polyacrylamide gel. However, due to the polyacrylamide gel deformation during the polymerization process, it was challenging

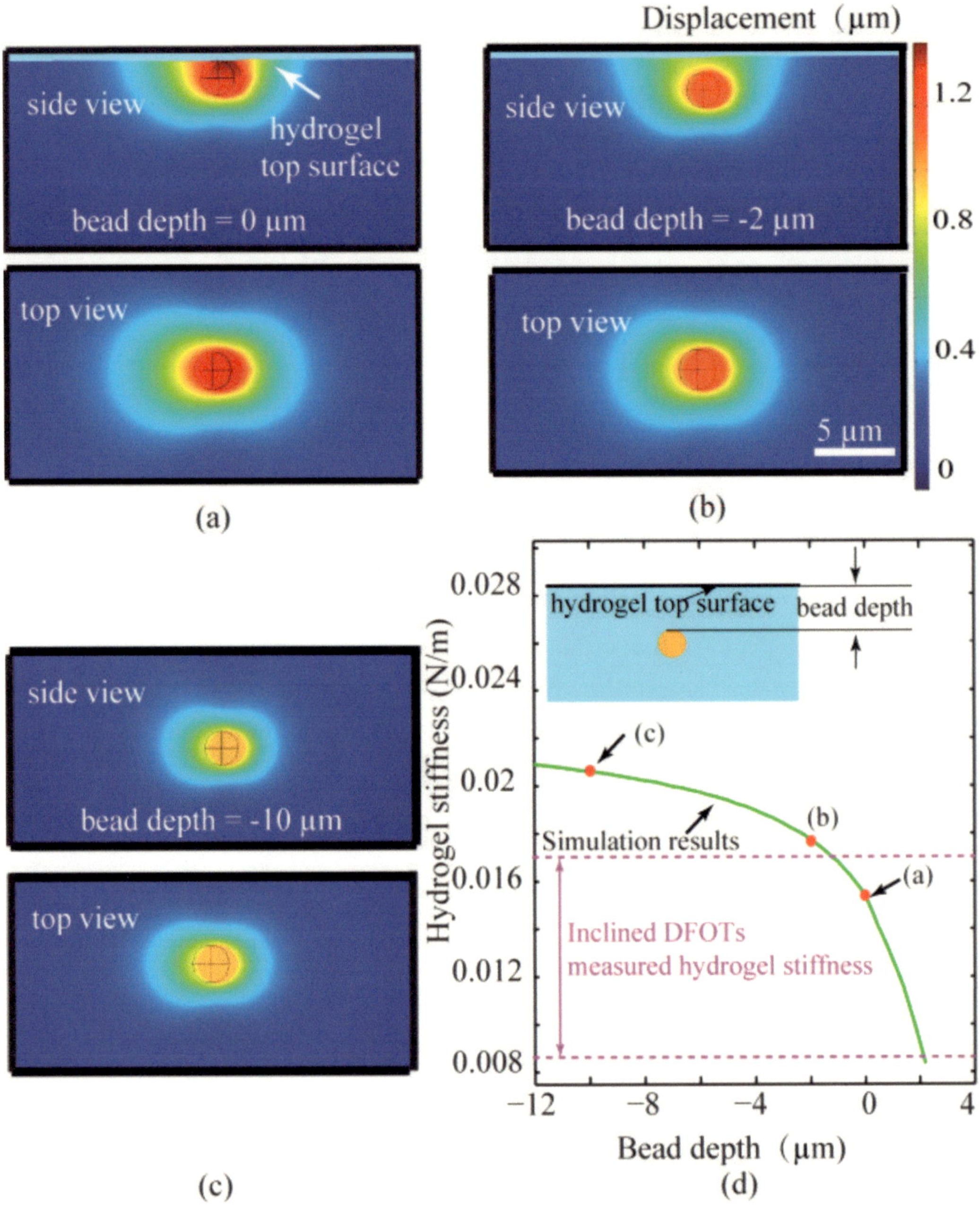

Figure 6. (a–c) COMSOL calculated the displacement field of a silica bead embedded in polyacrylamide gel when the bead depth is (a) 0, (b) −2 μm, and (c) −10 μm. The bead depth is defined to be 0 when the bead top surface is flush with the polyacrylamide gel top surface. The top and bottom figures are the side and top views, respectively. The total thickness of polyacrylamide gel used in the simulation is 50 μm. (d) Calculated dependence of the polyacrylamide gel stiffness on the bead depth based on simulation results. The three data points correspond to the results shown in (a–c), respectively. The inclined DFOTs measurements are also shown for comparison (two dashed lines). Inset: definition of bead depth.

http://dx.doi.org/10.5772/intechopen.71757

to determine the precise position of the bead with respect to the polyacrylamide gel top surface. Therefore, it was important to investigate the influence of the polyacrylamide gel stiffness on the depth of the bead in the simulation. In the simulation, we obtain the polyacrylamide gel stiffness shown in **Figure 6(a)–(c)** to be 0.0154 N/m, 0.0177 N/m, and 0.0206 N/m, respectively. It is noted that the deeper the bead, the larger the stiffness. In addition, due to the weaker polyacrylamide gels surface effects, the rate of change of the stiffness is smaller with deeper bead positions. The polyacrylamide gel displacement field is not influenced much by the surface when the bead is far away from the polyacrylamide gel surface, as shown in **Figure 6(c)**. When the bead is deeper than 10 μm from the surface, the stiffness variation becomes small, as shown in **Figure 6(d)**.

According to the simulation results shown in **Figure 6(d)**, we can see that the optical trapping measurement of polyacrylamide gel stiffness agrees with the simulation results based on AFM measurements when the bead top surface is in the range from −1 to 2 μm above the polyacrylamide gel surface. This proves that the inclined DFOTs can be a reliable tool to realize *in-situ*, non-invasive characterization of local material properties and forces in a 3D medium. In addition, since the spring constant of optical trapping is dependent on the optical restoring force applied on the embedded bead, we can vary the spring constant by tuning optical power. It demonstrates the capability of the inclined DFOTs to apply tunable forces in 3D compartments.

The ability of simultaneously applying and measuring forces enables the inclined DFOTs to be a potential tool for cell mechanics study in 3D compartments. The spring constant measurements bestows upon the inclined DFOTs capable of directly measuring forces on the bead if the displacement is monitored.

4. Conclusion

In this chapter, we show the inclined dual fiber optical tweezers (DFOTs) can be used for simultaneous applications and measurements of optical forces on particles in water and those in a 3D polyacrylamide gel compartment. Moreover, we demonstrate *in-situ* characterization of the polyacrylamide gel stiffness by the optical trapping measurements. The measured polyacrylamide gel stiffness agrees with finite element method (FEM) simulation results based on experimental AFM measurements. Since the optical trapping measurements do not require the medium to be mechanically homogeneous and linear, the inclined DFOTs can measure the mechanical properties of materials that are heterogeneous and nonlinear. The ability of simultaneous applying and measuring optical forces in a 3D compartment enables the inclined DFOTs to be a versatile tool that can be potentially useful for biomechanics and mechanobiology study.

Acknowledgements

This research is supported by the National Science Foundation under Grant No. CBET-1403257.

Author details

Chaoyang Ti[1]*, Minh-Tri Ho Thanh[2], Yao Shen[1], Qi Wen[2] and Yuxiang Liu[1]

*Address all correspondence to: chaoyang@wpi.edu

1 Department of Mechanical Engineering, Worcester Polytechnic Institute, Worcester, MA, USA

2 Department of Physics, Worcester Polytechnic Institute, Worcester, MA, USA

References

[1] Ashkin A. Acceleration and trapping of particles by radiation pressure. Physical Review Letters. 1970;**24**:156

[2] Svoboda K, Schmidt CF, Schnapp BJ, Block SM. Direct observation of kinesin stepping by optical trapping interferometry. Nature. 1993;**365**:721-727

[3] Finer JT, Simmons RM, Spudich JA. Single myosin molecule mechanics: Piconewton forces and nanometre steps. Nature. 1994;**368**:113-119

[4] Fabry B, Maksym GN, Butler JP, Glogauer M, Navajas D, Fredberg JJ. Scaling the microrheology of living cells. Physical Review Letters. 2001;**87**:148102

[5] Fredberg J, Fabry B. The cytoskeleton as a soft glassy material. In: Mofrad MRK, Kamm RD, editors. Cytoskeletal Mechanics: Models and Measurements. Cambridge, NY: Cambridge University Press; 2006

[6] Grier DG. Optical tweezers in colloid and interface science. Current Opinion in Colloid & Interface Science. 1997;**2**:264-270

[7] Dembo M, Wang Y-L. Stresses at the cell-to-substrate interface during locomotion of fibroblasts. Biophysical Journal. 1999;**76**:2307-2316

[8] Hickory WB, Nanda R. Effect of tensile force magnitude on release of cranial suture cells into S phase. American Journal of Orthodontics & Dentofacial Orthopedics. 1987;**91**:328-334

[9] Chen CS, Mrksich M, Huang S, Whitesides GM, Ingber DE. Geometric control of cell life and death. Science. 1997;**276**:1425-1428

[10] Roskelley C, Desprez P, Bissell M. Extracellular matrix-dependent tissue-specific gene expression in mammary epithelial cells requires both physical and biochemical signal transduction. Proceedings of the National Academy of Sciences. 1994;**91**:12378-12382

[11] Harris AK, Wild P, Stopak D. Silicone rubber substrata: A new wrinkle in the study of cell locomotion. Science. 1980;**208**:177-179

[12] Schoen I, Pruitt BL, Vogel V. The yin-yang of rigidity sensing: How forces and mechanical properties regulate the cellular response to materials. Annual Review of Materials Research. 2013;**43**:589-618

[13] Pelham RJ, Wang Y-l. Cell locomotion and focal adhesions are regulated by substrate flexibility. Proceedings of the National Academy of Sciences. 1997;**94**:13661-13665

[14] Gray DS, Tien J, Chen CS. Repositioning of cells by mechanotaxis on surfaces with micropatterned Young's modulus. Journal of Biomedical Materials Research Part A. 2003;**66**:605-614

[15] Legant WR, Miller JS, Blakely BL, Cohen DM, Genin GM, Chen CS. Measurement of mechanical tractions exerted by cells in three-dimensional matrices. Nature Methods. 2010;**7**:969-971

[16] Mahaffy R, Park S, Gerde E, Käs J, Shih C. Quantitative analysis of the viscoelastic properties of thin regions of fibroblasts using atomic force microscopy. Biophysical Journal. 2004;**86**:1777-1793

[17] Thomas G, Burnham NA, Camesano TA, Wen Q. Measuring the mechanical properties of living cells using atomic force microscopy. JoVE (Journal of Visualized Experiments). 2013;**76**:e50497-e50497

[18] Du Roure O, Saez A, Buguin A, Austin RH, Chavrier P, Siberzan P, Ladoux B. Force mapping in epithelial cell migration. Proceedings of the National Academy of Sciences of the United States of America. 2005;**102**:2390-2395

[19] Cukierman E, Pankov R, Stevens DR, Yamada KM. Taking cell-matrix adhesions to the third dimension. Science. 2001;**294**:1708-1712

[20] Liu Y, Yu M. Investigation of inclined dual-fiber optical tweezers for 3D manipulation and force sensing. Optics Express. 2009;**17**:13624-13638

[21] Liu Y, Yu M. Multiple traps created with an inclined dual-fiber system. Optics Express. 2009;**17**:21680-21690

[22] Neuman KC, Block SM. Optical trapping. Review of Scientific Instruments. 2004;**75**:2787-2809

[23] Svoboda K, Block SM. Biological applications of optical forces. Annual Review of Biophysics and Biomolecular Structure. 1994;**23**:247-285

[24] Ti C, Thomas GM, Wen Q, Liu Y. Fiber optical tweezers for simultaneous force exertion and measurements in a 3D hydrogel compartment. In: Lasers and Electro-Optics (CLEO). Washington, DC: Optical Society of America; 2015;**79**:1-2. JW2A.79

[25] Ti C, Thomas GM, Ren Y, Zhang R, Wen Q, Liu Y. Fiber based optical tweezers for simultaneous in situ force exertion and measurements in a 3D polyacrylamide gel compartment. Biomedical Optics Express. 2015;**6**:2325-2336

[26] Berg-Sørensen K, Flyvbjerg H. Power spectrum analysis for optical tweezers. Review of Scientific Instruments. 2004;**75**:594-612

[27] Hällström W, Lexholm M, Suyatin DB, Hammarin G, Hessman D, Samuelson L, Montelius L, Kanje M, Prinz CN. Fifteen-piconewton force detection from neural growth cones using nanowire arrays. Nano Letters. 2010;**10**:782-787

[28] Ananthakrishnan R, Ehrlicher A. The forces behind cell movement. International Journal of Biological Sciences. 2007;**3**:303-317

[29] Rodriguez ML. Review on cell mechanics: Experimental and modeling approaches. Applied Mechanics Reviews. 2013;**65**:510-518

[30] Seol Y, Carpenter AE, Perkins TT. Gold nanoparticles: Enhanced optical trapping and sensitivity coupled with significant heating. Optics Letters. 2006;**31**:2429-2431

[31] Chan BP. Biomedical applications of photochemistry. Tissue Engineering Part B: Reviews. 2010;**16**:509-522

[32] Mosiewicz KA, Kolb L, Van Der Vlies AJ, Lutolf MP. Microscale patterning of hydrogel stiffness through light-triggered uncaging of thiols. Biomaterials Science. 2014;**2**:1640-1651

Chapter 8

Nonlinearity-Tolerant Modulation Formats for Coherent Optical Communications

Keisuke Kojima, Toshiaki Koike-Akino,
Tsuyoshi Yoshida, David S. Millar and Kieran Parsons

Additional information is available at the end of the chapter

http://dx.doi.org/10.5772/intechopen.73031

Abstract

Fiber nonlinearity is the main factor limiting the transmission distance of coherent optical communications. We overview several modulation formats intrinsically tolerant to fiber nonlinearity. We recently proposed family of 4D modulation formats based on 2-ary amplitude 8-ary phase-shift keying (2A8PSK), covering the spectral efficiency of 5, 6, and 7 bits/4D symbol, which will be explained in detail in this chapter. These coded modulation formats fill the gap of spectral efficiency between DP-QPSK and DP-16QAM, showing superb performance both in linear and nonlinear regimes. Since these modulation formats share the same constellation and use different parity bit expressions only, digital signal processing can accommodate those multiple modulation formats with minimum additional complexity. Nonlinear transmission simulations indicate that these modulation formats outperform the conventional formats at each spectral efficiency. We also review DSP algorithms and experimental results. Their application to time-domain hybrid modulation for 4–8 bits/4D symbol is also reviewed. Furthermore, an overview of an eight-dimensional 2A8PSK-based modulation format based on a Grassmann code is also given. All these results indicate that the 4D-2A8PSK family show great promise of excellent linear and nonlinear performances in the spectral efficiency between 3.5 and 8 bits/4D symbol.

Keywords: high-dimensional modulation format, adaptive modulation and coding, spectral efficiency, nonlinearity tolerance, QPSK, 16QAM, 8PSK, 8QAM

1. Introduction

Optical fiber nonlinearity is the main factor limiting the transmission distance of coherent optical communications in many cases [1–3]. This is an especially critical issue for low dispersion fiber case, since the waveforms stay in the original shape for a longer time and the nonlinearity effect is enhanced. In addition, in the dense wavelength multiplex division (DWDM) systems, multiple wavelength travels nearly at the same speed, so there is less opportunity of averaging out the nonlinear effects. Typical examples are legacy submarine cable systems [4].

At the same time, more service providers prefer flexible networks where adaptive transceivers can operate multiple data rates, modulation formats, and forward error correction (FEC) overheads for the efficient use of network capacity [5–8]. For example, it has been shown that the mean loss in throughput per transceiver is nearly proportional to the granularity of the data rate [8]. In order to accommodate a wide range of channel conditions, many modulation formats with various spectral efficiencies have been studied extensively [1, 9–17].

In order to mitigate or minimize the penalty from the nonlinear effects, efforts have been made to optimize the modulation formats, which are intrinsically tolerant to optical fiber nonlinearity [4, 18–20]. In other words, the modulation format is constructed such that they are less susceptible to fiber nonlinearity. In Section 2, we discuss the so-called X-constellation, which is an eight-dimensional (8D) code and has higher nonlinearity tolerance than dual-polarization (DP)-binary phase shift keying (BPSK) format of the same spectral efficiency of 2 bits/4D symbol. This significantly reduces cross polarization modulation (XPolM) and increases the transmission distance. Note that we use bits/4D symbol as a unit of spectral efficiency throughout this chapter.

Another method for reducing the fiber nonlinearity is 4D constant modulus modulation. The power of combined x- and y-polarization is constant at each time slot. This very effectively suppresses self-phase modulation (SPM) and cross-phase modulation (CPM). We discuss two earlier 4D constant modulus modulation formats [19, 20] in Section 3. More recently proposed 4D-2A8PSK is another example of the 4D constant modulus modulation. It has been widely recognized that DP-Star-8QAM for 6 bits/symbol does not perform very well, and many formats have been investigated [21–25] for this spectral efficiency. In particular, 4D-2A8PSK with the spectral efficiency of 6 bits/symbol has been shown to have superior linear and nonlinear performance than many other formats, because of its large Euclidean distance, 4D constant modulus characteristics (i.e., constant power in each time slot), and Gray labeling [26–28]. By relating this to the block coding approach described in the context of high-dimensional modulation [14], 5, 6, and 7 bits/symbol modulation formats were described as a family of block-coded 4D-2A8PSK in a unified form in [29, 30]. In Section 4, we first describe the 4D-2A8PSK modulation format family for 5–7 bits/symbol spectral efficiencies. Transmission simulations in a nonlinear dispersion-managed (DM) link, as well as a dispersion-uncompensated link are performed, so that this proposed family of modulation formats can be confirmed to exhibit excellent linear and nonlinear transmission performances. An analysis on separated nonlinear components demonstrate that reduction of self-phase modulation (SPM) and cross-phase modulation (XPM) are the main causes of the improvement by the 4D constant modulus formats. We then review some items relevant to practical implementations in a digital signal processor (DSP), which need careful consideration since the constellation of the 2A8PSK family is different from that of the widely used QAM-based formats. Experimental verification, extension to TDH to seamlessly cover 4–8 bits/symbol spectral efficiency, as well as the combination of Grassmann code and 4D-2A8PSK for the spectral efficiency of 3.5 bits/symbol, will also be reviewed.

2. X-constellation

For coherent optical communications, block codes with 8–24 dimensions have been proposed to achieve coding gain compared to the conventional 2D modulation formats [14, 31–35]. For example, using eight bit block code (four information bits and four parity bits) achieved almost 3 dB asymptotic gain. However, they are not specifically designed for nonlinearity-tolerant modulation. Shiner et al. [4] used the 8D code to achieve the coding gain and also arranged the code such that the degree of polarization (DOP) over a symbol (two time slots) becomes zero, as shown in **Table 1**. This significantly reduces the impact of polarization change toward other channels, as well as receiving polarization effect from other channels. The authors call this modulation format "X-constellation," since the constellation of Slot-A and Slot-B are cross-polarized.

The authors conducted a transmission experiment using 5000 km of dispersion-managed high density wavelength division multiplexing (WDM) link and compared the Q-factor of DP-BPSK and X-constellation. **Figure 1** shows that X-constellation showed 2 dB improvement in the Q-factor demonstrating the benefit of the X-constellation.

Binary Value		0000	0001	0010	0011	0100	0101	0110	0111	1000	1001	1010	1011	1100	1101	1110	1111
Slot A	x-pol	−1-i	−1-i	−1-i	−1-i	−1+i	−1+i	−1+i	−1+i	1-i	1-i	1-i	1-i	1+i	1+i	1+i	1+i
	y-pol	1+i	−1+i	1-i	−1-i	−1-i	1-i	−1+i	1+i	−1-i	1-i	−1+i	1+i	1+i	−1+i	1-i	−1-i
	$S_A = (S_1,S_2,S_3)$	(0,-4,0)	(0,0,-4)	(0,0,4)	(0,4,0)	(0,0,4)	(0,-4,0)	(0,4,0)	(0,0,-4)	(0,0,-4)	(0,4,0)	(0,-4,0)	(0,0,4)	(0,4,0)	(0,0,4)	(0,0,-4)	(0,-4,0)
Slot B	x-pol	1+i	−1+i	1-i	−1-i	1+i	−1+i	1-i	−1-i	1+i	−1+i	1-i	−1-i	1+i	−1+i	1-i	−1-i
	y-pol	1+i	−1-i	−1-i	1+i	1-i	−1+i	−1+i	1-i	−1+i	1-i	1-i	−1+i	−1-i	1+i	1+i	−1-i
	$S_B = (S_1,S_2,S_3)$	(0,4,0)	(0,0,4)	(0,0,-4)	(0,-4,0)	(0,0,-4)	(0,4,0)	(0,-4,0)	(0,0,4)	(0,0,4)	(0,-4,0)	(0,4,0)	(0,0,-4)	(0,-4,0)	(0,0,-4)	(0,0,4)	(0,4,0)
DOP	$\|S_A + S_B\|$	0	0	0	0	0	0	0	0	0	0	0	0	0	0	0	0

Table 1. The optical field Jones vectors for the two consecutive time slots (Slot-A and Slot-B) that define the eight-dimensional X-constellation symbols, and their corresponding binary symbol labels [4].

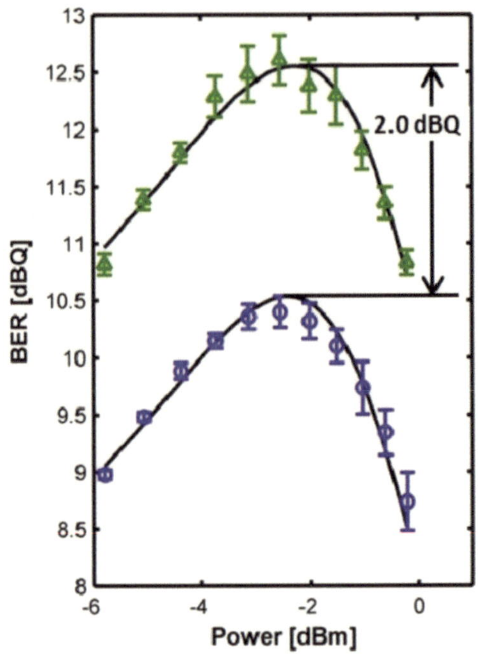

Figure 1. The measured Q-factor for DP-BPSK (circle) and X-constellation (triangle) following 5000 km of high density WDM propagation [4].

3. 4D constant modulus formats

DP-QPSK is a very robust modulation format for nonlinear transmission, and one of the main reasons is that it is 2D constant modulus, in that the power for each polarization is constant. 2D constant modulus property can also be achieved in DP-8PSK, DP-16PSK, etc.; however, their Euclidean distance is smaller than that of DP-Star-8PSK and DP-16QAM, and overall transmission characteristics are not as good. Instead, 4D constant modulus modulation formats (i.e., power of the combined X- and Y-polarizations is constant) were proposed. One of the 4D constant modulus format is 8PolSK-QPSK [19], in which eight polarization states in the Stokes space representation carry four different absolute phases, as shown in **Figure 2**. This gives 32 code words or 5 bits/symbol spectral efficiency. Compared to DP-Star-8QAM, 8PolSK-QPSK showed significantly reduced SPM and XPM.

Another example of 4D constant modulus format is POL-QAM 6–4, where six polarization states carry four different absolute phases [20]. This gives 24 code words.

In the next section, we review a family of 4D-2A8PSK modulation formats, which are also 4D constant modulus formats, covering multiple spectral efficiencies and having large coding gain.

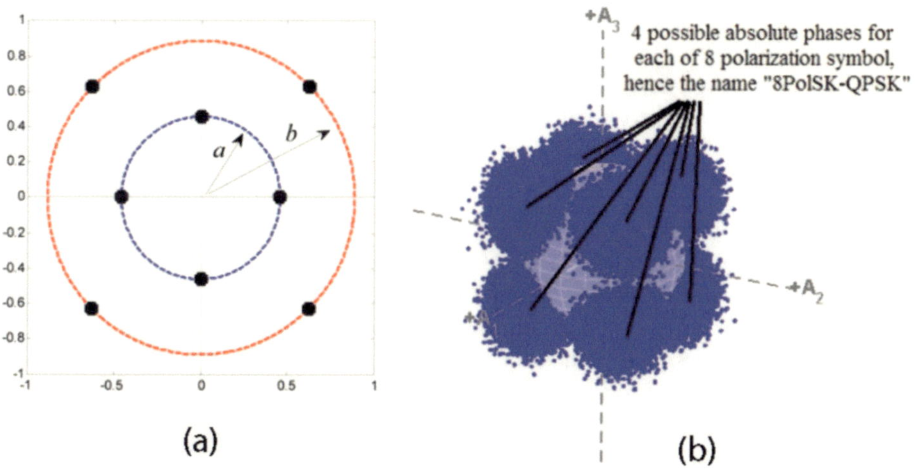

Figure 2. 8PolSK-QPSK modulation format [19]. (a) Star-8QAM format. In DP-8QAM, the two polarizations can both independently be at amplitude a or b. For 8PolSK-QPSK format, that independence is removed. (b) Stokes space representation of the 8PolSK-QPSK with added noise giving 16 dB of SNR.

4. 4D-2A8PSK

4.1. Generalized mutual information (GMI)

As a first step, we give an overview of a metric in order to compare the modulation formats under the most relevant condition. Pre-FEC bit error ratio (BER) has traditionally been used to predict post-FEC BER performance of hard decision (HD) FEC systems. However, pre-FEC BER is not directly applicable to modern long distance fiber-optic communications using soft decision (SD) FEC based on bit-interleaved coded modulation (BICM). As an alternative performance metric more suitable for SD-FEC systems, the BICM limit, called generalized mutual information (GMI), was introduced to the optical communications research community for comparing different modulation formats [36, 37]. This metric has been used to compare several modulation formats [23, 38]. The normalized GMI (i.e., GMI per bit) can be described from the log-likelihood ratio (LLR) outputs of the demodulator at the receiver as follows [39–41]:

$$I = 1 - \mathbb{E}_{L,b}\left[\log_2\left(1 + \exp\left((-1)^{b+1}L\right)\right)\right], \tag{1}$$

where b, L, and $\mathbb{E}[\cdot]$ denote the transmitted bit $b \in \{0,1\}$, the corresponding LLR, and an expectation (i.e., ensemble average over all LLR outputs L and transmitted bits b), respectively. We define "normalized" GMI as the mutual information per modulation (information) bit, not per modulation symbol. The normalized GMI can therefore set the upper limit of the possible code rate of SD-FEC coding for BICM systems. Therefore, multiplying the normalized GMI with the number of bits per symbol is equivalent to the achievable throughput per symbol.

The relationship between Q-factor calculated from pre-FEC BER and normalized GMI of four different modulation formats (DP-QPSK, D P-Star-8QAM, 6b4D-2A8PSK, and DP-16QAM) is shown in **Figure 3**. We will give a detailed explanation of the 6b4D-2A8PSK modulation format in Section 4.4. Here, the Q-factor is defined by

$$Q_{\mathrm{BER}}^2 = 2 \cdot \left\{ \mathrm{erfc}^{-1}(2 \cdot \mathrm{BER}) \right\}^2, \tag{2}$$

which is a classical measure to calculate the required signal-to-noise ratio (SNR) to achieve the BER for binary-input additive white Gaussian noise (AWGN) channels. Here, $\mathrm{erfc}^{-1}(\cdot)$ is an inverse complementary error function. **Figure 3** shows that the same pre-FEC BER (Q-factor) does not necessarily give the same BICM limit among various formats, especially at lower code rate regions. When the normalized GMI is 0.85, the Q-factor lies between 4.77 (BER = 4.16×10^{-2}) and 4.86 dB (BER = 4.01×10^{-2}), corresponding to the typical Q_{BER}^2 threshold of the state-of-the-art SD-FEC having a code rate of 0.8 [42, 43]. Accordingly, we will use 0.85 as the target of the normalized GMI throughout this chapter.

4.2. Generic 2A8PSK

The generic constellation of 4D-2A8PSK [26–29] is shown in **Figure 4**. It is similar to 8PSK, with two different amplitudes represented by the radii, r_1 and r_2 (suppose $r_1 \leq r_2$ without loss of generality). By combining the two polarizations (i.e., 4D space), 2^8 = 256 combinations (i.e., 8 bits per 4D symbol) are possible. With a condition that X- and Y-polarizations have complimentary radius, that is, if r_1 is used for X-polarization, then r_2 needs to be chosen for Y-polarization, and vice versa; we generate set-partitioned (SP) 4D codes achieving the property of 4D constant modulus, which leads to excellent nonlinear transmission performances. We define r_1/r_2 ($\leqslant 1$) as a ring ratio. When the ring ratio is equal to 1, the modulation format is reduced to regular DP-8PSK.

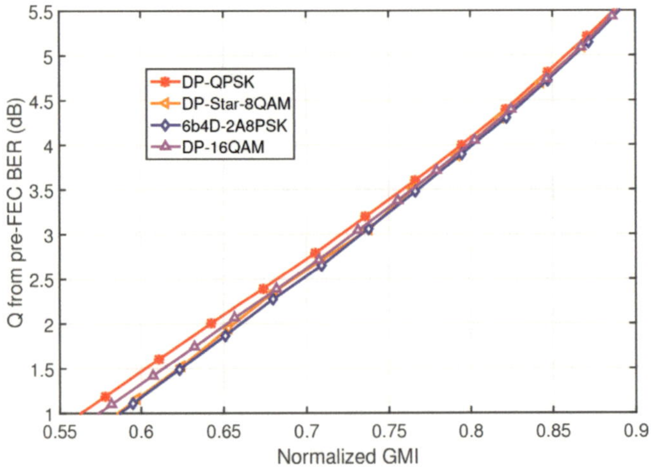

Figure 3. Q-factor calculated from pre-FEC BER vs. normalized GMI for four modulation formats [30].

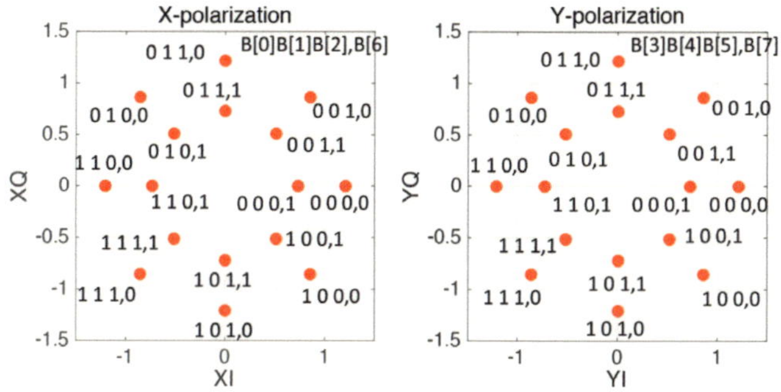

Figure 4. Constellation and bit-to-symbol mapping of 2A8PSK [30].

The mapping rule of 4D-2A8PSK is also included in **Figure 4** [44]. Let $B[0], \ldots, B[7]$ express eight modulation bits, and $B[0]$–$B[2]$ and $B[3]$–$B[5]$ denote the Gray-mapped 8PSK at X- and Y-polarizations, respectively. Whereas, $B[6]$ and $B[7]$ are used to express the amplitude in each polarization. By selecting the optimum 32, 64, and 128 point constellations out of 256 combinations, we can construct 32SP-, 64SP-, and 128SP-2A8PSK, for the spectral efficiency of 5, 6, and 7 bits/symbol, respectively. We also call these 5b4D-, 6b4D-, and 7b4D-2A8PSK for convenience.

4.3. 5b4D-2A8PSK

For the spectral efficiency of 5 bits/symbol, 32SP-2A8PSK (5b4D-2A8PSK) can be expressed by a linear code, with five information bits $B[0]$–$B[4]$, and three parity bits $B[5]$–$B[7]$. Since $B[5]$ and $B[6]$ can independently be represented as the linear combination of the five information bits, $2^{10} = 1024$ is the total number of possible linear codes to be designed. We chose the combination, which gives the least required SNR for the target GMI of 0.85, through Monte-Carlo simulations in AWGN.

In order to maintain a 4D constant modulus property, for each code word, the X-polarization ring size is always complementary to the Y-polarization ring size. Negating another parity bit $B[6]$ for $B[7]$ achieves this. As a whole, the parity-check equations for 5b4D-2A8PSK can be described as:

$$B[5] = B[0] \oplus B[1] \oplus B[2], \tag{3}$$

$$B[6] = B[2] \oplus B[3] \oplus B[4], \tag{4}$$

$$B[7] = \overline{B[6]}, \tag{5}$$

where $\oplus$ and $\overline{[\cdot]}$ denote the modulo-2 addition and negation, respectively.

4.4. 6b4D-2A8PSK

In 64SP-2A8PSK (6b4D-2A8PSK), which has 6 bits/symbol spectral efficiency, $B[6]$ is a parity bit of single-parity-check code protecting all the information bits, and can be expressed as an exclusive-or (XOR) of all the information bits $B[0]$–$B[5]$. Another parity bit $B[7]$ is the negation of $B[6]$ as used in 5b4D-2A8PSK. The optimum code for the target GMI of 0.85 is

$$B[6] = B[0] \oplus B[1] \oplus B[2] \oplus B[3] \oplus B[4] \oplus B[5], \quad (6)$$

$$B[7] = \overline{B[6]}. \quad (7)$$

4.5. 7b4D-2A8PSK

For the spectral efficiency of 7 bits/symbol, 128SP-2A8PSK (7b4D-2A8PSK) can be constructed simply as follows. In this code, $B[0]$–$B[6]$ are the information bits while there is only one parity bit at $B[7]$. In order to realize 4D constant modulus format, just like 5b4D- and 6b4D-2A8PSK, we can express the single parity bit $B[7]$ as:

$$B[7] = \overline{B[6]}. \quad (8)$$

4.6. Other modulation formats for comparison

In order to evaluate the performance of 5b4D-2A8PSK, we consider three other modulation formats having 5 bits/symbol spectral efficiency, that is, 8PolSK-QPSK [19], 32SP-16QAM [11], and time-domain hybrid (TDH) modulation. 8PolSK-QPSK [19] was briefly explained in Section 3. 32SP-16QAM is a 4D set-partitioned modulation format derived from DP-16QAM. To generate 32 code words, the parity rule shown in [11] is used. TDH modulation using a 1:1 mixture of DP-QPSK and 6b4D-2A8PSK to achieve an average of 5 bits/symbol spectral efficiency is also included.

For comparison with 6b4D-2A8PSK, three other modulation formats of 6 bits/symbol spectral efficiency were evaluated; specifically, DP-8PSK, DP-Star-8QAM, and DP-Circular-8QAM [23]. DP-8PSK and DP-Star-8QAM are conventional modulation formats. DP-Circular-8QAM has one center point and seven circular constellation points, and has larger Euclidean distance than DP-8PSK [23].

To compare with 7b4D-2A8PSK, two modulation formats of 7 bits/symbol spectral efficiency are evaluated. 128SP-16QAM is a 4D modulation format based on DP-16QAM, where 128 code words are generated using the parity rule described in [11]. We also included TDH modulation using 1:1 mixture of 6b4D-2A8PSK and DP-16QAM. Furthermore, we included DP-16QAM; however, to compare for the same data rate, we used the Baud rate of $(7/8) \times 34$ GBd.

4.7. Nonlinear transmission simulations

4.7.1. Simulation procedure

Nonlinear transmission simulations are conducted over a 2000 km DM link at a rate of 34 GBd per channel to evaluate the effect of modulation format on high fiber nonlinearity. At the transmitter, pulses were filtered by a root-raised-cosine (RRC) filter with a roll-off factor of 10%. Eleven DWDM channels of the same modulation format were combined with 37.5 GHz spacing without using any optical filtering. The link consists of 25 spans of 80 km nonzero dispersion shifted fiber (NZDSF) in which loss is compensated by Erbium-doped fiber amplifiers (EDFAs). The performance of each modulation format can be quantified by a span loss budget, which is defined as [45]

$$\text{Span Loss Budget} = 58 + P - \text{ROSNR} \\ -10\log_{10}(N) - \text{NF}, \tag{9}$$

where P is the launch power per channel expressed in dBm, ROSNR is the required OSNR to achieve the target GMI in dB, N is the number of spans, and NF is the noise figure of the EDFAs in dB.

The parameters for NZDSF were $\gamma = 1.6$ /W/km, $D = 3.9$ ps/nm/km, and $\alpha = 0.2$ dB/km. Other fiber effects such as polarization mode dispersion (PMD) and dispersion slope were not included. At the end of each span, 90% of the chromatic dispersion was compensated as a lumped linear dispersion compensator. Dispersion pre-compensation was applied at the transmitter side using 50% of the residual dispersion of the full link. The rest of the dispersion is compensated just before the receiver.

An ideal homodyne coherent receiver was used, with an RRC filter with a roll-off factor of 10%, followed by sampling at twice the symbol rate. For adaptive equalization, we used a time-domain data-aided least-mean-square equalizer utilizing the transmitted data directly as the training sequences for simplicity. A discussion on a more realistic equalizer will be given in Section 4.9. We did not use carrier phase estimation (CPE) in Sections 4.7 and 4.8.

All the optical noise due to the EDFAs are loaded just before the receiver. The calculated required OSNR at the target GMI is used to obtain the span loss budget as in (9). We used an EDFA noise figure of 5 dB to calculate the span loss budget.

4.7.2. 5 bits/symbol modulation formats

Four 5 bit/symbol formats are compared as shown in **Figure 5**. In this case, we use the ring ratio of $r_1/r_2 = 0.61$, optimal for 5b4D-2A8PSK for maximizing the span loss budget. Note that

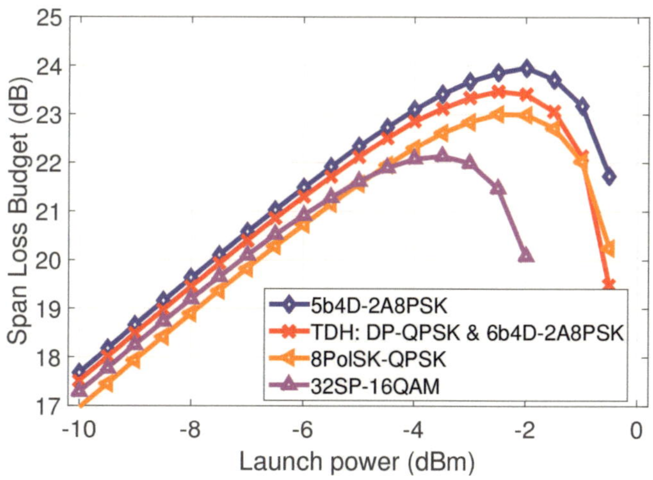

Figure 5. Span loss budget of four 5 bits/symbol modulation formats as a function of launch power for the DM link [30].

ring ratio is not a sensitive parameter, and in fact between 0.56 and 0.66, the peak span loss budget changed only by 0.03 dB.

As the launch power increases, the span loss budget for 32SP-16QAM decreases fast, because of large power variations at each time slot. On the other hand, 8PolSK-QPSK [19] has 0.65 dB worse OSNR for the linear case, while the saturation characteristics are very similar to 5b4D-2A8PSK due to constant power. TDH modulation with a 1:1 mixture of DP-QPSK and 6b4D-2A8PSK has 4D constant modulus property at each time slot. However, we used an optimized power allocation for TDH modulation (i.e., 6b4D-2A8PSK has 2.7 dB higher power than DP-QPSK), and there is a power variation between time slots generating some penalty due to the nonlinearity.

Overall, 5b4D-2A8PSK has the higher maximum span loss budget by 0.5 dB over the TDH modulation, by 0.9 dB over 8PolSK-QPSK, and by 1.8 dB over 32SP-16QAM.

4.7.3. 6 bits/symbol modulation formats

Four 6 bits/symbol modulation formats are compared as in **Figure 6**. The optimal ring ratio is $r_1/r_2 = 0.65$ for 6b4D-2A8PSK for maximum span loss budget. The maximum span loss budget for 6b4D-2A8PSK is shown to be higher than DP-Circular-8QAM, DP-8PSK, and DP-Star-8QAM by 0.6, 0.5, and 1.6 dB, respectively.

4.7.4. 7 bits/symbol modulation formats

Figure 7 shows performance comparison among three 7 bit/symbol formats at 34 GBd and DP-16QAM of the same data rate ($7/8 \times 34$ GBd). Here, the ring ratio of $r_1/r_2 = 0.59$ is chosen for 7b4D-2A8PSK to maximize the span loss budget. Here, TDH modulation uses a 1:1 mixture of 6b4D-2A8PSK and 128SP-QAM, whose optimized power ratio was 0.1 dB. We can see that 7b4D-2A8PSK outperformed THD modulation, 128SP-16QAM, and $(7/8) \times 34$ GBd DP-16QAM by 0.7, 1.4, and 2.2 dB, respectively.

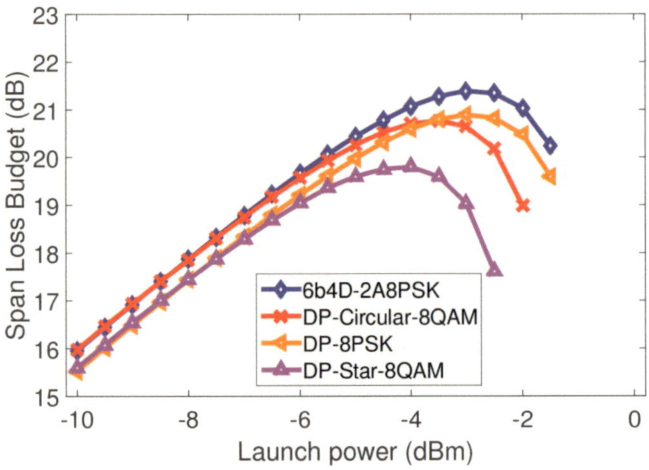

Figure 6. Span loss budget of four 6 bits/symbol modulation formats for the DM link [30].

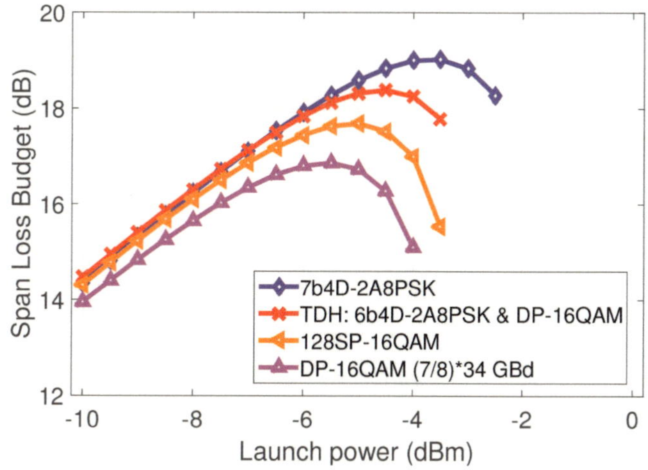

Figure 7. Span loss budget of three 7 bits/symbol modulation formats at 34 GBd and DP-16QAM with the same data rate (7/8 × 34 GBd) as a function of launch power for the DM link [30].

4.7.5. Summary for the dispersion managed link results

The peak span loss budget for the DM link is summarized in **Figure 8**. The circles connected by the dashed lines include DP-QPSK, 5b4D-, 6b4D-, 7b4D-2A8PSK, and DP-16QAM, all at 34 GBd. Squares are taken from TDH modulation formats, and triangles are from other (conventional) modulation formats in **Figures 5–7**. This shows that the 4D-2A8PSK family fills the gap between DP-QPSK and DP-16QAM almost linearly (in the dB scale), and each one offers a good improvement from the conventional modulation formats at the same spectral efficiency.

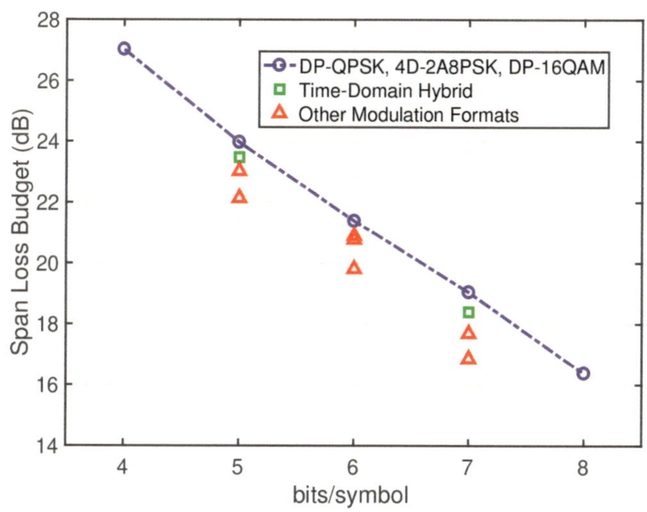

Figure 8. Peak span loss budget for the 2000 km dispersion managed link. Circles connected with blue dotted lines are for the three 4D-2A8PSK formats, DP-QPSK, and DP-16QAM. Green squares are for TDH modulation formats, and red triangles are for other modulation formats appeared in **Figures 5–7** [30].

4.7.6. 5 bits/symbol under dispersion uncompensated link

For evaluating the transmission characteristics of various modulation formats under a reduced nonlinearity situation, representing terrestrial cases, we also simulated the link with 50 spans of 80 km standard single-mode fiber (SSMF) without inline dispersion compensation or dispersion pre-compensation. SSMF parameters are $\gamma = 1.2$ /W/km, $D = 17$ ps/nm/km, $\alpha = 0.2$ dB/km. We used the same 0.85 as the target GMI. The span loss budget of the four modulation formats for the spectral efficiency of 5 bits/symbol are shown in **Figure 9**, as an example. The overall differences among the modulation formats are smaller than the case of DM-NZDSF link. 5b4D-2A8PSK shows the highest performance with the peak span loss budget outperforming those of TDH of DP-QPSK and 6b4D-2A8PSK, 8PolSK, and 32SP-8QAM by 0.2, 1.0, and 0.8 dB, respectively. TDH and 32SP-16QAM in the dispersion uncompensated link case did not suffer as much as they did in the DM case. The reason is the weaker nonlinear distortion in the uncompensated SSMF links compared to DM-NZDSF links.

4.8. Separated nonlinearity

For better understanding, the reason of the outperformance of 4D constant modulus modulation, additional simulations are conducted with separated nonlinear components, using the method proposed in [46]. With this method, the nonlinear transmission performance with nonlinear effects of SPM, XPM, and XPolM can be evaluated individually.

Figure 10 shows the simulated Q-factor as a function of OSNR for 6b4D-2A8PSK and DP-Star-8QAM in the DM link, where the simulation parameters are kept the same as in Section 4.7.1. Here, we use the recently proposed Q-factor definition based on GMI and not on pre-FEC BER as follows [44]:

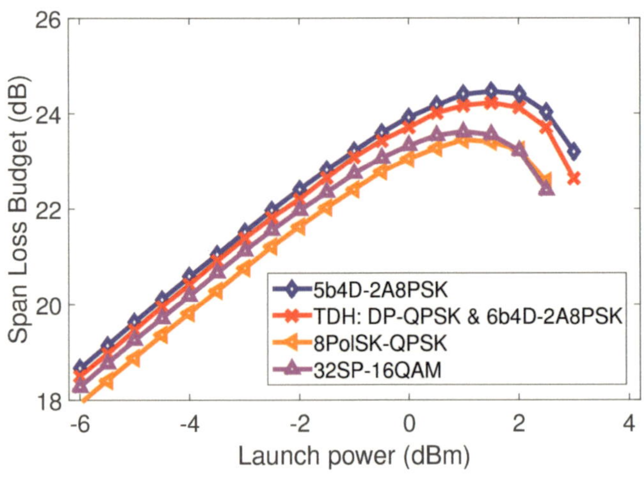

Figure 9. Span loss budget of four 5 bits/symbol modulation formats as a function of launch power for the dispersion unmanaged link [30].

$$Q_{\mathrm{GMI}}^{2}=\left\{0.5\cdot J^{-1}(\mathrm{GMI})\right\}^{2}, \tag{10}$$

where $J^{-1}(\cdot)$ is the inverse J function, widely used in extrinsic information transfer chart analysis [39]. The inverse J function is well approximated by

$$\begin{aligned}
&J^{-1}(I)\simeq\begin{cases} a_1 I^2+b_1 I+c_1\sqrt{I}, & 0\le I\le I^{*},\\ -a_2\ln\left[b_2(1-I)\right]-c_2 I, & I^{*}\le I\le 1,\end{cases}\\
&I^{*}=0.3646,\\
&a_1=1.09542,\qquad b_1=0.214217,\qquad c_1=2.33727,\\
&a_2=0.706692,\qquad b_2=0.386013,\qquad c_2=-1.75017.
\end{aligned} \tag{11}$$

The Q-factor defines the above based on GMI in Eq. (10) is a generalized extension from the conventional Q-factor based on BER in (2). With this new Q-factor, the effective SNR to achieve same post-FEC BER performance with SD-FEC systems can be evaluated. Even though both definitions provide identical Q performance in binary-input AWGN channels, the generalized Q-factor can predict SNR gain more accurately than the conventional BER-based Q-factor for BICM systems using SD-FEC coding and/or high-order high-dimensional modulation.

The curves marked with "AWGN" in **Figure 10** indicate the case in which the nonlinear effects are fully ignored, and the curves depicted with "SPM," "XPM," and "XPolM" show that these nonlinear components are individually added. The curve with "SPM + XPM + XPolM" shows the situation when all of these nonlinear effects are taken into account. The launch power is set to be −4 dBm, giving the peak span loss budget for DP-Star-8QAM. At this launch power, OSNR of 15.2 dB gives a normalized GMI of 0.85.

Figure 11 is a re-plot of the simulated Q versus separated nonlinear effects when the OSNR is 15.2 dB. Q under the linear condition (AWGN) for 6b4D-2A8PSK is higher than DP-Star-8QAM by 0.4 dB. The contributions from SPM and XPM in 6b4D-2A8PSK are much smaller

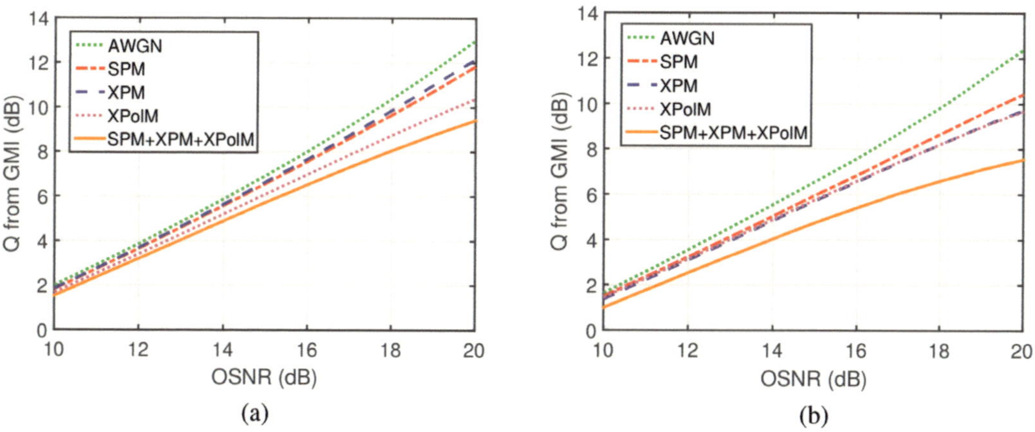

Figure 10. Generalized Q-factor as a function of OSNR with separated nonlinear effects at a launch power of 4 dBm [30]. (a) 6b4D-2A8PSK (b) DP-Star-8QAM (Curves for XPM and XPolM overlapping).

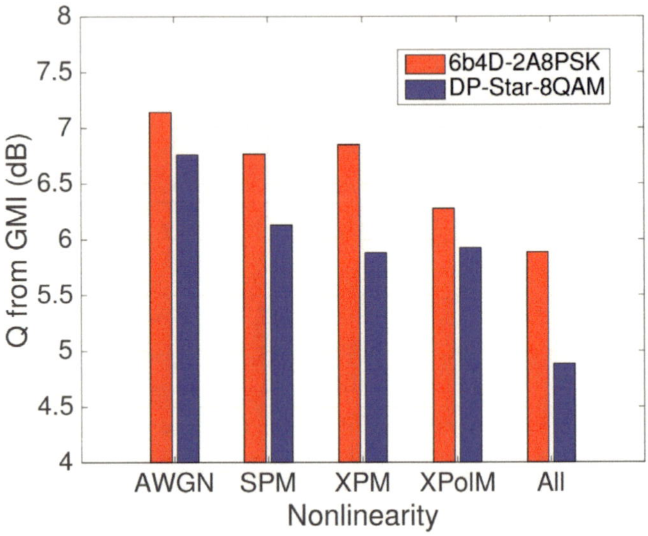

Figure 11. Generalized Q-factor for 6b4D-2A8PSK and DP-Star-8QAM with separated nonlinear effects at a launch power of −4 dBm and OSNR of 15.2 dBm [30].

than those in DP-Star-8QAM. This confirms that 4D-2A8PSK family can be robust against XPM and SPM nonlinearity. However, the contribution of XPolM is similar in 6b4D-2A8PSK and DP-Star-8QAM. This is due to the fact that individual polarization power in 6b4D-2A8PSK fluctuates over symbol time even though the combined power of both polarizations is constant. This is consistent with a report in which another 4D constant modulus modulation format 8PolSK shows a significant reduction in SPM and XPM, but not necessarily in XPolM [19].

4.9. DSP algorithm

4.9.1. Adaptive equalizer

To understand the fundamental benefit of the 4D-2A8PSK family, we used an idealized data-directed least-mean-square equalizer up to this point. In this section, we address the performance impact when more realistic equalizers [47, 48] are used, considering practical implementations into account.

We first consider a conventional radius-directed equalizer (RDE) [48] for 6b4D-2A8PSK, where the decision on the ring radii is performed at each polarization separately. In this case, we observe 0.12 and 0.10 dB degradation in the span loss budget, in comparison to the idealized least-mean-square equalizer at a launch power of −10 and −4 dBm, respectively.

We then take advantage of the 4D constant modulus property, by using the relative power of two polarizations for soft decision of the ring radii. For soft decision information, we use a heuristic sigmoid function $S(x) = 1/\left(1 + e^{-x/a}\right)$, where a is a softness parameter, and x is a relative power of two polarizations. In this manner, we can compensate for the degradation by 0.07 dB from the conventional RDE. The overall degradation due to the realistic adaptive equalizer compared to the ideal one is no worse than 0.05 dB.

http://dx.doi.org/10.5772/intechopen.73031

4.9.2. LLR computation

For SD-FEC, it is necessary to calculate log-likelihood ratio (LLR) with moderate circuit complexity. A fast-decoding algorithm and LLR computation for high-order set-partitioned 4D–QAM formats [49] is now extended to 6b4D-2A8PSK to use two lookup tables [44]. The schematic of the soft-demapping circuit is shown in **Figure 12**, and also used the asymmetry between the radial and the axial LLR and the offline processing of the experimental data showed only a small power penalty [44]. It also used the LLR calculation method robust against residual phase noise [50].

4.10. Experiment

We have also conducted a transmission experiment comparing 6b4D-2A8PSK and DP-Star-8QAM [44]. The signals were either 6b4D-2A8PSK or DP-Star-8QAM modulated at 32 GBd and filtered with a root-raised cosine filter with a roll-off factor of 0.15. Seventy channels were spaced at 50 GHz spacing. The transmission line was 1260 km, having an average span length of 70 km. Chromatic dispersion was managed inline by the mixture of nonzero dispersion shifted fiber (NZDSF) having negative local CD of −3 ps/nm and standard single-mode fiber (SSMF). In the receiver side, the signal stored by 64 GS/s analog-to-digital converters (ADCs) was processed offline, which included CD compensation, adaptive equalization with constant modulus algorithm for initial convergence, and radius directed equalization afterward, carrier recovery (CR) with multipilot algorithm [47] having an window size of 63, pilot-aided phase-slip recovery, and the proposed soft-demapping as described in Section 4–9.

Figures 13 and **14** show the experimental results, **Figure 14(a)** is Q from GMI as a function of launched power. In the case of ideal soft-demapping (only 16 level quantization for SD-FEC decoding was applied), we observed 0.6 dB improvement at maximum Q by 4D-2A8PSK compared to DP-Star-8QAM. The proposed technique had performance degradation of 0.15 and 0.06 dB for 4D-2A8PSK and DP-Star-8QAM, respectively, compared to the ideal LLR.

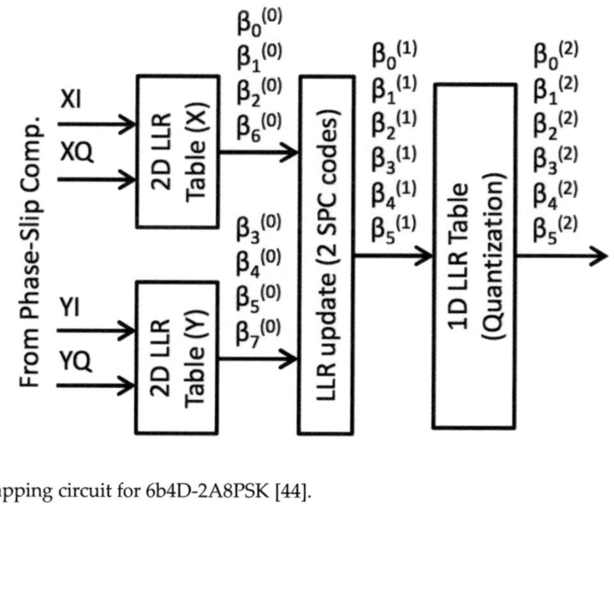

Figure 12. Soft-demapping circuit for 6b4D-2A8PSK [44].

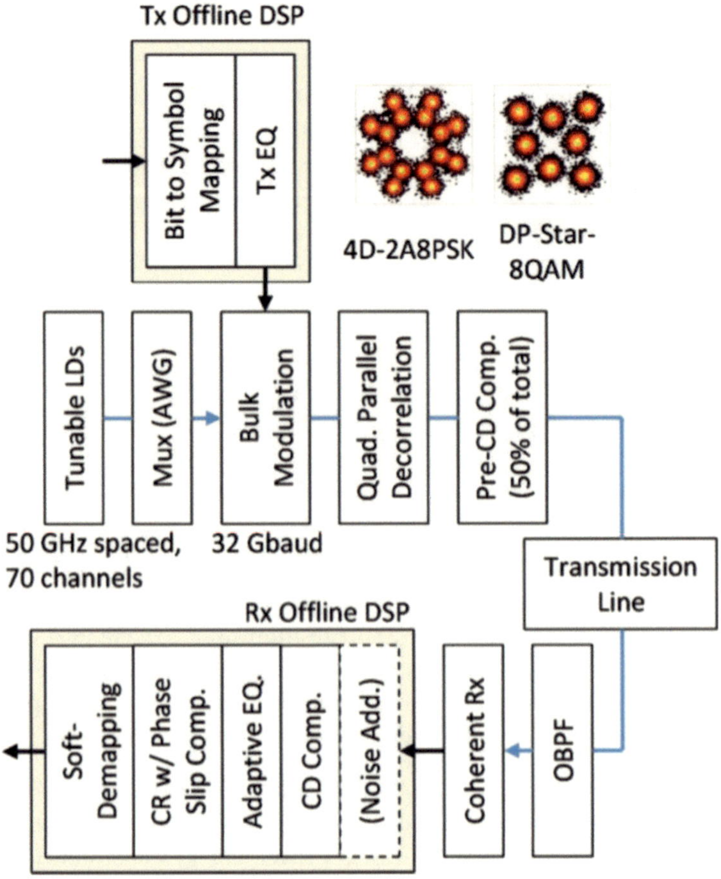

Figure 13. Experimental setup [44].

The overall performance gain of 0.5 dB was still significant in the highly nonlinear transmissions. **Figure 14(b)** shows required OSNR, which was calculated by loading noise at the receiver DSP to emulate OSNR decrease. The target normalized GMI was set to 0.92, which was close to 20.5% SD-FEC limit [53]. The proposed soft-demapping worked even at such low OSNR conditions and 4D-2A8PSK outperformed DP-Star-8QAM as the launched power increase.

4.11. Time domain hybrid modulation

TDH modulation has been studied considerably to cover a wide range of channel conditions, due to its flexibility in choosing the nearly arbitrary spectral efficiency [12, 51, 52]. As the constituent modulation formats, we use DP-QPSK (4 bits/symbol) and QP-16QAM (8 bits/symbol) in conjunction with 5b4D, 6b4D, and 7b4D-2A8PSK to widen the range of TDH [54]. For a comparison, we also use TDH modulation using conventional modulation formats, that is, DP-QPSK, 32SP-QAM, DP-Star-8QAM (S8QAM), 128SP-QAM, and DP-16QAM. The benefit of the 4D-2A8PSK family is the 4D constant modulus property. In other words, there is no

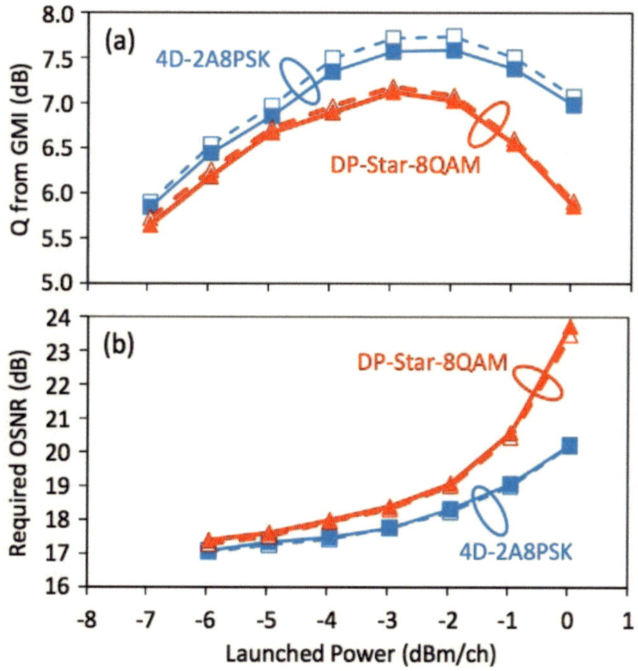

Figure 14. Experimental result of (a) Q from GMI and (b) required OSNR for two types of LLR calculation: Ideal (dotted line) and the proposed in **Figure 12** (solid line) [44].

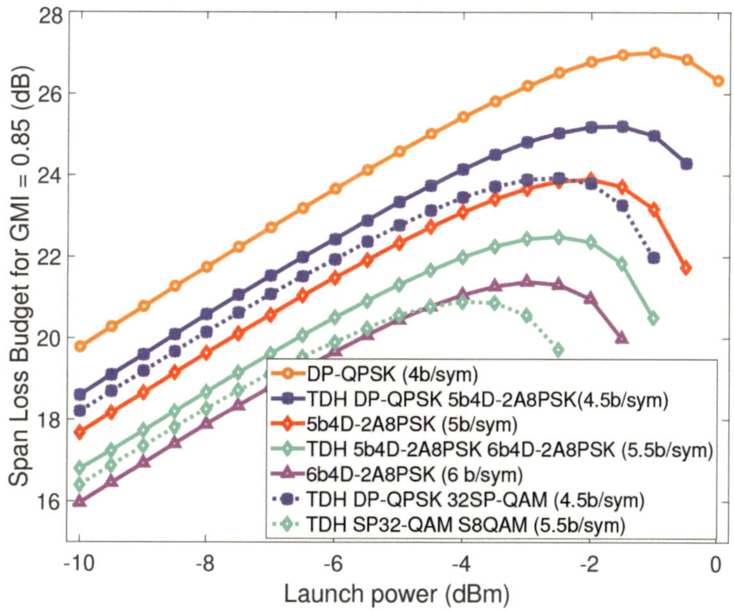

Figure 15. Span loss budget for various modulation formats in the range of 4–6 bits/symbol [54].

compromise in choosing the power ratio (ratio between the two modulation formats). On the other hand, conventional formats experience power fluctuations, causing compromise in the power ratio [29, 54].

We simulated transmission performance over the same link condition as described in Section 4.7. For 5b4D, 6b4D, and 7b4D-2A8PSK formats, we choose the ring ratio of 0.60, 0.65, and 0.59 for the best nonlinear performance. For all the THD modulation, we use 1:1 ratio with alternating formats; however, in actual systems, any arbitrary ratio can be used. The important parameter for TDH is the power ratio, that is, how much power will be allocated for each time slot. We optimize the power ratio for the best nonlinear performance.

Figure 15 shows the calculated span loss budget for 4–6 bits/symbol modulation formats, including the 2A8PSK-based and the conventional TDH modulation. DP-QPSK and 6b4D-2A8PSK data are also included as a reference. **Figure 16** shows the span loss budget for 6.5–8 bits/symbol modulation formats. From these figures, we can see that the TDH modulation based on 2A8PSK has much better nonlinear performance than that based on the conventional modulation formats, due to their constant modulus property.

The peak span loss budget for various spectral efficiencies is shown in **Figure 17**. Here, 4.5, 5.5, 6.5, and 7.5 bits/symbol TDH based on 2A8PSK used DP-QPSK, 5b4D-2A8PSK, 6b4D-2A8PSK, 7b4D-2A8PSK, and DP-16QAM. TDH based on the conventional formats used DP-QPSK, 32SP-QAM, S8QAM, 128SP-QAM, and DP-16QAM. We observed 1.3, 1.6, 1.6, and 0.6 dB increase in peak span loss budget, when TDH used 2A8PSK, at 4.5, 5.5, 6.5, and 7.5 bits/symbol, respectively. This shows the versatility of the 4D-2A8PSK family.

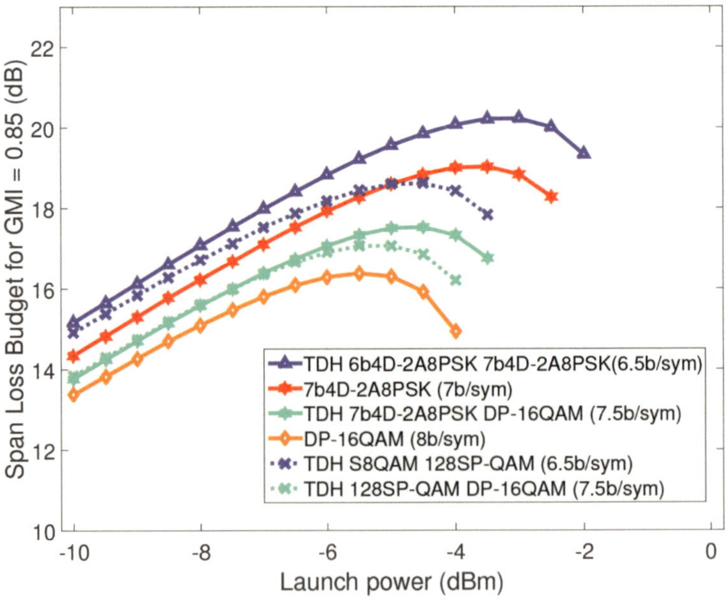

Figure 16. Span loss budget for various modulation formats in the range of 6.5–8 bits/symbol [54].

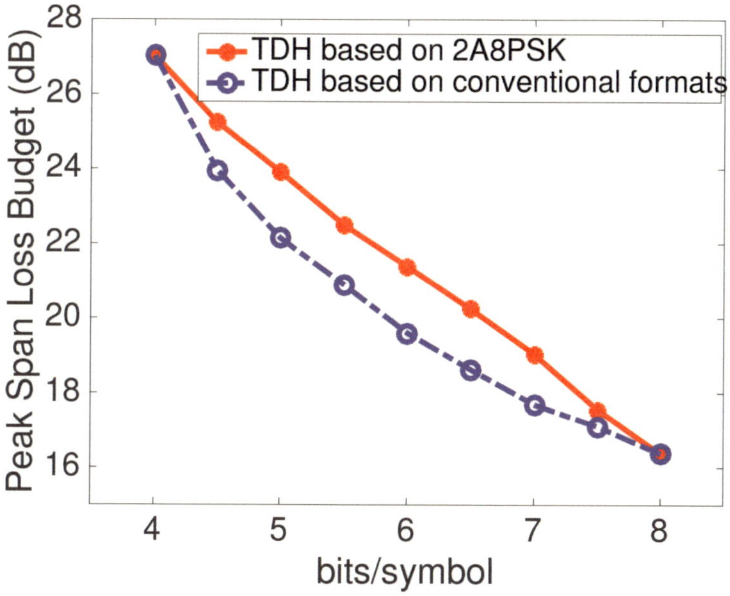

Figure 17. Span loss budget for TDH modulation based on the 4D-2A8PSK formats, and that on the conventional modulation formats [54].

4.12. 3.5 bits/symbol modulation format

Grassmann code [4, 55] is known to be robust against state of polarization (SOP) rotation including cross polarization modulation (XPolM) as described in Section 2. We investigated a Grassmann code-based 7-bit 8D code [56], whose schematic is shown in **Figure 18**. 2-ary amplitude QPSK (2AQPSK) and 2A8PSK are used for the first and the second time slots, respectively. x_1, x_2, y_1, y_2 are x- and y-polarization component of the first and second time slot, respectively. Let b_0–b_6 be the information bits. In a similar manner as described in Sec 4.7, for the 2AQPSK part, b_0 and b_1 are used for the angle of x_1, and b_2, b_2 are used for the angle of y_1, respectively. For x_2, b_4–b_6 are used for the angle representation of 2A8PSK. All use Gray coding for the angle. The radius of x_1 is expressed as $XOR(b_4, b_5, b_6)$, where "0" means the larger radius and "1" means the smaller radius. The ratio of the radii is called the ring ratio. The radius of y_1 is expressed as $\overline{XOR(b_4, b_5, b_6)}$. Both 2AQPSK and 2A8PSK share the same ring ratio of 0.70, which was optimized for the nonlinear performance. The radius of x_2 is expressed as $XOR(b_0, b_1, \ldots, b_6)$. y_2 is calculated from the Grassmannian condition $x_1\, y_1^* + x_2\, y_2^* = 0$. This guarantees the 4D constant modulus condition for both time slots.

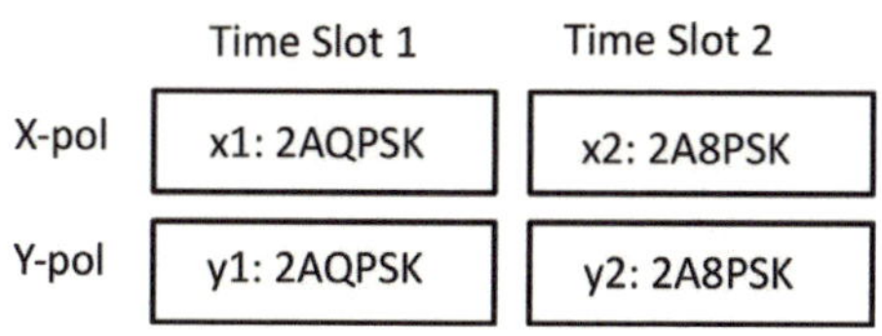

Figure 18. Structure of the 7b8D-Grassmann code.

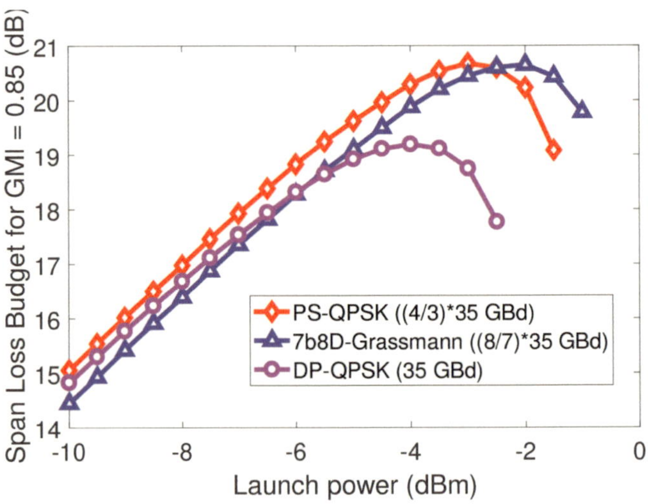

Figure 19. Span loss budget of three modulation for the same data rate, as a function of launch power for the target normalized GMI = 0.85.

We compared 7b8D-2A8PSK (3.5 bits/symbol), PS-QPSK (3bits/symbol), and DP-QPSK (4 bits/symbol) of the same data rate. We also chose the channel spacing as 1.15 times of the Baud rate. Simulation procedures and parameters are nearly identical to that described in Section 4–71, except that we used nine channels. The simulated results are shown in **Figure 19**. 7b8D-Grassmann format exhibits almost the same span loss budget as PS-QPSK, which has higher Baud rate and broader spectrum. On the other hand, 7b8D-Grassmann format shows much larger span loss than DP-QPSK, although the latter has narrower spectrum. Therefore, depending on the application, 7b8D-Grassmann format may be an alternative to PS-QPSK and DP-QPSK.

5. Conclusion

We reviewed nonlinearity-tolerant modulation formats, including the recently proposed 5, 6, and 7 bits/symbol 4D modulation format family based on 2A8PSK. A series of transmission simulation results show that this 2A8PSK family shows better nonlinear performance than the conventional modulation formats at each corresponding spectral efficiency, especially for dispersion-managed links, which are known to have high fiber nonlinearity. It is also determined that the primary benefits of the 4D constant modulus property comes from reduced effects of SPM and XPM. Since these modulation formats in the 4D-2A8PSK family differ just in the parity bits, they can be realized with very similar hardware over different spectral efficiency between DP-QPSK and DP-16QAM. Furthermore, this modulation format family can be the components of time-domain hybrid modulation, where almost arbitrary spectral efficiency can be realized between 4 and 8 bits/symbol, when combined with DP-QPAK and DP-16QAM.

http://dx.doi.org/10.5772/intechopen.73031

Author details

Keisuke Kojima[1]*, Toshiaki Koike-Akino[1], Tsuyoshi Yoshida[2,3], David S. Millar[1] and Kieran Parsons[1]

*Address all correspondence to: kojima@merl.com

1 Mitsubishi Electric Research Laboratories (MERL), Cambridge, MA, USA

2 Information Technology R&D Center, Mitsubishi Electric Corporation, Kamakura, Kanagawa, Japan

3 Chalmers University of Technology, Gothenburg, Sweden

References

[1] Essiambre R-J, Kramer G, Winzer PJ, Foschini GJ, Goebel B. Capacity limits of optical fiber networks. Journal of Lightwave Technology. 2010;**28**(4):662-701

[2] Secondini M, Forestieri E, Prati G. Achievable information rate in nonlinear WDM fiber-optic systems with arbitrary modulation formats and dispersion maps. Journal of Lightwave Technology. 2013;**31**(23):3839-3852

[3] Agrell E, Alvarado A, Durisi G, Karlsson M. Capacity of a nonlinear Optical Channel with finite memory. Journal of Lightwave Technology. 2015;**32**(16):2862-2876

[4] Shiner AD, Reimer M, Borowiec A, Gharan SO, Gaudette J, Mehta P, Charlton D, Roberts K, O'Sullivan M. Demonstration of an 8-dimensional modulation format with reduced inter-channel nonlinearities in a polarization multiplexed coherent system. Optics Express. 2014;**22**(17):20366-20374

[5] Jinno M, Kozicki B, Takara H, Watanabe A, Sone Y, Tanaka T, Hirano A. Distance-adaptive spectrum resource allocation in spectrum-sliced elastic optical path network. IEEE Communications Magazine. 2010;**48**(8):138-145

[6] Nag A, Tornatore M, Mukherjee B. Optical network design with mixed line rates and multiple modulation formats. Journal of Lightwave Technology. 2010;**28**(4):466-475

[7] Alvarado A, Ives DJ, Savory SJ, Bayvel P. On the impact of optimal modulation and FEC overhead on future optical networks. Journal of Lightwave Technology. 2016;**34**(9):2339-2352

[8] Ives DJ, Alvarado A, Savory SJ. Adaptive transceivers in nonlinear flexible networks. In: European Confrence on Optical Communication. 2016. Düsseldorf, Germany. Paper M.1.B.1

[9] Agrell E, Karlsson M. Power-efficient modulation formats in coherent transmission systems. Journal of Lightwave Technology. 2009;**27**(22):5115-5126

[10] Bosco G, Curri V, Carena A, Poggiolini P, Forghieri F. On the performance of Nyquist-WDM terabit superchannels based on PM-BPSK, PM-QPSK, PM-8QAM or PM-16QAM subcarriers. Journal of Lightwave Technology. 2011;**29**(1):53-61

[11] Renaudier J, Voicila A, Bertran-Pardo O, Rival O, Karlsson M, Charlet G, Bigo S. Comparison of set-partitioned two-polarization 16QAM formats with PDM-QPSK and PDM-8QAM for optical transmission systems with error-correction coding. In: European Conf. Optical Communications. 2012. Amsterdam, The Netherlands. Paper We.1.C.5

[12] Zhuge Q, Xu X, Morsy-Osman M, Chagnon M, Qiu M, Plant DV. Time domain hybrid QAM based rate-adaptive optical transmissions using high speed DACs. In: Optical Fiber Commun. Conf., 2013 Anaheim, CA. Paper OTh4E.6

[13] Fischer JK, Alreesh S, Elschner R, Frey F, Nölle M, Schubert C. Bandwidth-variable transceivers based on 4D modulation formats for future flexible networks. In: European Conf. Optical Communications. 2013. London, UK. Paper Tu.3.C.1

[14] Millar DS, Koike-Akino T, Arik SÖ, Kojima K, Yoshida T, Parsons K. High-dimensional modulation for coherent optical communications systems. Optics Express. 2014;**22**(7): 8798-8812

[15] Reimer M, Gharan SO, Shiner AD, O'Sullivan M. Optimized 4 and 8 dimensional modulation formats for variable capacity in optical networks. In: Optical Fiber Commun. Conf. 2016. Anaheim, CA, Paper M3A. 4

[16] Koike-Akino T, Kojima K, Millar DS, Parsons K, Yoshida T, Sugihara T. Pareto-efficient set of modulation and coding based on RGMI in nonlinear fiber transmissions. In: Optical Fiber Commun. Conf., 2016. Anaheim, CA. Paper Th1D.4

[17] Koike-Akino T, Kojima K, Millar DS, Parsons K, Yoshida T, Sugihara T. Pareto optimization of adaptive modulation and coding set in nonlinear fiber-optic systems. Journal of Lightwave Technology. 2017;**35**(3):1-9

[18] Liu X, Chraplyvy AR, Winzer PJ, Tkach RW, Chandrasekhar S. Phase-conjugated twin waves for communication beyond the Kerr nonlinearity limit. Natue Photonics. 2013; **7**:560-568

[19] Chagnon M, Osman M, Zhuge Q, Xu X, Plant DV. Analysis and experimental demonstration of novel 8PolSK-QPSK modulatoin at 5 bis/symbol for passive mitigation of nonlinear impairments. Optics Express. 2013;**21**(25):30204-30220

[20] Bülow H. Polarization QAM modulation (POL-QAM) for coherent detection schemes. In: Optical Fiber Commun. Conf. 2009. San Diego, CA. Paper OWG2

[21] Sjödin M, Agrell E, Karlsson M. Subset-optimized polarization-multiplexed PSK for fiber-optic communications. IEEE Communication Letter. 2013;**17**(5):838-840

[22] Bülow H, Lu X, Schmalen L, Klekamp A, Buchali F. Experimental performance of 4D optimized constellation alternatives for PM-8QAM and PM-16QAM. In: Optical Fiber Commun. Conf. 2014, San Francisco, CA, Paper M2A.6

[23] Rios-Mueller R, Renaudier J, Schmalen L, Charlet G. Joint coding rate and modulation format optimization for 8QAM constellations using BICM mutual information. In: Optical Fiber Commun. Conf. 2015, Los Angeles, CA, Paper W3K.4

[24] Zhang S, Nakamura K, Yaman F, Mateo E, Inoue T, Inada Y. Optimized BICM-8QAM formats based on generalized mutual information. In: European Conf. Optical Communications. 2015, Valencia, Spain, Paper Mo.3.6.5

[25] Nakamura T, de Gabory ELT, Noguchi H, Maeda W, Abe J, Fukuchi K. Long haul transmission of four-dimensional 64SP-12QAM signal based on 16QAM constellation for longer distance at same spectral efficiency as PM-8QAM. In: European Conf. Optical Communications. 2015, Valencia, Spain, Paper Th.2.2.2

[26] Kojima K, Millar DS, Koike-Akino T, Parsons K. Constant modulus 4D optimized constellation alternative for DP-8QAM. In: European Conf. Optical Communications. 2014, Cannes, France, Paper P.3.25

[27] Kojima K, Koike-Akino T, Millar DS, Parsons K. BICM capacity analysis of 8QAM-alternative modulation formats in nonlinear fiber transmission. Tyrrhenian Int'l Workshop on Digital Comm. 2015, Florence, Italy, Paper P.4.5

[28] Kojima K, Koike-Akino T, Millar DS, Yoshida T, Parsons K. BICM capacity analysis of 8QAM-alternative 2D/4D modulation formats in nonlinear fiber transmission. In: Conf. Lasers and Electro-Optics. 2016, Anaheim, Paper SM2F.5

[29] Kojima K, Yoshida T, Koike-Akino T, Millar DS, Parsons K, Arlunno V. 5 and 7 bit/symbol 4D Modulation Formats Based on 2A8PSK. In: European Conf. Optical Communications. 2016, Düsseldorf, Germany, Paper W.2.D.1

[30] Kojima K, Yoshida T, Koike-Akino T, Millar DS, Parsons K, Pajovic M, Arlunno V. Nonlinearity-tolerant four-dimensional 2A8PSK family for 5-7 bits/symbol spectral efficiency. Journal of Lightwave Technology. 2015;**33**(10):1993-2003

[31] Millar DS, Koike-Akino T, Kojima K, Parsons K. A 24-Dimensional Modulation Format Achieving 6 dB Asymptotic Power Efficiency. In: Signal Processing in Photonic Communications. 2013, Puerto Rico, Paper SPM3D.6

[32] Eriksson TA, Johannisson P, Sjödin M, Agrell E, Andrekson PA, Karlsson M. Frequency and polarization switched QPSK. European Conf. Optical Communications. 2013, London, UK, Paper Th.2.D.4

[33] Koike-Akino T, Millar DS, Kojima K, Parsons K. Eight-dimensional modulation for coherent optical communications. European Conf. Optical Communications. 2013, London, UK, Paper Tu.3.C.3

[34] Millar DS, Koike-Akino T, Maher R, Lavery D, Paskov M, Kojima K, Parsons K, Thomsen BC, Savory SJ, Bayvel P. Experimental demonstration of 24-dimensional extended Golay coded modulation with LDPC. In: Optical Fiber Commun. Conf. 2014, San Francisco, CA, Paper M2I.2

[35] Millar DS, Koike-Akino T, Arik ., Kojima K, Parsons K. Comparison of quaternary block-coding and sphere-cutting for high-dimensional modulation. In: Optical Fiber Commun. Conf. 2014, San Francisco, CA, Paper M3A.4

[36] Alvarado A, Agrell E, Lavery D, Bayvel P. LDPC codes for optical channels: Is the "FEC Limit" a good predictor of post-FEC BER. In: Optical Fiber Commun. Conf. 2015, Los Angeles, CA, Paper Th3E.5

[37] Alvarado A, Agrell E. Four-dimensional coded modulation with bit-wise decoders for future optical communications. Journal of Lightwave Technology. 2017;**35**(8):1383-1391

[38] Maher R, Alvarado A, Lavery D, Bayvel EP. Modulation order and code rate optimisation for digital coherent transceivers using generalised mutual information. In: European Conf. Optical Communications. 2015, Valencia, Spain, Paper M.3.3.4

[39] ten Brink S, Kramer G, Ashikhmin A. Design of low-density parity-check codes for modulation and detection. IEEE Transactions on Communications. Apr. 2004;**52**(4):670-678

[40] Bennatan A, Bushtein D. Design and analysis of nonbinary LDPC codes for arbitrary discrete-memoryless channels. IEEE Transactions on Information Theory. 2006;**52**(2):549-583

[41] Szczecinski L, Alvarado A. Bit-Interleaved Coded Modulation: Fundamentals, Analysis, and Design. UK: Wiley; 2015. p. 228

[42] Zhang S, Arabaci M, Yaman F, Djordjevic IB, Xu L, Wnag T, Inada Y, Ogata T, Aoki Y. Experimental study of non-binary LDPC coding for long-haul coherent optical QPSK transmissions. Optics Express. 2011;**19**(20):19042-19049

[43] Sugihara K, Miyata Y, Sugihara T, Kubo K, Yoshida H, Matsumoto W, Mizuochi T. A spatially-coupled type LDPC code with an NCG of 12 dB for optical transmission beyond 100 Gb/s. In: Optical Fiber Commun. Conf. 2013, Anaheim, CA, Paper OM2B.4

[44] Yoshida T, Matsuda K, Kojima K, Miura H, Dohi D, Pajovic M, Koike-Akino T, Millar DS, Parsons K, Sugihara T. Hardware-efficient precise and flexible soft-demapping for multi-dimensional complementary APSK signals. In: European Conf. Optical Communications. 2016, Düsseldorf, Germany, Paper Th.2.P2.SC3.27

[45] Poggiolini P, Bosco G, Carena A, Curri V, Forghieri F. Performance evaluation of coherent WDM PS-QPSK (HEXA) accounting for non-linear fiber propagation effects. Optics Express. 2010;**18**(11):11360-11371

[46] Bononi A, Serena P, Rossi N, Sperti D. Which is the dominant nonlinearity in long-haul PDM-QPSK coherent transmissions? In: European Conf. Optical Communications. 2010, Torino, Italy, Paper Th.10.E.1

[47] Pajovic M, Millar DS, Koike-Akino T, Maher R, Lavery D, Alvarado A, Paskov M, Kojima K, Parsons K, Thomsen BC, Savory SJ, Bayvel P. Experimental demonstration of multi-pilot aided carrier phase estimation for DP-16QAM and DP-256QAM. In: European Conf. Optical Communications. 2015, Valencia, Spain, Paper Mo.4.3.3

[48] Ready MJ, Gooch RP. Blind equalization based on radius directed adaptation. Proceedings of ICASSP. 1990;**3**:1699-1702

[49] Ishimura S, Kikuchi K. Fast decoding and LLR-computation algorithms for high-order set-partitioned 4D–QAM constellations. In: European Conf. Optical Communications. 2015, Valencia, Paper P.3.01

[50] Koike-Akino T, Millar DS, Kojima K, Parsons K. Phase noise-robust LLR calculation with linear/bilinear transform for LDPC-coded coherent communications. Conf. Lasers and Electro-Optics. 2015, San Diego, Paper SW1M.3

[51] Zhou X, Nelson LE, Isaac R, Magill PD, Zhu B, Borel P, Carlson K, Peckham DW. 12,000km transmission of 100GHz spaced, 8x495-Gb/s PDM time-domain hybrid QPSK-8QAM signals. In: Optical Fiber Commun. Conf. 2012, Los Angeles, CA, Paper OTu2B.4

[52] Yan L et al. Sensitivity comparison of time domain hybrid modulation and rate adaptive coding. Proc. OFC, W1I.3, Anaheim. 2013

[53] Ishii K, Dohi K, Kubo K, Sugihara K, Miyata Y, Sugihara T. A study on power-scaling of triple-concatenated FEC for optical transport networks. In: European Conf. Optical Communications. 2015, Valencia, Paper Tu3.4.2

[54] Kojima K, Yoshida T, Parsons K, Koike-Akino T, Millar DS, Matsuda K. Nonlinearity-tolerant time domain hybrid modulation for 4-8 bits/symbol based on 2A8PSK. In: Optical Fiber Commun. Conf. 2017, Los Angeles, CA, Paper W4A.5

[55] Koike-Akino T, Kojima K, Parsons K. Trellis-coded high-dimensional modulation for polarization crosstalk self cancellation in coherent optical communications. In: Signal Processing in Photonic Communications. 2015, Boston, Paper SpS3D.6

[56] Kojima K, Yoshida T, Koike-Akino T, Millar DS, Matsuda K, Parsons K. Nonlinearity-tolerant modulation formats at 3.5 bits/symbol. In: Conf. Lasers and Electro-Optics. 2017, San Jose, Paper STu4M.3

Chapter 9

Fiber Laser for Phase-Sensitive Optical Time-Domain Reflectometry

Vasily V. Spirin, Cesar A. López-Mercado,
Patrice Mégret and Andrei A. Fotiadi

Additional information is available at the end of the chapter

http://dx.doi.org/10.5772/intechopen.72553

Abstract

We have designed a new fiber laser configuration with an injection-locked DFB laser applicable for phase-sensitive optical time-domain reflectometry. A low-loss fiber optical ring resonator (FORR) is used as a high finesse filter for the self-injection locking of the DFB (IL-DFB) laser. By varying the FORR fidelity, we have compared the DFB laser locking with FORR operating in the under-coupled, critically coupled, and over-coupled regimes. The critical coupling provides better frequency locking and superior narrowing of the laser linewidth. We have demonstrated that the locked DFB laser generates a single-frequency radiation with a linewidth less than 2.5 kHz if the FORR operates in the critically coupled regime. We have employed new IL-DFB laser configuration operating in the critical coupling regime for detection and localization of the perturbations in phase-sensitive OTDR system. The locked DFB laser with a narrow linewidth provides reliable long-distance monitoring of the perturbations measured through the moving differential processing algorithm. The IL-DFB laser delivers accurate localization of the vibrations with a frequency as low as ~50 Hz at a distance of 9270 m providing the same signal-to-noise ratio that is achievable with an expensive ultra-narrow linewidth OEwaves laser (OE4020–155000-PA-00).

Keywords: fiber laser, self-injection locking, phase-sensitive OTDR

1. Introduction

Distributed fiber optic sensors are widely used for variety of applications such as structural health monitoring, perimeter and pipeline security, temperature, pressure, strain, and vibration measurements due to its lightweight, ease of installation, and immunity to electromagnetic fields [1–18]. One of the modern forward-looking fiber optic techniques, the so-called phase-sensitive optical time-domain reflectometry (φ-OTDR), enables detection of acoustical perturbations along sensing optical fibers of several kilometers length [19–22].

Such optical sensor analyzer operates as a conventional OTDR, where a light pulse is injected into the optical fiber and Rayleigh backscattering radiation originating from the natural refractive index inhomogeneities frozen in the fiber core is recorded as time-dependent traces. However, in contrast to the conventional OTDR utilizing low coherence laser sources and hence based on recording of Rayleigh backscattering intensity, the φ-OTDR systems require highly coherent lasers with a coherence length exceeding the pulse duration and employ a difference between consequent time-dependent traces as a readout signal. For proper operation of φ-OTDR systems, the allowed optical frequency shift between two neighboring pulses should be low enough to keep Rayleigh backscattering interference pattern recorded as a result of pulse reflections from multiple scattering fiber centers unchangeable. Under these conditions, two consecutive traces recorded in an undisturbed fiber are identical. Meanwhile, any change in geometry of the frozen distribution of the refractive index in the fiber core caused by stress, strain, or temperature variations applied to some fiber points affects the difference between successive traces and, therefore, can be detected and localized with φ-OTDR systems [19–22]. Typically for long-distance measurements, a coherent laser source with a few kHz linewidth and frequency drift less than 1 MHz/min is required [2].

It is well known that self-injection locking of conventional telecom DFB lasers could significantly improve their spectral performance [23–34]. In our previous works, we have demonstrated substantial narrowing of the laser linewidth due to a spectrally selective feedback realized with FORR built from low-cost standard fiber telecom components [35–37]. To provide the DFB laser locking, a part of the optical radiation emitted by the DFB laser is passed through the filtering in a fiber ring resonator and returned into the laser cavity. This low-cost all-fiber solution allows achieving the laser linewidth as narrow as 500 Hz [36].

In this chapter, we demonstrate application of the DFB laser locked through a fiber ring cavity for φ-OTDR systems. In particular, we show that the proposed laser solution in combination with the moving differential processing algorithm enables accurate detection and localization of vibrations applied to the sensing fiber at a distance up to 10 km.

2. Experimental results and discussion

Figure 1 shows the experimental configuration of the DFB laser locked through a fiber optic ring resonator (FORR). The MITSUBISHI FU-68PDF-V520M27B DFB laser with a built-in optical isolator operates a linewidth of ~2 MHz at the wavelength of ~ 1534.85 nm. The output DFB laser radiation is passed through an optical circulator (OC), optical coupler (C1), and polarization controller (PC1) and then introduced into a fiber optic ring resonator. The FORR consists of a variable ratio coupler VRC, 95/5 coupler C2 and comprises ~4 m length of a standard SMF-28 fiber. The operation of FORR is similar to the Fabry-Perot interferometer

http://dx.doi.org/10.5772/intechopen.72553

with the reflected power detected in port B and transmitted power directed to the coupler C3 [20]. The radiation at the port A is used as an output of the injection-locked DFB (IL-DFB) laser, while port B and C are connected to detectors for the monitoring of the reflected and transmitted powers, respectively. Optical isolators prevent reflections from the fiber ends that potentially could affect the DFB laser behavior. The polarization controller (PC2) and the optical switcher (OS) are used to adjust the feedback strength and activate or deactivate the optical feedback loop that returns transmitted through the FORR power back into the DFB laser cavity.

Once the DFB laser frequency gets a resonance with the FORR, the DFB laser is locked in frequency to one of the FORR frequency modes. This effect could be observed as a suppression of the temporal power fluctuations recorded at ports C and B. **Figure 2** shows typical oscilloscope traces of transmitted and reflected powers recorded at ports C and B, respectively. Single

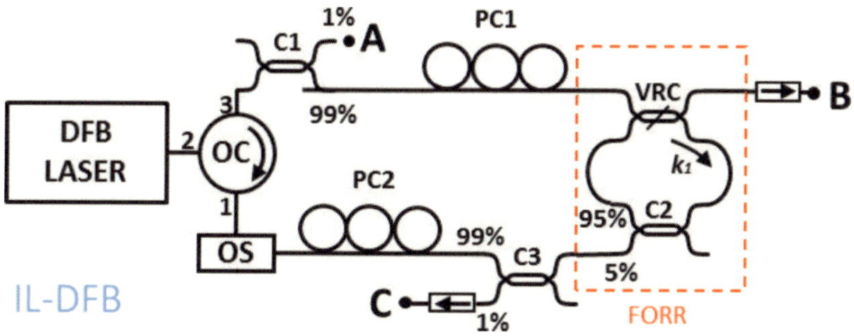

Figure 1. The experimental configuration of the DFB laser locked through FORR. OC—Optical circulator, VRC—Variable ratio coupler, C—Optical coupler, PC—Polarization controller, OS—Optical switcher.

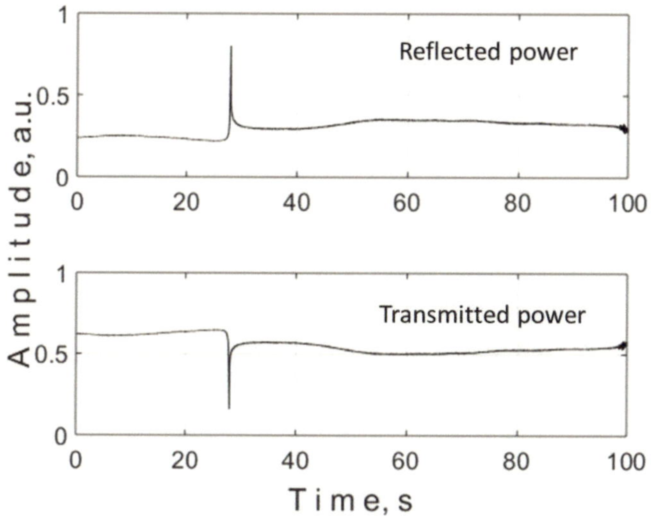

Figure 2. Typical oscilloscope traces for transmitted and reflected powers.

frequency regime is observed during some time intervals that are interrupted by short-time mode hopping events. The frequency drift during the stable time interval strongly depends on environmental conditions and can be less than 1–2 MHz/min, if the FORR is placed in an isolation box used for environmental protection.

Stabilization of the laser operation is accompanied by stabilization of the polarization state of the locked DFB laser radiation recorded at the port A and the transmitted power recorded at the port C (see **Figure 3**). However, during the mode hopping, the polarization state can be changed. In our previous work, we have reported two possible regimes of mode hopping [35]. The first regime, we have attributed to the hopping between FORR modes of nearest orders, but of the same polarization state, and the second one to the hopping between orthogonal polarizations modes of the same order. These two regimes have significantly different time which the system takes to recover steady-state operation after the mode hopping event. In the first most common regime, the recovery time is typically ~5 ms, but in the second regime, the typical recovery time is significantly less and equal to ~100 μs [35].

The locking performance strongly depends on the parameters of the VRC coupler used in the FORR. The coupling regime is defined by a relation between the VRC coupling coefficient k_1 and the total losses inside the cavity α [38]:

$$\begin{array}{ll} k_1 < 1 - \alpha(1 - \gamma_1) & \text{under-coupling regime,} \\ k_1 = 1 - \alpha(1 - \gamma_1) & \text{critical-coupling regime,} \\ k_1 > 1 - \alpha(1 - \gamma_1) & \text{over-coupling regime,} \end{array} \quad (1)$$

where the power transmission coefficient α includes all losses inside the fiber loop and γ_1 is the intensity loss in the variable ratio coupler.

Figure 4 shows the normalized reflected power versus the coupling coefficient k_1 at port B. In the critical coupling regime, the reflected power reaches the minimum, and therefore, all nearby input power are transmitted through the FORR (see **Figure 5**). For that reason, the critical coupling provides higher feedback strength leading to better locking.

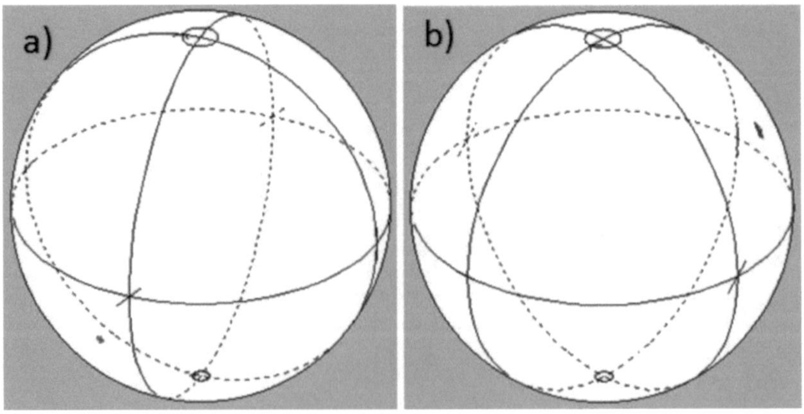

Figure 3. Polarization behavior during the stable interval: a) at port a and b) at port C.

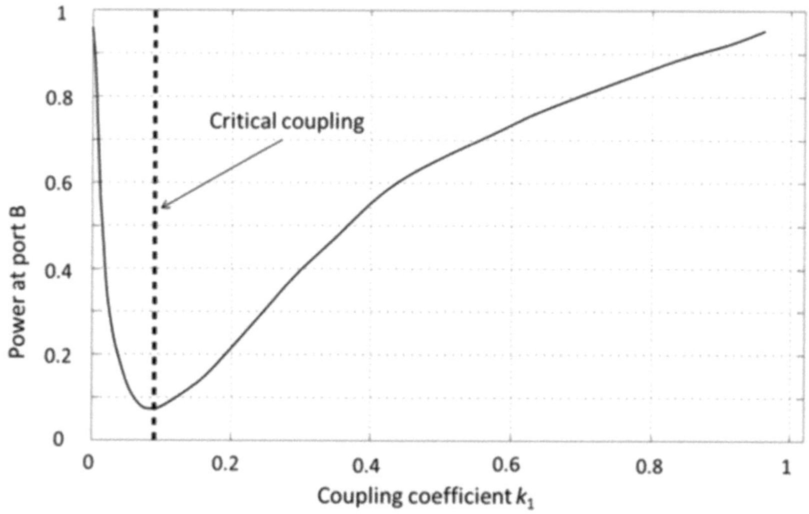

Figure 4. Normalized reflected power versus coupling coefficient k_1 at port B.

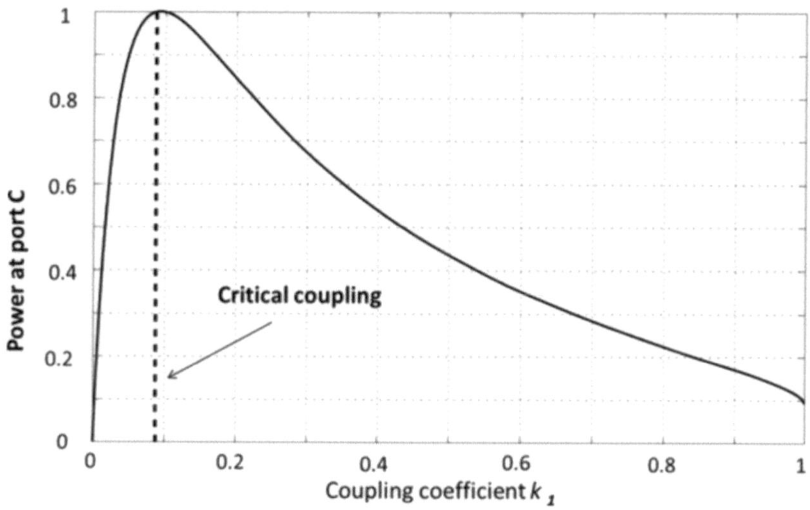

Figure 5. Normalized transmitted power versus coupling coefficient k_1 at port C.

Let us stress out that for the DFB laser locked at the resonance frequency of the FORR, the transmitted and reflected powers from FORR are in good agreement with the theoretical estimations.

The finesse of the resonance peak of the FORR is defined as a ratio of the FORR intermode interval to the spectral peak width:

$$\mathcal{F} = FSR/\Delta f \tag{2}$$

where $FSR = c/nL \approx 50$ MHz is the free spectral range of the FORR and Δf is the full-width at half-maximum (FWHM) of the feedback loop transmission peak.

For the FORR configuration with an additional coupler C2, Δf could be found following the same procedure as it is described in Ref. 26 for a simple optical ring resonator. For our resonator, the FWHM of the cavity mode can be estimated as [37]:

$$\Delta f = \frac{1}{2\pi\tau}\cos^{-1}\left(2 - \frac{1+\alpha(1-\gamma_1)(1-k_1)}{2\sqrt{\alpha(1-\gamma_1)(1-k_1)}}\right) \tag{3}$$

In the critical coupling regime at $k_1 = 0.08$, Eq. (3) gives $\Delta f = 0.77$ MHz and the finesse $\mathcal{F} = 65.8$.

The finesse of the resonator strongly depends on the total cavity losses and the coupling coefficient. The finesse is higher for smaller losses and lower coupling coefficients (see **Figure 6**).

The delayed self-heterodyne spectra of the IL-DFB laser measured with unbalanced Mach-Zehnder interferometer [39] during the time intervals of stable laser operation at the critical coupling regime are shown in **Figure 7**. The Mach-Zehnder interferometer comprises 35 km of the delay fiber line in one arm and 20 MHz phase modulator in the other. The beat signal of two interferometer arms is detected by 125 MHz photodiode and RF spectrum analyzer. One can see that the DFB laser linewidth $\Delta\nu$ decreases from approximately 2 MHz for a free running laser to 2.4 kHz for the IL-DFB laser operating in critical coupling regime.

Assuming the Lorentzian shape of the laser line, the laser coherence length L_c [40] is described as:

$$L_c = \frac{c}{\pi N_{eff}\Delta\nu} \tag{4}$$

where c is the speed of the light in vacuum, N_{eff} is the effective group refraction index in the sensing fiber equal to 1.468 [41], and $\Delta\nu$ is the laser linewidth.

Therefore, the coherence length is increased from approximately 26 m for a free running DFB laser up to 27.2 km for the IL-DFB laser operating in the critical coupling regime.

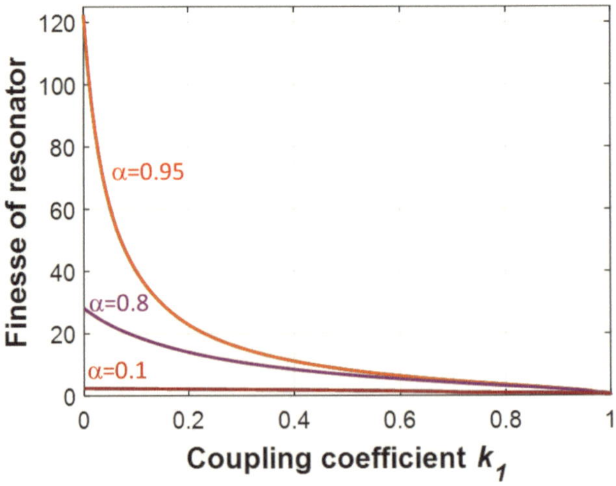

Figure 6. The finesse of resonator versus coupling factor k_1.

Figure 8 shows the experimentally measured linewidth of the locked DFB laser versus the coupling coefficient k_1. The minimal linewidth of about 2.5 kHz is achieved for the laser operating in the under-coupling and critical coupling regimes. In the over-coupling regime, the laser linewidth significantly increases.

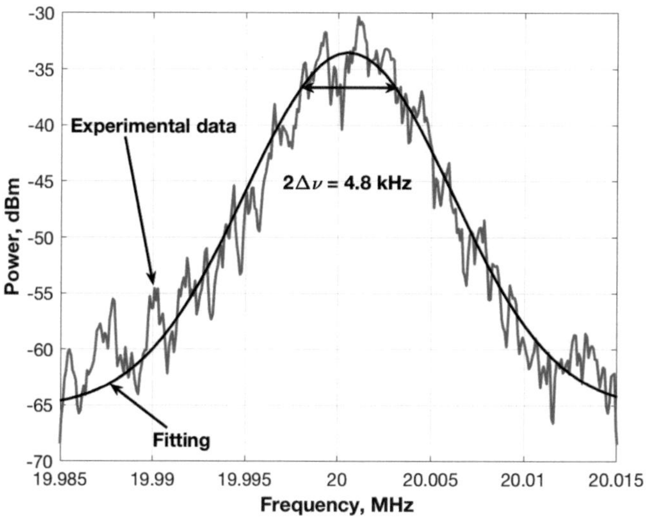

Figure 7. Delayed self-heterodyne spectra of the IL-DFB laser recorded with the laser stabilized for operation in the critical coupling regime.

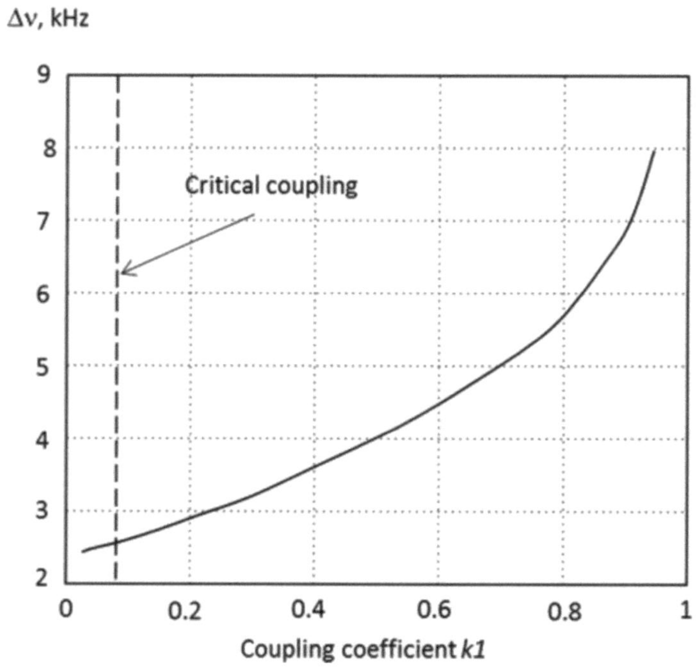

Figure 8. Experimental linewidth of the locked DFB laser versus the coupling coefficient k_1.

Thus, the critical coupling provides the best IL-DFB laser stability and narrowest linewidth making the laser applicable in φ-OTDR systems.

The configuration of the φ-OTDR system with IL-DFB laser operating in the critical coupling regime is shown in **Figure 9**. The output radiation of the IL-DBF laser is amplified by an EDFA up to 23 mW and passes through a bandpass filter (BPF) to filter out spontaneous emission noise.

An optical intensity modulator (OIM) provides generation of optical pulses with the width of 350 ns and repetition rate of 10 kHz introduced into the sensing fiber through 1/99 coupler and an optical circulator. The sensing fiber comprises 8400 m of standard SMF-28e and 950 m of Raman Optical Fiber (OFS) with the loss coefficient of about 0.33 dB/km. Acoustic perturbations at the frequency of 50 Hz are generated at the sensing fiber distance of 9270 m by a loudspeaker.

The Rayleigh backscattered signal traces have been detected at the port E by 125 MHz photodetector and recorded by an oscilloscope. Each from about 1024 similar traces of 100 ms length has been digitized with a speed of 20 mega samples per second.

Since a low-frequency perturbation causes very small modification of the Rayleigh scattering traces, a simple calculation of the differences between two neighboring successive traces is not enough to localize the perturbed fiber segment. **Figure 10** shows superposition of the absolute values of differences between all neighboring successive traces. One can see that the signal obtained for the distance of perturbations is the same as for any other positions.

However, the signal-to-noise ratio can be significantly improved with applying a special averaging procedure. For these purposes, we have processed the storage traces by employing three-step algorithm which is similar to moving differential method [19].

First, we determine the partial sums by averaging of *N* consequent traces. Each next partial sum is shifted by a number *m* of traces from the previous one. Then, we calculate the absolute value of the differences between every consequent partial sum.

The last step of the algorithm includes an analysis of the superposition of the absolute values for all differences.

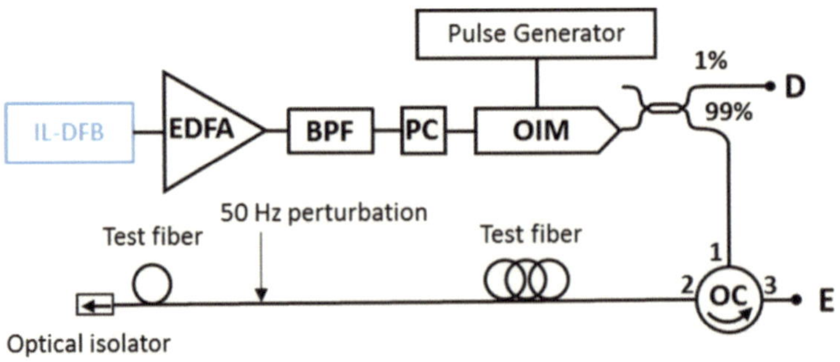

Figure 9. Experimental scheme for detection of the perturbation in the sensing fiber. BPF—Bandpass filter, PC—Polarization controller, OIM—Optical intensity modulator, OC—Optical circulator.

Figure 11 shows the superposition of all differences for $N = 75$ and $m = 10$. The highest peak denotes the point of the perturbation at 9270 m along the fiber under test. The signal peak exceeds the maximum of the noise value in 1.6 times and the average value of the noise approximately by 9 dB. Let us stress out that with the unlocked DFB laser, we never register the perturbation even at distances less than 0.1 km.

It is experimentally demonstrated that the relation between the signal and average value of the noise strongly depends on the shift between the averaging data arrays. However, a choice of

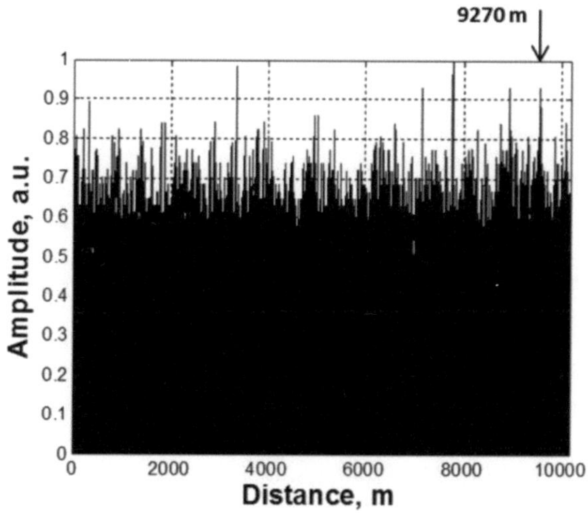

Figure 10. Superposition of the absolute values of differences between neighboring successive traces recorded with IL-DFB laser.

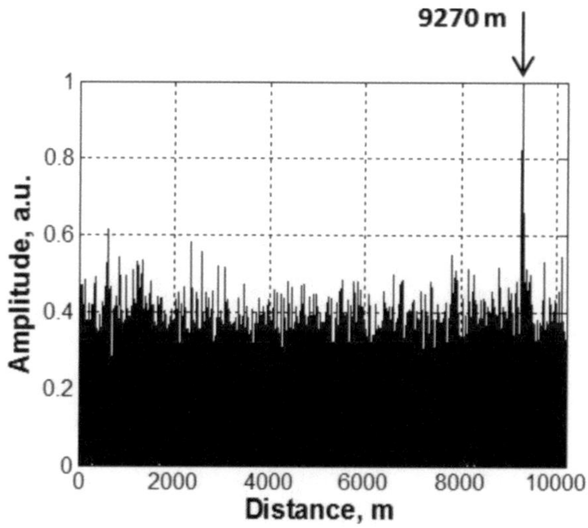

Figure 11. Superposition of the absolute values of differences between averaged traces for IL-DFB laser, $m = 10$ and $N = 75$.

the number of averaging not so strongly affects the signal-to-noise ratio and could be selected arbitrary within the interval between 40 and 120. **Figure 12** shows superposition of the absolute values of differences between the averaged traces for the shift m = 23 and averaging number N = 75.

The signal peak exceeds the maximum of the noise value in 2.1 times and the average noise value approximately by 10 dB.

Figure 13 shows a superposition of the absolute values of differences between the averaged traces for the shift of m = 45 and averaging number N = 75. The signal peak exceeds the maximum of the noise value in 1.4 times and the average noise value by approximately 8 dB.

A number of the performed experiments allow us to conclude that in our experimental conditions, the maximum of the signal-to-noise ratio is achieved with the shift m approximately equal to 20.

The obtained value of the signal-to-noise ratio allows correct localization of the perturbations with the IL-DFB laser system at distances of about 10 km and resolution of around 10–15 m but does not allow comparing a capacity of the proposed solution with commercially available techniques.

In order to fulfill this gap, we have performed the same measurements, under the same experimental conditions, but utilizing a commercially available ultra-narrow linewidth (~ 300 Hz) laser OEwaves laser OE4020–155000-PA-00. **Figure 14** shows the superposition of the subtractions of averaged traces for the measurements utilizing OEwaves laser for the shift m = 23 and averaging number N = 75. The signal peak at the distance of ~9270 m exceeds the highest noise signal in about 2.25 times that is nearly the same result as obtained with our IL-DFB laser.

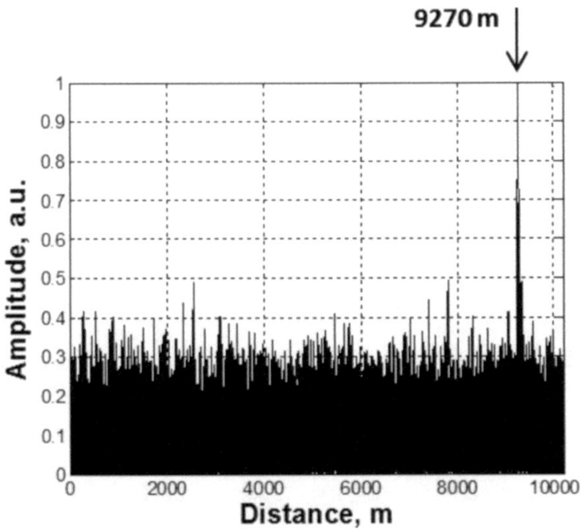

Figure 12. Superposition of the absolute values of differences between averaged traces for IL-DFB laser, m = 23 and N = 75.

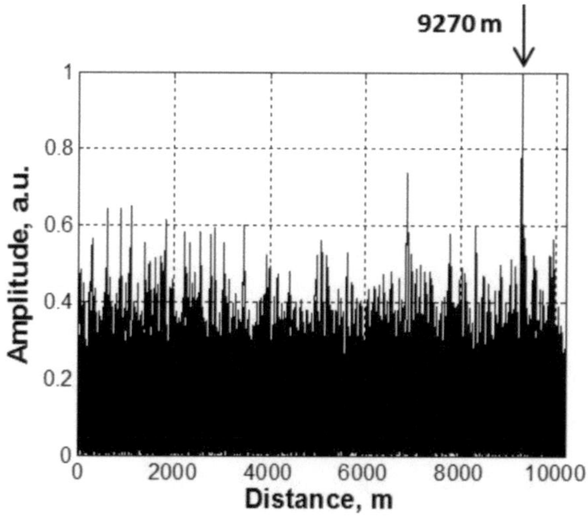

Figure 13. Superposition of the absolute values of differences between the averaged traces for IL-DFB laser, $m = 45$ and $N = 75$.

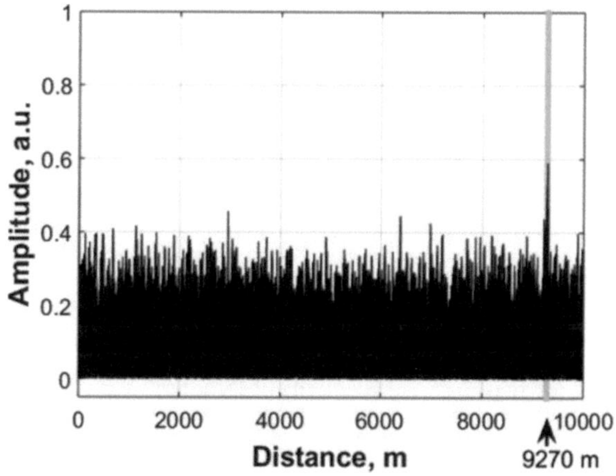

Figure 14. Superposition of the absolute values of the differences between average traces for ultra-narrow linewidth OEwaves laser OE4020–155000-PA-00.

3. Conclusion

We have employed IL-DFB laser configuration at the critical coupling regime for detection and localization of the perturbations by φ-OTDR system.

At the critical coupling regime, the laser power is practically totally accumulated inside the cavity providing strong feedback for laser locking and resulting in the best laser stability and significant narrowing of the laser generation spectrum. The locked DFB laser with the linewidth of about 2.4 kHz provides the long-distance measurements of the perturbations as the moving differential processing algorithm is applied. The IL-DFB laser delivers accurate localization of vibrations at the frequency as low as low 50 Hz at the distance of ~9270 m with the same signal-to-noise ratio that is achieved with an expensive ultra-narrow linewidth laser OEwaves laser OE4020–155000-PA-00.

We believe that proposed solution can be useful for applications in cost-effective φ-OTDR system for the measurement of the perturbations at distances up to ten kilometers.

Acknowledgements

This work was supported by project N 265517 CONACYT, Mexico, Ministry of Education and Science of Russian Federation (14.Z50.31.0015, "State Assignment" 3.3889.2017), Russian Fund of Basic Research (16-42-732135 R-OFIM), and the IAP program VII/35 of the Belgian Science Policy. C.A. López-Mercado was sponsored by the CONACYT, Mexico as Postdoctoral Fellow at the University of Mons, Belgium.

Author details

Vasily V. Spirin[1*], Cesar A. López-Mercado[2], Patrice Mégret[2] and Andrei A. Fotiadi[2,3,4]

*Address all correspondence to: vaspir@cicese.mx

1 División de Física Aplicada, CICESE, Ensenada, B.C., México

2 Electromagnetism and Telecommunication Department, University of Mons, Mons, Belgium

3 Ioffe Physico-Technical Institute of the RAS, St. Petersburg, Russia

4 Ulyanovsk State University, Ulyanovsk, Russia

References

[1] Shi Y, Feng H, Zeng Z. Distributed fiber sensing system with wide frequency response and accurate location. Optics and Lasers in Engineering. 2016;**77**:219-224

[2] Peng F, Wu H, Jia X-H, Rao YJ, Wang Z-N, Peng Z-P. Ultra-long high-sensitivity Φ-OTDR for high spatial resolution intrusion detection of pipelines. Optics Express. 2014;**22**:13804-13810

[3] Spirin VV, Swart PL, Chtcherbakov AA, Miridonov SV, Shlyagin MG. Distributed fibre-optic loss sensor with chirped Bragg grating based on transmission-reflection analysis. Electronics Letters. 2003;**39**:895-897

[4] López RM, Spirin VV, Shlyagin MG, Miridonov SV, Beltrán G, Kuzin EA, Márquez Lucero A. Coherent optical frequency domain reflectometry for interrogation of bend-based fiber optic hydrocarbon sensors. Optical Fiber Technology. 2004;**28**:79-90

[5] Spirin VV, Mendieta FJ, Miridonov SV, Shlyagin MG, Chtcherbakov AA, Swart PL. Localization of a loss-inducing perturbation with variable accuracy along a test fiber using transmission-reflection analysis. IEEE Photonics Technology Letters. 2004;**16**:569-571

[6] Spirin VV, Swart PL, Chtcherbakov AA, Miridonov SV, Shlyagin MG. 20-km length distributed fiber-optical loss sensor based on transmission - reflection analysis. Optical Engineering. 2005;**44**:040501

[7] Spirin VV. Autonomous Measurement System for Localization of Loss-Induced Perturbation Based on Transmission-Reflection Analysis. In: Kr Sharma M, editor. Advances in Measurement Systems. InTech; 2010 ISBN 978–953–307-061-2, 81–104

[8] Spirin VV, Shlyagin MG, Miridonov SV, Swart PL. Transmission/reflection analysis for distributed optical fibre loss sensor interrogation. Electronics Letters. 2002;**38**:117-118

[9] Spirin VV, Shlyagin MG, Miridonov SV, Swart PL. Alarm-condition detection and localization using Rayleigh scattering for a fiber optic bending sensor with an unmodulated light source. Optics Communications. 2002;**205**:37-41

[10] López RM, Spirin VV, Miridonov SV, Shlyagin MG, Beltrán G, Kuzin EA. Fiber optic distributed sensor for hydrocarbon leak localization based on transmission/reflection measurement. Optics & Laser Technology. 2002;**34**:465-469

[11] Spirin VV. Transmission/reflection analysis for localization of temporally successive multi-point perturbations in distributed fiber-optic loss sensor based on Rayleigh backscattering. Applied Optics-OT. 2003;**42**:1175-1181

[12] Bueno Escobedo JL, Spirin VV, López-Mercado CA, Lucero AM, Mégret P, Zolotovskii IO, Fotiadi AA. Self-injection locking of the DFB laser through an external ring fiber cavity: Application for phase sensitive OTDR acoustic sensor. Results in Physics. 2017;**7**:641-643

[13] Wang C, Wang C, Shang Y, Liu X, Peng G. Distributed acoustic mapping based on interferometry of phase optical time-domain reflectometry. Optics Communication. 2015;**346**:172-177

[14] Faustov A, Gussarov A, Wuilpart M, Fotiadi AA, Liokumovich LB, Kotov OI, Zolotovskiy IO, Tomashuk AL, Deschoutheete T, Mégret P. Distributed optical fibre temperature measurements in a low dose rate radiation environment based on Rayleigh backscattering. In: Berghmans F, Mignani AG, De Moor P, editors. Proceedings of the SPIE 8439Optical Sensing and Detection II. 2012 pp. 84390C-84390C-8

[15] Faustov AV, Gusarov AV, Mégret P, Wuilpart M, Zhukov AV, Novikov SG, Svetukhin VV, Fotiadi AA. The use of optical frequency-domain Reflectometry in remote distributed measurements of the γ-radiation dose. Technical Physics Letters. 2015;**41**(5):412-415

[16] Faustov AV, Gusarov AV, Mégret P, Wuilpart M, Zhukov AV, Novikov SG, Svetukhin VV, Fotiadi AA. Application of phosphate doped fibers for OFDR dosimetry. Results in Physics. 2016;**6**:86-87

[17] Faustov AV, Gusarov A, Wuilpart M, Fotiadi AA, Liokumovich LB, Zolotovskiy IO, Tomashuk AL, de Schoutheete T, Megret P. Comparison of gamma-radiation induced attenuation in Al-doped, P-doped and Ge-doped fibres for Dosimetry. IEEE Transactions on in Nuclear Science. 2013;**60**(4):2511-2517

[18] Faustov AV, Andrei G, Liokumovich LB, Fotiadi AA, Wuilpart M, Mégret P. Comparison of simulated and experimental results for distributed radiation-induced absorption measurement using OFDR reflectometry. Proc. SPIE. 2013;**8794**:87943O

[19] Lu Y, Zhu T, Chen L, Bao X. Distributed vibration sensor based on coherent detection of phase-OTDR. Lightwave. 2010;**28**:3243-3249

[20] Li Q, Zhang C, Li L, Zhong X. Localization mechanisms and location methods of the disturbance sensor based on phase-sensitive OTDR. Optik (Stuttg). 2014;**125**:2099-2103

[21] Li Q, Zhang C, Li C. Fiber-optic distributed sensor based on phase-sensitive OTDR and wavelet packet transform for multiple disturbances location. Optik (Stuttg). 2014;**125**: 7235-7238

[22] Zhan Y, Yu Q, Wang K, Yang F, Zhang B. Optimization of a distributed optical fiber sensor system based on phase sensitive OTDR for disturbance detection. Sensor Review. 2015;**35**:382-388

[23] Spirin VV, Kellerman J, Swart PL, Fotiadi AA. Intensity noise in SBS with injection locking generation of stokes seed signal. Optics Express. 2006;**14**:8328-8335

[24] Spirin VV, Kellerman J, Swart PL, Fotiadi AA. Intensity noise in SBS with seed signal generated through injection locking. Conference Digest: CLEO-Europe'2007, IEEE

[25] Spirin VV, Castro M, López-Mercado CA, Mégret P, Fotiadi AA. Optical locking of two semiconductor lasers through high order Brillouin stokes components in optical Fiber. Laser Physics. 2012;**22**(4):760-764

[26] Fotiadi AA, Kinet D, Mégret P, Spirin VV, Lopez-Mercado CA, Zolotovskii I. Brillouin Fiber Laser Passively Stabilized at Pump Resonance Frequency. Symposium: IEEE Photonics Benelux Chapter; 2012. pp. 365-368

[27] Spirin VV, López-Mercado CA, Kinet D, Mégret P, Zolotovskiy IO, Fotiadi AA. Single longitudinal-mode Brillouin fiber laser passively stabilized at pump resonance frequency with dynamic population inversion grating. Laser Physics Letters. 2013;**10**:015102

[28] Lopez-Mercado CA, Spirin VV, Zlobina EA, Kablukov SI, Mégret P, Fotiadi AA. Doubly-resonant Brillouin fiber cavity: Algorithm for cavity length adjustment. IEEE Photonics Benelux Chapter. 2012:369-372

[29] Spirin VV, López-Mercado CA, Kinet D, Mégret P, Zolotovskiy IO, Fotiadi AA. Passively stabilized Brillouin fiber lasers with doubly resonant cavities. Proceedings of SPIE. 2013; **8601**:860135

[30] Spirin VV, López-Mercado CA, Kinet D, Zlobina EA, Kablukov SI, Mégret P, Zolotovskiy IO, Fotiadi AA. Double-frequency Brillouin fiber lasers. Proc. SPIE 8772. 2013 87720U

[31] Fotiadi A, Spirin V, López-Mercado C, Kinet D, Preda E, Zolotovskii I, Zlobina E, Kablukov S, Mégret P. Recent progress in passively stabilized single-frequency Brillouin fiber lasers with doubly-resonant cavities. CLEO/Europe and EQEC. CLEO-Europe: Conference Digest; 2013

[32] Spirin VV, López-Mercado CA, Bueno-Escobedo JL, Lucero AM, Zolotovskii IO, Mégret P, Fotiadi AA. Self-injection locking of the DFB laser through ring fiber optic resonator. Proc. SPIE 9344. 2015 93442B

[33] Lopez-Mercado CA, Spirin VV, Nava-Vega A, Patrice M, Andrei F. Láser de Brillouin con cavidad corta de fibra estabilizado pasivamente en la resonancia de bombeo por fenómeno de auto-encadenamiento por inyección óptica. Revista Mexicana de Física. 2014;**60**:53-58

[34] Spirin VV, Mégret P, Fotiadi AA. Passively Stabilized Doubly Resonant Brillouin Fiber Lasers. In: Paul M, Kolkata J, editors. Fiber Lasers. InTech; 2016

[35] Bueno Escobedo JL, Spirin VV, López-mercado CA, Mégret P, Zolotovskii IO, Fotiadi AA. Self-injection locking of the DFB laser through an external ring fiber cavity : Polarization behavior. Results in Physics. 2016;**6**:59-60

[36] Spirin VV, López-Mercado CA, Mégret P, Fotiadi AA. Single-mode Brillouin fiber laser passively stabilized at resonance frequency with self-injection locked pump laser. Laser Physics Letters. 2012;**9**:377-380

[37] López-Mercado CA, Spirin VV, Bueno Escobedo JL, Lucero AM, Mégret P, Zolotovskii IO, Fotiadi AA. Locking of the DFB laser through fiber optic resonator on different coupling regimes. Optics Communications. 2016;**359**:195-199

[38] Faramarz E. Seraji, steady-state performance analysis of fiber-optic ring resonator. Progress in Quantum Electronics. 2009;**33**:1-16

[39] Derickson D. Fiber Optic Test and Measurement. Prentice Hall PTR, c: Upper Saddle River, N.J; 1998

[40] Muanenda Y, Oton CJ, Faralli S, Di Pasquale F. A cost-effective distributed acoustic sensor using a commercial off-the-shelf DFB laser and direct detection phase-OTDR. IEEE Photonics Journal. 2016;**8**:1-10

[41] http://www.corning.com/opticalfiber/index.asp

Chapter 10

Optoelectronic Design of a Closed-Loop Depolarized IFOG with Sinusoidal Phase Modulation for Intermediate Grade Applications

Ramón José Pérez Menéndez

Additional information is available at the end of the chapter

http://dx.doi.org/10.5772/intechopen.72592

Abstract

A depolarized fiber optic gyroscope (DFOG) prototype with closed-loop configuration, sinusoidal-bias, and serrodyne-feedback electrooptic phase modulations was designed. A complete optoelectronic design is realized by using computational simulation tools (optical subsystem: Synopsys®-Optsim™ software and electronic subsystem: National Instruments®-MultiSim™ software). The design presented here includes both optical and electronic circuits, being the main innovation, is the use of an analogical integrator provided with reset and placed in the feedback of the electrooptic phase-modulation chain that produces a serrodyne-shaped voltage ramp signal for obtaining the interferometric signal phase cancellation. The performance obtained for this model (threshold sensitivity ≤0.052°/h; dynamic range = ± 78.19°/s) does reach the IFOG intermediate grade (tactical and industrial applications) and does demonstrate the suitability and reliability of simulation-based software tools for this kind of optoelectronic design.

Keywords: interferometric-fiber-optic-gyroscope (IFOG), depolarized-fiber-optic-gyroscope (DFOG), super-luminescent-laser-diode (SLD), single-mode-fiber (SMF), phase-modulator (PM), closed-loop configuration, bias phase-modulation, feedback phase-modulation, serrodyne-wave, IFOG intermediate grade, IFOG navigation grade, phase-sensitive-demodulator (PSD), Lyot depolarizer

1. Introduction

In all the electro-optical engineering areas, particularly in the design of high-cost devices like IFOGs, the computational simulation resources can provide a powerful and inestimable advance. It stems from the rapidity, the reproducibility, and the reliability of this kind of hardware to obtain the ultimate design of a preconceived model. Furthermore, it is possible to obtain substantial cost savings in components and time consuming for the model assembly in optical bench. Only after having obtained an ideal design so much for the performance characteristics all that for the adaptation to a specific application, it is suitable to initiate the laboratory manufacture stage for the prototype designed previously. In this article, it is shown to the reader an aspect that is not usually in the literature, namely: how to realize the simulation of a classical IFOG system without having to make the real model in the laboratory. For this proposal, three classical electrooptic simulation tools: Synopsys™ OptSim®, National Instruments™ MultiSim® and MathWorks™ Matlab-Simulink® will be used. In the present decade, the design trends on interferometric fiber optic gyroscope (IFOG) are focused on devices with very high performance (navigation-grade, sensitivity ≤0.001°/h), mainly targeting aeronautics and spacecraft applications. Nevertheless, it is also possible to realize designs for certain applications that do not need such a high grade of performance (intermediate-grade, sensitivity ≤0.01°/h or industrial-grade, sensitivity ≤1°/h). The latter mentioned will constitute the objective of the model presented. What continues next is a brief overview of the basis of IFOG performance.

The nonreciprocal phase shift between the two waves in counter-propagation (clockwise and counterclockwise) induced by rotation when both propagate across the sensing coil of optical-fiber, also known as Sagnac effect, is usually given by the expression (see, for instance, [1–10]):

$$\phi_S = \frac{2\,\pi\,L\,D}{\lambda\;c}\,\Omega \qquad (1)$$

being L the total length (m) of the sensing coil, D its diameter (m), Ω the rotation rate (rad/s), and ϕ_S is the phase shift difference (rad), λ and c are the wavelength (m) and the speed of light (m/s) in free space, respectively, of the radiation emitted by the laser source. The proportionality factor that precedes the rotation-ratio is known as the scale-factor (SF) of the gyroscope, and it is a basic constructive constant that depends on geometric and optical parameters of the device. Taking the following initial values for the design: $L = 300$ m, $D = 0.08$ m, and $\lambda = 1310$ nm, a value of 1.86 μrad/(°/h) is obtained for SF. A detailed study of the depolarization mechanism of optical counter-propagated waves within the fiber optic sensing coil can be consulted in [11–17]. The main advantage of the depolarization technique is that this approach allows using a single-mode optical fiber for the sensing coil, with the consequent economic savings on optical components costs of the gyroscope. This design is based on a conventional IFOG structure (interferometric fiber optic gyroscope) with a sinusoidal electrooptic phase modulation and a closed-loop feedback phase modulation realized with classic analogical electronic components, which provides a better stability and linearity of the gyroscope's SF, while using cost-competitive components. The rest of the paper is organized as follows. The next section (Section 2) is focused on the design of optical and electronic sub-systems of the model. Section 3 provides some important calculations and estimations of the performance of the design, and Section 4 shows

http://dx.doi.org/10.5772/intechopen.72592

the simulation results (optical and electronic subsystems). Finally, Section 5 includes a discussion on simulation results, and Section 6 collects the main conclusions of this paper.

2. Sensor design

2.1. Design of the optical system

The components of the optical system of this gyroscope are depicted in **Figure 1**. The light source is a 1310 nm superluminescent diode (SLD) with a Gaussian low ripple spectral profile. For this unit, the commercial reference SLD1024S of Thorlabs was used, with DIL-14 pin assembly package, with FC/APC fiber pigtailing and realized in standard single-mode optical fiber. This unit provides an adjustable optical power up to 22 mW maximum level, although only 5 mW maximum level is needed for the present model. This unit takes an integrated thermistor to perform the temperature control, so that it is possible to obtain the stabilization of the power source on the spectral range. Accordingly with the temperature stabilization, the chip package must not exceed a maximum temperature of 65°C. The directional optical coupler is four ports (2 × 2 configuration), with 50/50 output ratio, realized with fiber-optic side-polished technique, and an insertion loss of 0.60 dB. The linear polarizer placed at the output of the directional input-output coupler is featured in polarization-maintaining fiber (PMF) with a 2.50 m length, insertion loss of 0.1dB, and a polarization extinction ratio (PER) > 50 dB. The integrated optical circuit IOC (integrated optical chip) performs the function of optical directional coupler at the input of the sensing fiber-optic coil (Y-Junction) and also the function of electro-optic phase modulator (PM). In a more advanced design, the linear fiber-optic polarizer

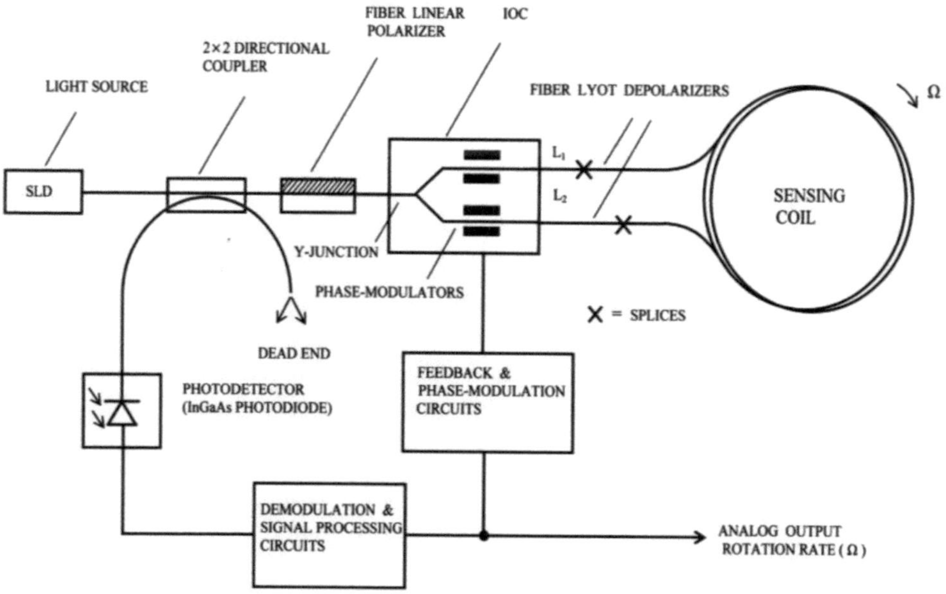

Figure 1. Electro-optical system configuration.

can be replaced by an integrated approach, so that the former remains joined at the input of the IOC wave-guide [18]. This way, the bulk optic polarizer is avoided, which is one important contribution to reduce the whole space occupied by the optical system of the gyroscope.

The chosen PM is electro-optical class. Its electrodes remain parallel to the wave-guide channels obtained by diffusion of Ti on a lithium-niobate ($LiNbO_3$) substrate. The PM zone of the IOC includes two pairs of electrodes placed symmetrically with regard to the central axis of the integrated block. The output ports of the IOC remain connected, respectively, to the heads of the two Lyot depolarizers (both made on PM-fiber), with lengths L_1 and L_2, respectively. These Lyot depolarizers are realized in polarization-maintaining optical fiber (PMF), connecting two segments appropriate lengths, so that the axes of birefringence of both form angles of 45°.

Calculations of Lyot depolarizer lengths are shown next. Calculated lengths L_1 and L_2 of the Lyot depolarizers summarize 26.20 and 52.40 cm, respectively.

$$\acute{c}\text{alculo despolarizadores Lyot}$$

$$\frac{L_D}{L_c} \cong \frac{L_b}{\lambda} \Rightarrow L_D \cong \frac{L_c L_b}{\lambda} \approx \frac{(20\lambda)\left(\frac{\lambda}{B}\right)}{\lambda} = \frac{(20\lambda)\left(\frac{\lambda}{|n_X - n_Y|}\right)}{\lambda}$$

$$L_c \approx 20\lambda = 20 \times \left(1310 \times 10^{-9}\right) = 26.20 \times 10^{-6}[m] = 26.20\,[\mu\text{m}] \quad (= \text{coherence source length})$$

$$L_b = \frac{\lambda}{|n_X - n_Y|} = \frac{1310 \times 10^{-9}}{1 \times 10^{-4}} = 10000 \times \left(1310 \times 10^{-9}\right) = 1.310 \times 10^{-2}[m]$$

$$(= fiber\text{-optical beat length})$$

$$L_D \cong \frac{\left(26.20 \times 10^{-6}\right)\left(1.310 \times 10^{-2}\right)}{1310 \times 10^{-9}} = 26.20 \times 10^{-2}[m](= fiber\text{-optical depolarization length})$$

$$L_1 = L_D = 26.20\,[\text{cm}]$$

$$L_2 = 2L_1 = 2 \times 26.20\,[\text{cm}] = 52.40\,[\text{cm}]$$

$$\theta = 45^{\circ} \ (= \text{angle between main birefringence axis of two fibers at splices})$$

These calculations were realized taking into account a 26.20 μm value for the coherence length of a broadband light source (emitting at 1310 nm wavelength) and a 13.10 mm value for the beat length of optical fiber. The two optical waves CW (clockwise) and CCW (counterclockwise) coming from the sensing coil gather together on the Y-Junction placed at the input of the IOC. The sensing coil consists of 300 m of optical standard single-mode fiber (commercial type SMF28), made by quadrupolar winding on a spool of 8 cm average-diameter, which provides 1194 turns. This optical fiber presents the following structural characteristics: step refractive index, basis material = fused-silica, external coating = acrylate, core diameter = 8.2 μm, cladding diameter = 125 ± 0.7 μm, and external coating diameter = 245 ± 5 μm, with the following optical parameters: n_{core} = 1.467, $n_{cladding}$ = 1.460, NA = 0.143, maximum attenuation = 0.35 dB/km at 1310 nm, h-parameter = 2×10^{-6} m^{-1}, dispersion coefficient ≤ 18.0 ps/(nm × km) at 1550 nm, polarization dispersion coefficient ≤ 0.2 ps/$km^{½}$, birefringence: B = 1.0×10^{-6}.

The chosen PM is electro-optical class. Its electrodes remain parallel to the wave-guide channels obtained by diffusion of Ti on a lithium-niobate ($LiNbO_3$) substrate. The PM zone of the

http://dx.doi.org/10.5772/intechopen.72592

IOC includes two pairs of electrodes placed symmetrically with regard to the central axis of the integrated block. The output ports of the IOC remain connected, respectively, to the heads of the two Lyot depolarizers (both made on PM-fiber), with lengths L_1 and L_2, respectively. These Lyot depolarizers are realized in polarization-maintaining optical fiber (PMF), connecting two segments of appropriate lengths, so that the axes of birefringence of both form angles of 45°.

2.2. Design of the electronic system

In absence of rotation (Ω = 0 rad/s), the transit-time of the two counter-propagated waves across the sensing coil is τ seconds, being its value:

$$\tau = \frac{L}{\left(c/n_{core}\right)} = \frac{n_{core}\, L}{c} \tag{2}$$

With the values of parameters adopted previously for the model design, assuming 1467 for n_{core} value and using 1194 turns of optical-fiber wrapped on standard fiber-optic coil, the resultant value for the transit time is τ = 1.467 μs. On the other hand, the transit time value also determines the value of modulation frequency f_m that must be applied to Phase-Modulator (PM), given by the expression:

$$f_m = \frac{1}{2\,\tau} \tag{3}$$

resulting, for the present design in a calculated value of 340.83 kHz. Equation (3) comes from the condition of maximum amplitude of the bias phase-difference modulation for the optical wave, which is possible to formulate by the following expression:

$$\Delta\phi_{bias}(t) = 2\phi_0 \sin\left(\frac{2\pi f_m \tau}{2}\right) \cos\left[2\pi f_m\left(t - \frac{\tau}{2}\right)\right] \tag{4}$$

The condition of maximum amplitude needs the $2\pi f_m \tau = \pi$ relation to be satisfied (and then, Eq. (3) is accomplished). The block diagram of the electronic scheme for phase modulation and demodulation circuits is represented in **Figure 2**. A closed-loop configuration has been adopted with sinusoidal bias phase modulation and serrodyne feedback phase modulation, taking as initial reference the state-of-the-art of demodulation circuits reported till now [19–24].

However, and this is the novelty, it has been changed the structure of feedback chain, adding now a new design of analogical integrator which incorporates one FET transistor (2N3848) as it is depicted in **Figure 3**. The function of this transistor is realizing periodically the shortcut of the capacitor therefore nulling instantaneously the voltage on feedback branch of integrator OPAMP. The time period for shortcut FET transistor is driving by the value of V_{gate} voltage, which, in turn, is controlled by an unstable Flip-Flop circuit.

Referring to **Figure 3**, block #7 generates a linear ramp voltage V_γ on its output, and this ramp resets each time period driving by V_{gate} voltage. In this way, a resultant serrodyne-wave voltage is easily generated at the output of this integrator circuit, obtaining finally the same

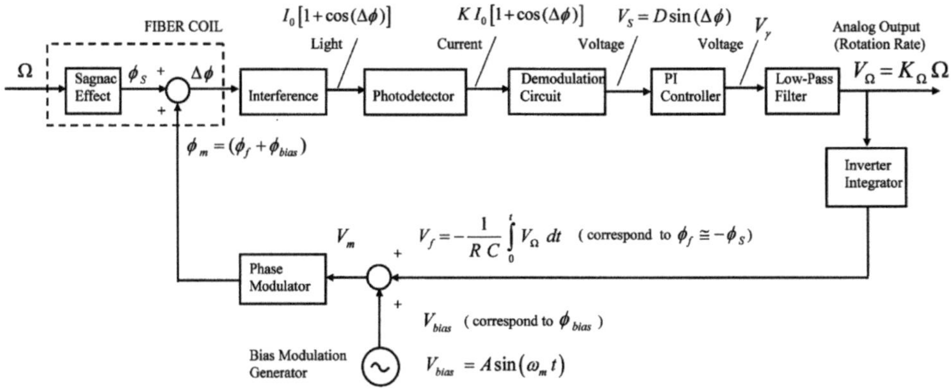

Figure 2. Analog closed-loop scheme for feedback phase-modulation configuration.

intended sawtooth voltage on feedback phase modulation chain as reported on previous designs by literature [25–27]. Working as feedback phase modulation signal, the analogical serrodyne wave presents two important advantages with regard to the sinusoidal-one: (a) it is possible to generate the serrodyne wave easily by means of a simple integrator circuit (Miller integrator) with simple and low-cost electronic components; (b) the phase cancellation process inside the control loop becomes simpler and more efficient.

In accordance with the interference principle, the light intensity at the photodetector optical input presents the following form (for sinusoidal phase-modulation):

$$I_d(t) = \frac{I_0}{2}\left[1 + \cos(\Delta\phi)\right] = \frac{I_0}{2}\left\{1 + \left[J_0(\phi_m) + 2\sum_{n=1}^{\infty} J_{2n}(\phi_m)\cos(2\,n\,\omega_m\,t)\right]\right.$$
$$\left.\cos\phi_S - 2\sum_{n=1}^{\infty} J_{2n-1}(\phi_m)\sin\left[(2\,n-1)\omega_m\,t\right]\sin\phi_S\right\} \tag{5}$$

being J_n the Bessel-function of the first kind of nth order. Here, $\Delta\phi$ represents the effective phase-difference of the two counter-propagating optical waves on sensing coil. This value results from the combined action of the phase-modulation process ($\phi_m = \phi_{bias} + \phi_f$) and the Sagnac phase shift induced by the rotation-rate (ϕ_S). The output signal of the photodetector, in photocurrent form, is proportional to the light intensity at its optical input. This photocurrent signal is converted to voltage with a transimpedance amplifier that is placed at the entry of demodulation circuit. The demodulation circuit takes the task of extracting the information of the Sagnac rotation-induced phase shift (ϕ_S). The corresponding voltage signal at its output (V_S) scales as sine-function of the effective Sagnac phase difference. The PI controller performs an integration of the V_S signal in time domain, so that a voltage signal (V_γ) is obtained; this signal growing almost linearly with the time. This latter signal is filtered by means of a low pass filter so that the corresponding output signal on voltage form (V_Ω) is a DC voltage value that is possible to consider to be almost proportional to the gyroscope rotation-rate Ω (when $\sin\phi_S \approx \phi_S$). Therefore, the analogical output voltage signal V_Ω constitutes the measurement of

http://dx.doi.org/10.5772/intechopen.72592

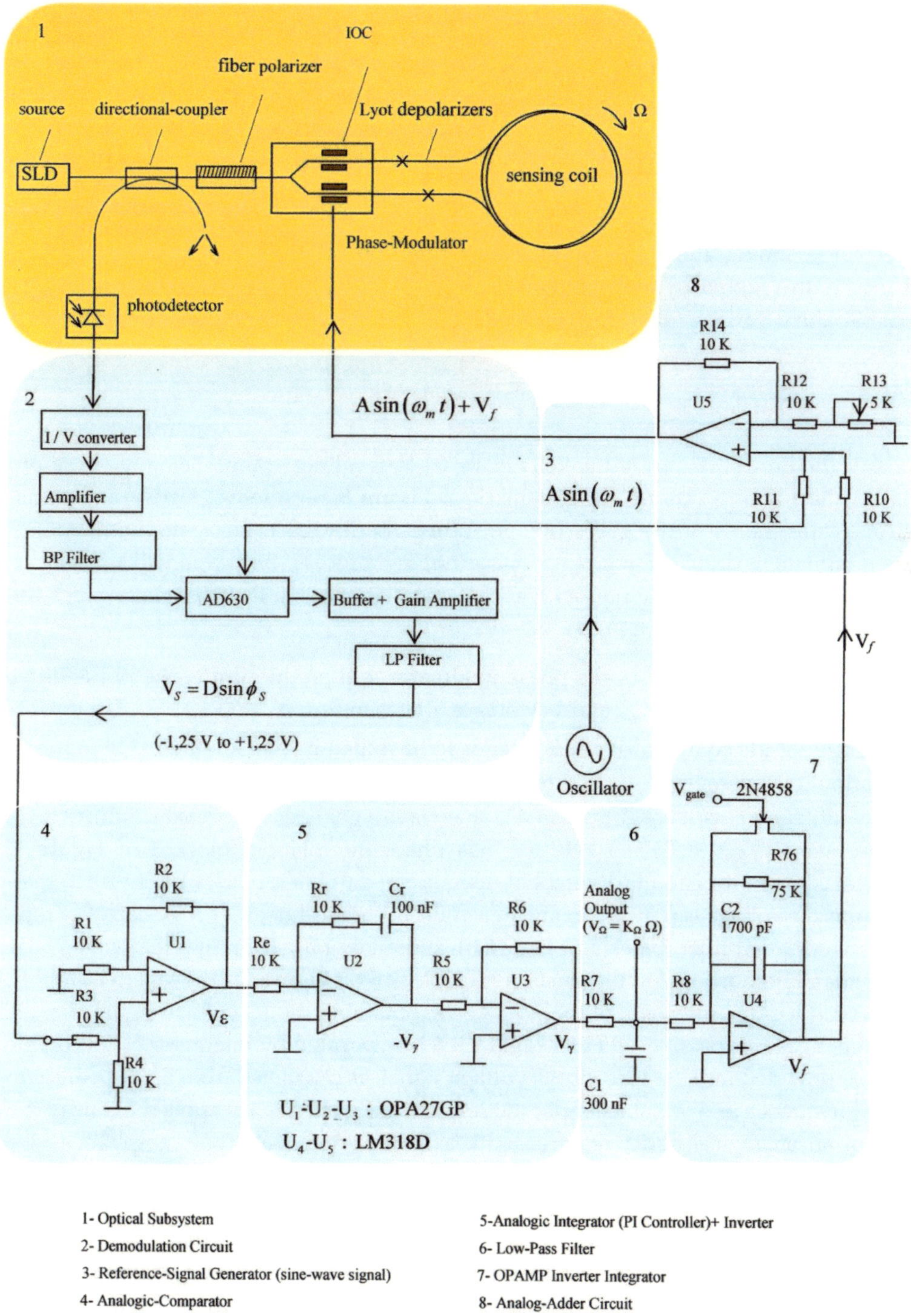

Figure 3. Block-diagram for electro-optical system and phase-sensitive-demodulation (PSD). The IFOG model is closed-loop configuration with sinusoidal-bias and serrodyne-feedback phase-modulations.

the rotation rate of the system. The control system, as a whole, acts as the principle of phase nulling. The phase-nulling process consists of generating a phase displacement ($\phi_m = \phi_{bias} + \phi_f$) in such a way that the phase-difference ϕ_f associated with the voltage output signal (V_f) is equal and with opposite sign with regard to the Sagnac phase-shift induced by the rotation rate, i.e., $\phi_f = -\phi_S$. To achieve this, the feedback phase modulation circuit holds a sample of the output signal V_Ω. Note that this voltage signal is obtained at the end of low pass filter (Block 6 on **Figure 3**) and is proportional to rotation-rate Ω. An integration operation is needed for obtaining a linear ramp voltage to apply on phase modulator. Then, it integrates and inverts this signal by means of an operational integrator-inverter circuit, turning this signal into the following form:

$$V_f = -\frac{1}{RC}\int_0^t V_\Omega \, dt \tag{6}$$

This way, the time variation of the voltage signal V_f is a linear ramp, being its slope proportional to the rotation rate of the system (V_Ω). **Figure 3** represents clearly the optical and electronic subsystems of the gyroscope, including the feedback phase-modulation and bias phase-modulation circuits for getting phase nulling process, both applied together to PM (Phase-Modulator). Referring now to **Figure 3**, then latter being the reference voltage for bias phase-modulation, see **Figure 2**), i.e., $V_m = V_{bias} + V_f$.

Therefore, the output signal of the phase modulator will be the sum of the phase-difference signals associated with the V_{bias} and V_f voltages, that is to say: $\phi_m = \phi_{bias} + \phi_f$. The error signal at the output of the comparator (V_ε $\Delta\phi$) tends to be nulled in average-time, due to the phase-cancellation (the average-time of the reference bias phase-modulation ϕ_{bias} is 0). The feedback phase-modulation circuit consists of an AC sine-wave signal generator that produces a voltage reference signal V_{bias} at 340.83 kHz for bias phase modulation (block 3 of **Figure 3**), an analogical comparator circuit (differential-operational-amplifier, block 4 of **Figure 3**) that generates an error voltage signal V_ε, an analogical Proportional-Integral (PI) Controller followed by one inverter-amplifier (block 5 of **Figure 3**), and a low-pass (LP) filter that yields a DC V_Ω voltage signal proportional to the rotation-rate (block 6 on **Figure 3**). The inverter-amplifier on block 5 produces the inversion of the $-V_\gamma$ signal, obtaining the V_γ voltage signal. The DC V_Ω output voltage after passive LP Filter on block 6 is integrated by integrator circuit on block 7 and then converted into the V_f feedback voltage signal, as calculated from Eq. (6), consisting on constant frequency and variable amplitude serrodyne wave which is applied to one of the two inputs of an analogical adder featured with a noninverter operational amplifier (the other input is connected to AC signal generator, block 8 on **Figure 3**). Therefore, the voltage output signal of this analogical adder is the V_m voltage signal that realizes the sum of the V_{bias} and V_f voltage signals, as described previously. **Figure 4** represents the detail block diagram of electronic scheme for detection and phase sensitive demodulation (PSD) circuits. It consists basically 12 functional blocks: (1) photodetector simulated output current, (2) transimpedance amplifier (current to voltage converter), (3) low-pass Filter (LP-Filter, f_c = 800 kHz), (4) band-pass-filter (BP Filter, f_{center} = 340.83 kHz), (5) analogical multiplier (AD630), (6) sinusoidal Oscillator (f = 340.83 kHz), (7) analogical inverter amplifier, (8) low-noise adjustable-gain

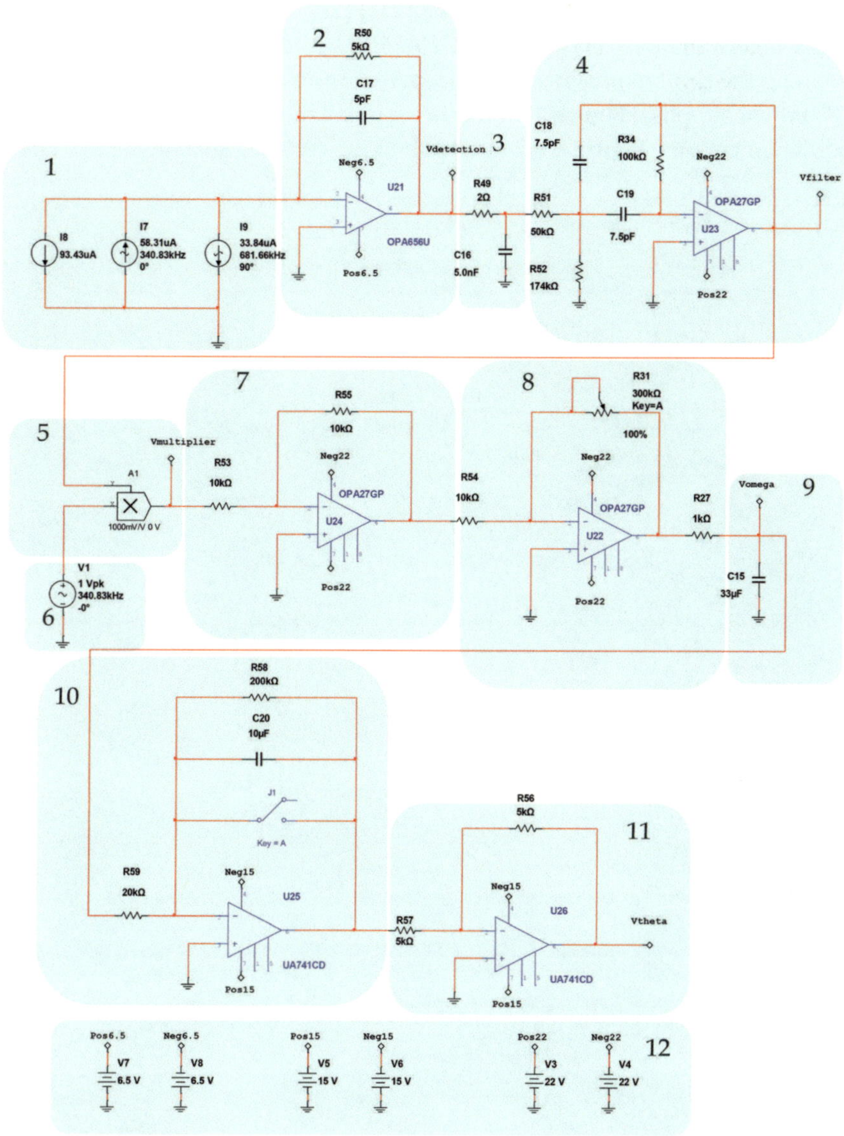

1- Photodetector current simulation
2- Transimpedance amplifier
3- LP Filter (fcutting = 800 kHz)
4- BP Filter (fcenter = 340.83 kHz)
5- Analog Multiplier (AD630)
6- Oscillator (sinusoidal generator f = 340.83 kHz)
7- Analog-inverter amplifier
8- Low noise adjustable gain amplifier
9- Passive LP Filter (fcuttting = 4,82 Hz)
10- Analog Inverter-Integrator
11- Analog-Inverter Amplifier
12- DC Power Supply

Figure 4. Detection and phase-sensitive-demodulation (PSD) circuits.

amplifier, (9) Low-Pass-Filter (LP Filter, f_c = 4.82 Hz) [28], (10) analogical integrator filter (for rotation-angle determination), (11) inverter OPAMP; the output voltage V_{theta} of this inverter allows obtaining the draft experienced by the system from a certain time (initialization time); and (12) DC power supplies. **Figure 5** represents in detail the analog PI controller and feedback phase modulation circuits. **Figure 6** represents the V_{gate} voltage signal generator circuit.

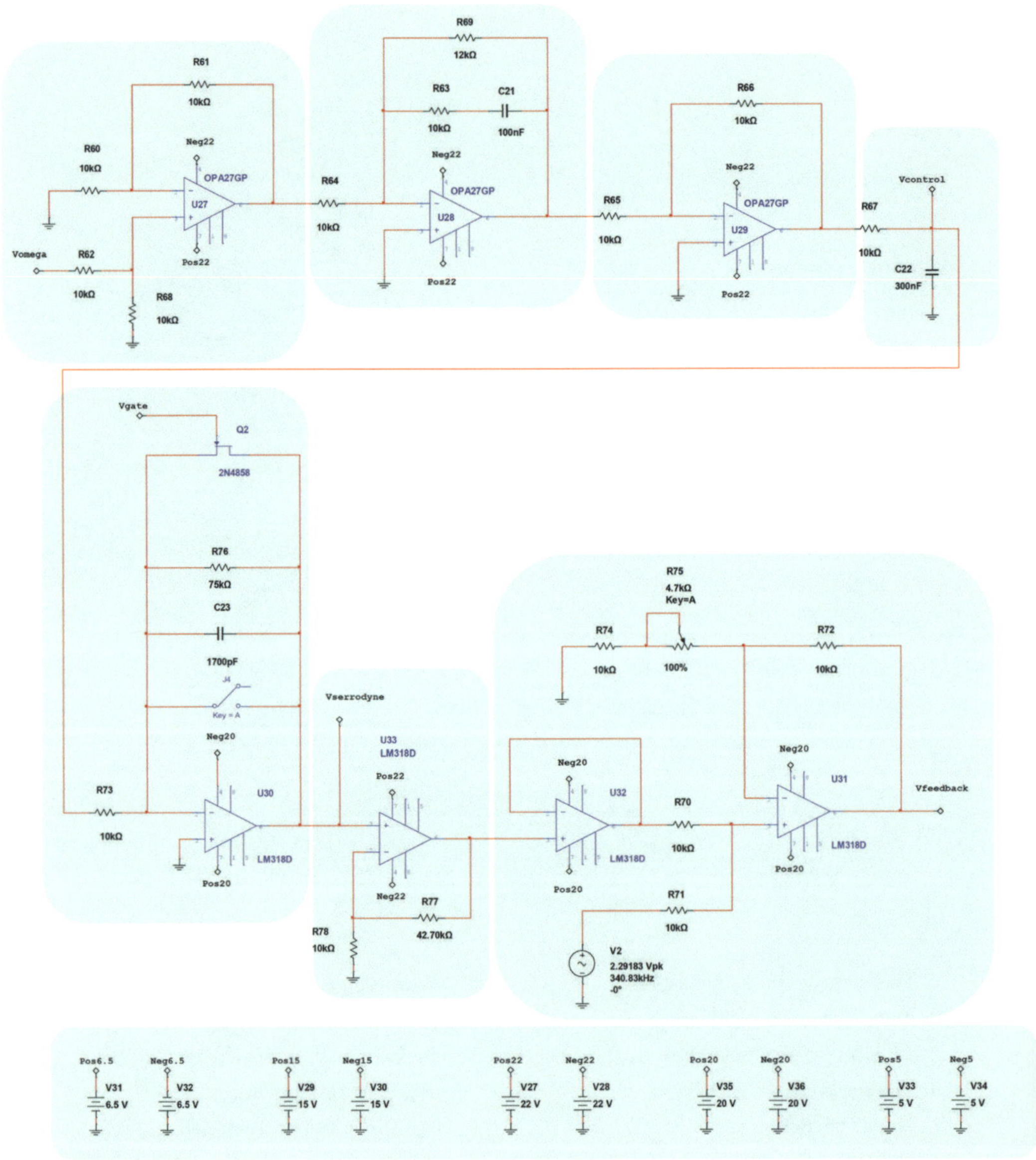

Figure 5. Analog controller circuit (includes blocks #1, #2, #3, and #4) and Serrodyne feedback phase-modulation circuit (includes blocks # 5, #6, and #7).

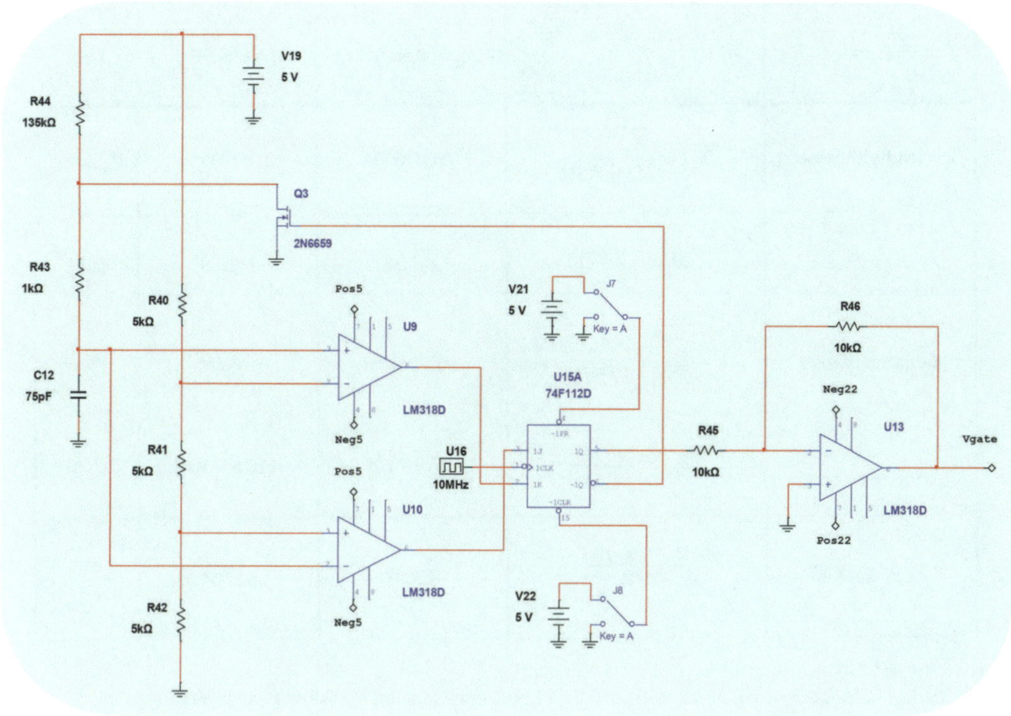

Figure 6. V_{gate} voltage signal generator (Astable pulse generator + J-K Flip-flop + analog inverter).

3. Calculations and estimations

This design has been simulated using Matlab-Simulink™ The MathWorks® and MultiSim™ National Instruments®. (Note that Synopsys® OptSim™ original version software only allows implementing APD-type photodetectors on optical circuit design, consequently an APD-PIN equivalent current-conversion will be necessary for connecting the simulation results to IFOG prototype designed in this article, which owes PIN photodetector). The open-loop scale factor K_0 can be calculated (being $c \approx 3 \times 10^8$ m/s the speed of light in vacuum) as:

$$K_0 = \frac{LD}{\lambda\, c} \tag{7}$$

The parameters of the model were chosen as fiber coil length L = 300 m, fiber coil diameter D = 80 mm, number of turns in the coil N = 1194, and light source wavelength λ = 1310 nm. The average optical-power at detector optical-input is P_d = 145.61 μW, and the responsivity of the InGaAs photodetector is R = 0.68678 μA/μW. The beat length of the optical fiber, L_b, can be calculated from its optical birefringence (B) as:

$$L_b = \frac{\lambda}{B} = \frac{\lambda}{|n_x - n_y|} \tag{8}$$

Parameter	Calculation Formula	Calculated Value	Estimated Value	Unit
Sensitivity threshold	$\Delta\Omega = \frac{2}{K_0}\sqrt{\frac{e}{P_d R t}}$	0.05193796	0.05193936	[°/hour]
Dynamic Range	$20\log\left(\frac{\Omega_{max}}{\Omega_{min}}\right)$	101.38	101.38	[dB]
	$\Omega_{max} = \frac{\lambda c}{12LD}$	±78.185	±78.185	[°/s]
	$\Omega_{min} \approx \frac{\sqrt{h L_b}}{LD}$	$\pm 1.164\times10^{-5}$	$\pm 1.164\times10^{-5}$	[°/s]
Scale Factor	$SF = \frac{2\pi LD}{\lambda c}$	0.3837	0.3664	$\left[\frac{rad}{\left(rad/s\right)}\right]$

Table 1. Performance parameters of the designed IFOG prototype (analog closed-loop configuration).

where n_x and n_y are the refractive indexes of the two orthogonally polarized modes along the x and y directions. For this model, the following performance parameters have been analyzed: sensitivity threshold [29], dynamic range, and scale factor (SF) [30]. The values calculated (using the formulae) and estimated (by the results of the simulations) for such parameters are shown in **Table 1**. In this table, the third column shows the value calculated directly by the formula, and the fourth shows estimated results from the optical and electronic simulations.

The threshold sensitivity considers the SNR at photodetector optical input provided by the optical simulation, and the dynamic range and scale factor are determined by the sine function nonlinearity (assuming the maximum value $\phi_S = \pm 7/6$). In the formulae, h is the h-parameter of the optical-fiber and t is the average integration time.

4. Simulation results

Three different kinds of computer simulations have been realized. First, the control system simulation has been realized using Matlab-Simulink™ for determining the 2% settling-time t_s of the complete electro-optic system. Second, an optical system simulation has been realized using Synopsys® OptSim™ for obtaining the optical interference signal at the PIN photodetector optical input and its main and representative values: average optical power and Signal-to-Noise-Ratio (SNR). Third and finally, the electronic circuit simulation made with MultiSim™ National Instruments® to obtain the V_Ω DC voltage as image of the rotation rate of the system and then for obtaining the output graph-response of gyroscope unit.

Figure 7 represents the parametric model of IFOG′ electro-optical system. It is depicted as a parameterized block diagram corresponding to electro-optical system equivalent to the gyroscopic sensor. Here, it is taking into account the values of the parameters identified on its optical and electronics circuits. The system's step-response curve (obtained with Matlab-Simulink®) is shown in **Figure 8**. A settling time t_s (2%) of 1.39 ms is obtained. This value can be used to estimate a value for the initialization time of the final gyroscope unit. Results of optical subsystem simulation (realized by means of OptSim™ software) are represented on **Figures 9–14**. **Figure 9** represents the optical schematic circuit of the designed model for

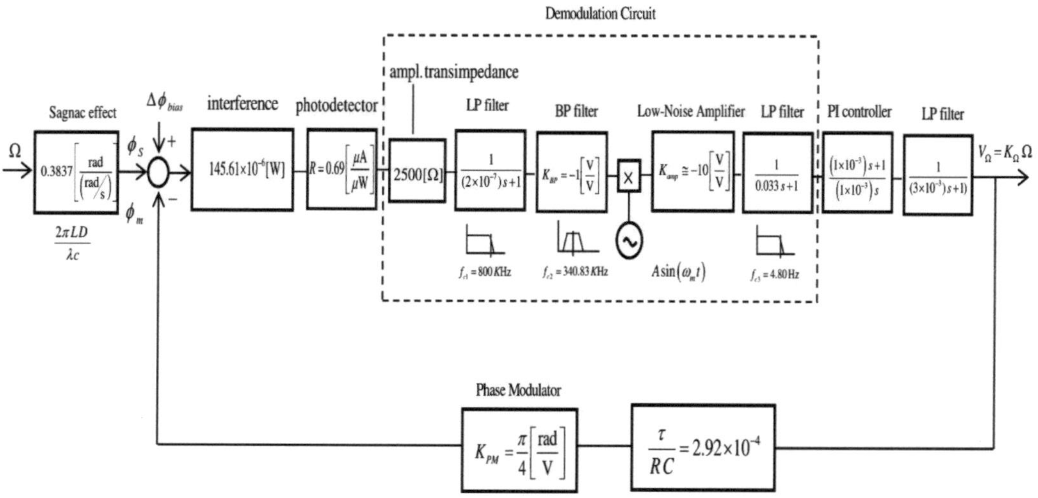

Figure 7. Parameterized block-diagram of the designed IFOG model considered as a continuous linear system (parameters: L = 300 m, D = 0.08 m, λ = 1310 nm, P_d = 145.61 μW, R = 0.69 μA/μW).

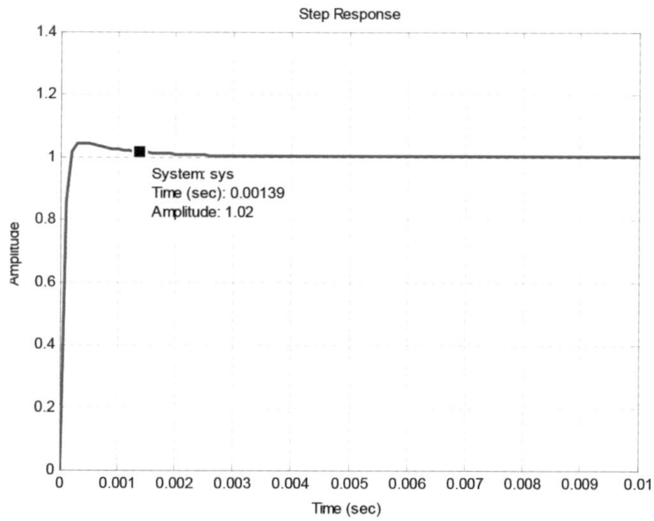

Figure 8. Time-response of the complete closed-loop system obtained for step-stimulus input signal (for step-stimulus input the t_s 2% settling time obtained is 1.39 ms).

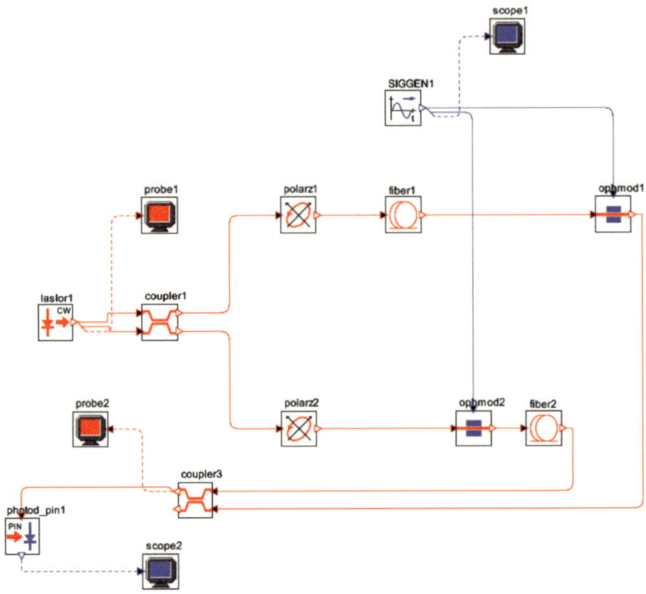

Figure 9. Optical circuit setup of the designed IFOG gyroscope for computer simulation (OptSim™).

obtaining its optical performance. **Figure 10** represents the sinusoidal electrical signal provided by the AC signal generator and applied to the PM (Phase-Modulator) as bias phase-modulation signal. **Figure 11** represents the optical-power spectrum at the photodetector optical input (central frequency is 288.844 THz).

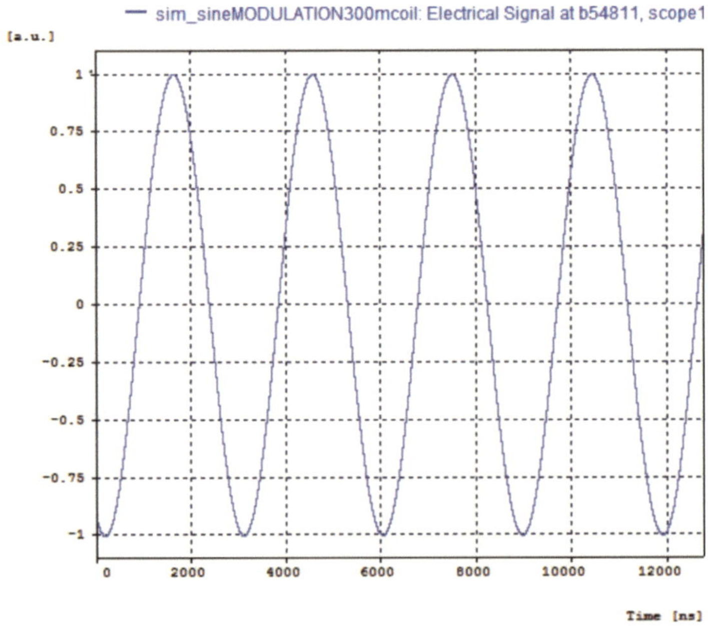

Figure 10. Sinusoidal voltage signal provided by AC signal generator and applied to PZT phase-modulator (PiezoZeramicTube).

http://dx.doi.org/10.5772/intechopen.72592

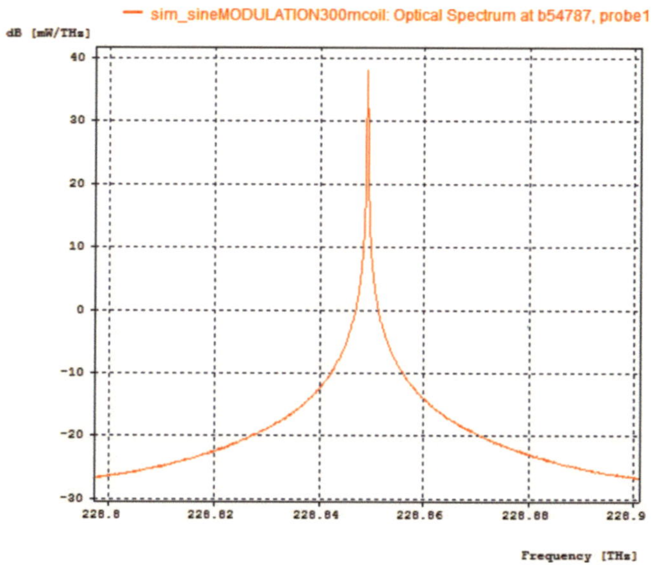

Figure 11. Optical-power spectrum obtained at photodetector optical-input.

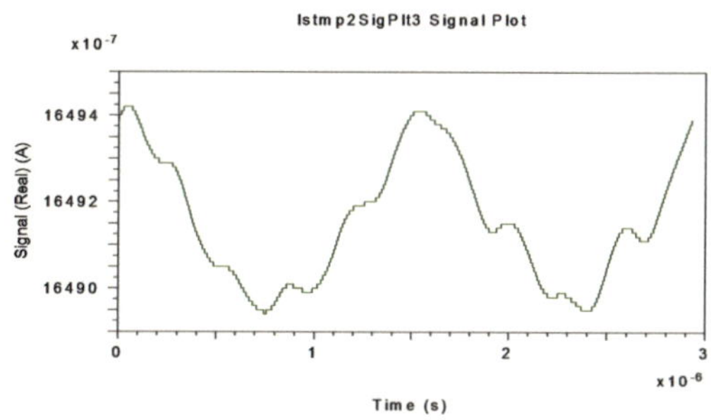

Figure 12. Interferometric signal at electrical output of APD-equivalent-photodetector (after electrical BP filter, f_{center} = 340.83 kHz) when $\Omega = \pm\, 10°/s$ is applied to system.

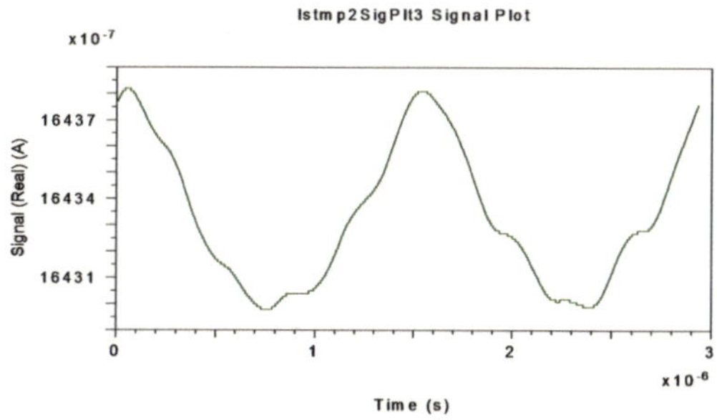

Figure 13. Interferometric signal at electrical output of APD-equivalent-photodetector (after electrical BP filter, f_{center} = 340.83 kHz) when $\Omega = \pm\, 20°/s$ is applied to system.

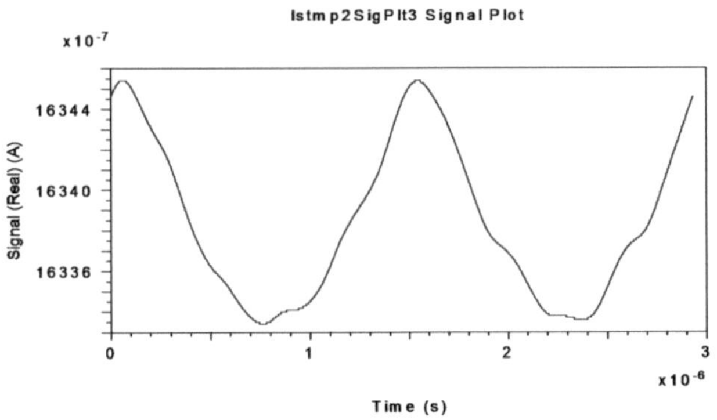

Figure 14. Interferometric signal at electrical output of APD-equivalent-photodetector (after electrical BP filter, f_{center} = 340.83 kHz) when Ω = ± 30°/s is applied to system.

Considering an input value of 210 μW as the average optical power value providing by light source, 145.61 μW were obtained at the optical input of photodetector, which means a power loss of −9.837808 dBm. Calculation of photon-shot-noise photocurrent at photodetector, taking into account 100 μA for average real value of photocurrent at its electrical output, is the following:

$$I_{sn} = \sqrt{\frac{e^2\,q\,\lambda}{h\,c} P_{\text{max-detector}} \Delta f} = \sqrt{\frac{\left(1.6\times10^{-19}\right)^2\times0.65\times\left(1310\times10^{-9}\right)}{\left(6.624\times10^{-34}\right)\times\left(3\times10^{8}\right)}\times\left(100\times10^{-6}\right)\times1}$$

$$= \sqrt{1.096940\times10^{-23}} = 3.312008\times10^{-12}\quad \text{A}$$

Note that lower the photon-shot-noise photocurrent value, lower is the threshold sensitivity of IFOG sensor and therefore also higher is its accuracy. On the other hand, it is needed to say that for low level of optical power coupled into photodetector, the main optical noise source of IFOG-sensor is photon shot noise (excess RIN can be neglected). This way, in accordance with photon-shot-noise photocurrent above calculated, the threshold sensitivity of gyro sensor (that is to say, the minimum rotation-rate which is able to measure) can be calculated as shown next (this value is collected in **Table 1**):

$$\Omega_{\text{lim}} \cong \left(\frac{h c^2}{\pi e q L D P_{\text{max}}}\right) I_{sn} = \left[\frac{6.624\times10^{-34}\times\left(3\times10^{8}\right)^2}{\pi\times\left(1.6\times10^{-19}\right)\times0.65\times300\times0.08\times\left(100\times10^{-6}\right)}\right]$$

$$\times\,3.312\times10^{-12} \approx 2.5180234\times10^{-7}\ [\text{rad/sec}]$$

$$= 2.5180234\times10^{-7}\times\left(\frac{180^{\circ}}{\pi}\right)\times\left(\frac{3600\ \text{sec}}{1\ \text{hour}}\right) = 0.05193796\ \left[{}^{\circ}/_{\text{hour}}\right]$$

Figures 12–14 represents the electrical interferometric signal (in APD photo-current form, after electrical BP filtering, f_{center} = 340.83 kHz) detected by an APD equivalent photodetector, when Ω = ± 10°/s, Ω = ± 20°/s, and Ω = ± 30°/s, respectively, are applied to the system. This is because the block-mode simulation only offers measurements realized by an APD equivalent photodetector as optical output of the system. The average mean values of APD photocurrents are, respectively, 1649.20 μA, 1643.30 μA and 1633.80 μA, which correspond to 99.873 μA, 99.515 μA and 98.940 μA for PIN-equivalent photodiode. Note that in this interval, the average current decreases almost linearly as rotation-rate increases linearly. These curves agree with theoretical interferometric curves as calculated on optical input photodetector.

The results of electronic circuits' simulation (realized by MultiSim™ software) collect the waveforms voltage on the following test-point voltage: $V_{detection}$, V_{filter}, $V_{multiplier}$, V_{theta}, $V_{serrodyne}$, and V_{gate} (referring to **Figures 4–6**). All these values are obtained on electronic circuits when $\Omega = +30°/s$ rotation-rate is applied to system and are gathered on **Figures 15–20. Figure 15** shows the detected output voltage after transimpedance amplifier ($V_{detection}$, see **Figure 4**). **Figure 16** represents the output voltage after BP Filter (V_{filter}, see **Figure 4**). **Figure 17** represents output voltage after analog Multiplier ($V_{multiplier}$, **Figure 4**). **Figure 18** represents output voltage after Angle analog integrator (V_{theta}, **Figure 4**). **Figure 19** represents output voltage after analog integrator ($V_{serrodyne}$: a sawtooth-voltage with constant frequency and variable amplitude, this amplitude depending on V_{Ω} voltage value). Finally, **Figure 20** represents V_{gate}

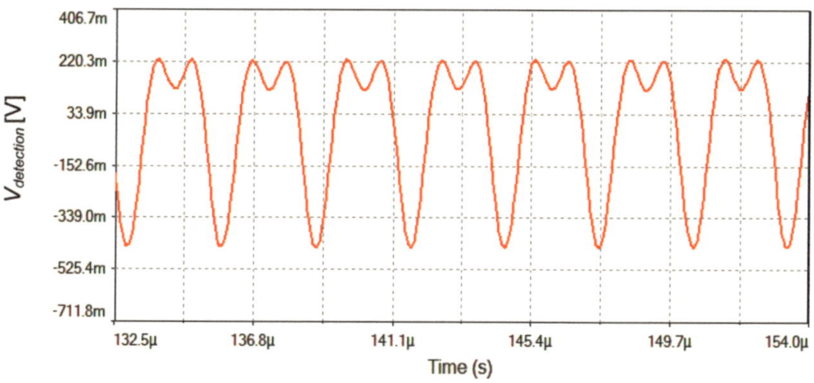

Figure 15. $V_{detection}$ voltage signal (after transimpedance amplifier) for $\Omega = +30°/s$ when $\Omega = \pm 30°/s$ is applied to system.

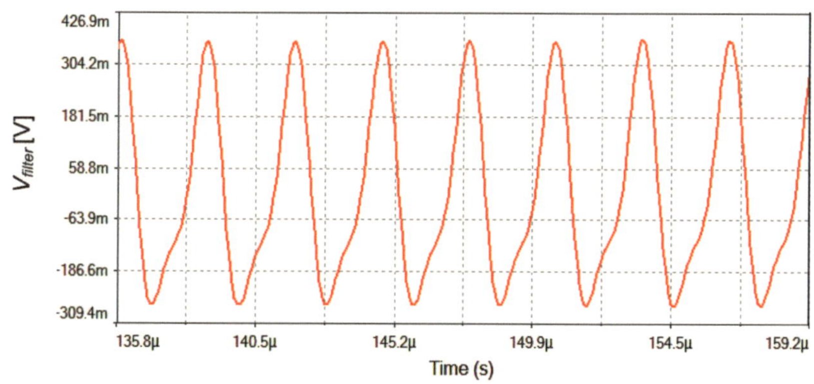

Figure 16. V_{filter} output voltage after BP filter for $\Omega = +30°/s$.

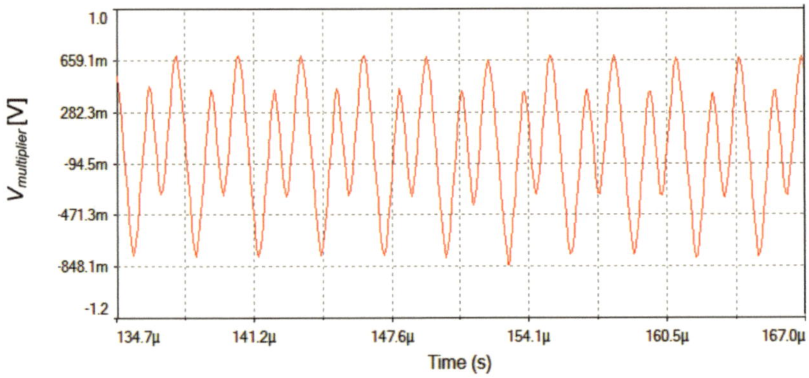

Figure 17. $V_{multiplier}$ output voltage after AD630 analog multiplier for Ω = + 30°/s.

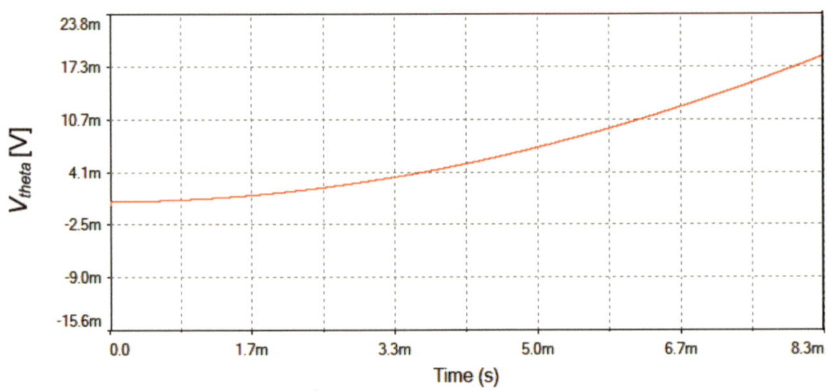

Figure 18. V_{theta} output voltage after angle analog integrator for Ω = + 30°/s.

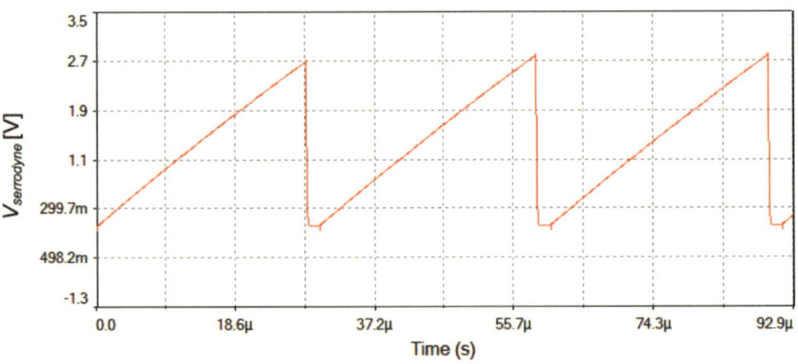

Figure 19. $V_{\text{serrodyne}}$ at analog integrator output (feedback-voltage signal to phase-modulator) for Ω = + 30°/s.

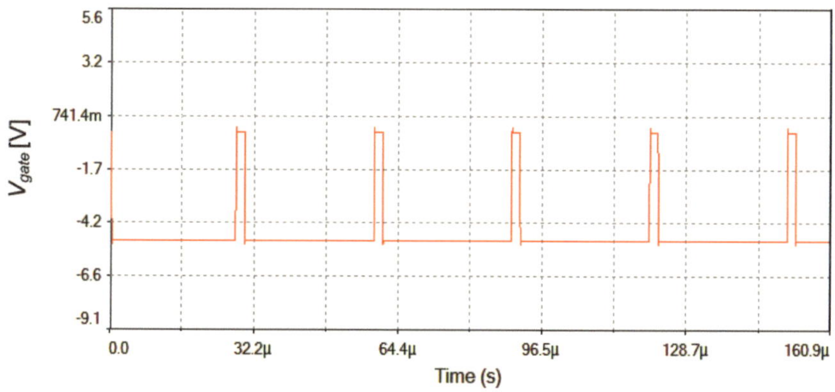

Figure 20. V_{gate} voltage generated by pulse generator circuit (fixed frequency $f = 32.59$ kHz).

generated by pulse generator circuit and applied to gate of J2 N4858 FET transistor (see the circuit on **Figure 6**).

The expansion of Eq. (5) with only the contribution of first two time-component harmonics allows obtaining an approximate value for detected $I_d(t)$ photo-current. The result of this approximation is:

$$I_d(t) \cong \frac{I_0}{2}\left[1 + J_0(\phi_m)\ \cos\phi_S\right] + I_0 J_2(\phi_m)\ \cos(2\omega_m t)\ \cos\phi_S - I_0 J_1(\phi_m)\ \sin(\omega_m t)\ \sin\phi_S \quad (9)$$

being I_0 the maximum value of detected photo-current and ϕ_m the amplitude of differential phase-modulation. Assuming the $\phi_m = 1.80$ value, this value corresponding to the maximum value of function $J_1(\phi_m)$, the following Bessel functions calculations are obtained:

$$J_0(1.80) \cong 0.33999, J_1(1.80) \cong 0.58150 \text{ and } J_2(1.80) \cong 0.30611.$$

then, taking into account 100 μA as the DC average detected photodetector-current and after some numerical adjusts, Eq. (9) yields the following analytical value:

$$I_d(t) \cong 74.63\left[1 + 0.34\ \cos\phi_S\right] + 45.69\ \cos(2\omega_m t)\ \cos\phi_S - 86.79\ \sin(\omega_m t)\ \sin\phi_S \quad [\mu A] \quad (10)$$

This analytical expression allows to calculate for every rotation-rate Ω value (i.e., ϕ_S Sagnac phase shift) the DC term and the first and second harmonics terms. These terms can later be introduced as current DC and AC generators on MultiSim™ circuit simulation program (block 1 on **Figure 4**). By this means, V_Ω can be measured on simulated circuit (see **Figure 4**), so that a table with V_Ω value versus Ω [°/s] value can be made. **Table 2** lists the correlation data obtained from the electronic demodulation circuit for the measured output-voltage signal V_Ω [mV] versus input rotation rate Ω [°/s] of the system for full dynamic range (0 to ±78.19 ≈ ±80 [°/s]) with a step of 10°/s. **Figure 21** shows the graphic relationship between both variables corresponding to mentioned data table. After performing the appropriate calculations, taking into account the theoretical value of the SF (Scale Factor) of the gyroscope that appeared on **Table 1**, a linear

Ω [°/s]	0	±10	±20	±30	±40	±50	±60	±70	±80
V_Ω [mV]	0	±305	±613	±931	±1280	±1609	±1970	±2337	±2708
$(V_\Omega)_{lin}$ [mV]	0	±338.7	±677.5	±1016	±1355	±1694	±2032	±2371	±2710
$\|\Delta(V_\Omega)\|$ [mV]	0	33.73	64.46	85.19	74.92	84.65	62.38	34.11	1.834
$\|\Delta(V_\Omega)/(V_\Omega)_{lin}\|$ %	0	9.959	9.514	8.385	5.529	4.997	3.070	1.439	0.068

Table 2. Output data and linear fitting for 0 ÷ ±80°/s (full dynamic range), step = 10°/s.

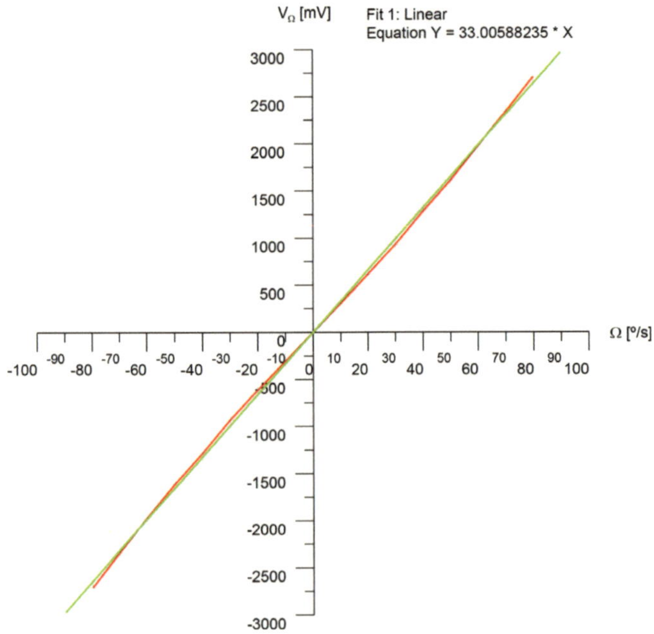

Figure 21. Output response curve V_Ω [mV] versus Ω [°/s] (in red color) and linear fitting (in green color) for 0 to ±80°/s (full dynamic range).

function can be obtained for the best fitting of the output response curve. This way, it is possible to evaluate the linearity of this graphic and then assuming this value as scale factor linearity of the IFOG-model. **Tables 3** and **4** are constructed to obtain detailed scale-factor linearity analysis in more restricted dynamic ranges.

Ω [°/s]	0	±1	±2	±3	±4	±5	±6	±7	±8	±9	±10
V_Ω [mV]	0	±30	±59	±90	±120	±150	±180	±210	±240	±270	±305
$(V_\Omega)_{lin}$ [mV]	0	±30.13	±60.26	±90.39	±120.52	±150.65	±180.78	±210.91	±241.04	±271.17	±301.30
$\|\Delta(V_\Omega)\|$ [mV]	0	0.13	1.26	0.39	0.52	0.65	0.78	0.91	1.04	1.17	3.70
$\|\Delta(V_\Omega)/(V_\Omega)_{lin}\|$ %	0	0.431	2.091	0.431	0.431	0.431	0.431	0.431	0.431	0.431	1.228

Table 3. Output data and linear fitting for 0 ÷ ±10°/s (restricted dynamic range), step = 1°/s.

Ω [°/s]	0	±0.1	±0.2	±0.3	±0.4	±0.5	±0.6	±0.7	±0.8	±0.9	±1.0
V_Ω [mV]	0	±3	±6	±9	±12	±15	±18	±21	±24	±27	±30
$(V_\Omega)_{lin}$ [mV]	0	±3	±6	±9	±12	±15	±18	±21	±24	±27	±30
$\lvert\Delta(V_\Omega)\rvert$ [mV]	0	0.0	0.0	0.0	0.0	0.0	0.0	0.0	0.0	0.0	0.0
$\lvert\Delta(V_\Omega)/(V_\Omega)_{lin}\rvert$ %	0	0.0	0.0	0.0	0.0	0.0	0.0	0.0	0.0	0.0	0.0

Table 4. Output data and linear fitting for 0 ÷ ±1 °/s (restricted dynamic range), step = 0.1°/s.

Table 3 shows data obtained for 0 to ±10 [°/s] dynamic range with a step of 1°/s, and **Figure 22** shows the corresponding V_Ω versus Ω graphical representation. **Table 4** shows data obtained for 0 to ±1 [°/s] dynamic range with a step of 0. 10°/s, and **Figure 23** shows the corresponding V_Ω versus Ω graph. **Tables 2–4** also includes the values $(V_\Omega)_{lin}$ [mV] of the correspondent linear fitting, the module of the differential values $\Delta(V_\Omega)$ [mV], and the module $\lvert\Delta(V_\Omega)/(V_\Omega)_{lin}\rvert$ of the ratio values.

The $\Delta(V_\Omega)$ [mV] value is defined as:

$$\Delta V_\Omega = V_\Omega - (V_\Omega)_{lin} \ [mV] \tag{11}$$

from correlation values of both curves (output data curve and linear fitting curve), it can be determined the non-linearity percentage coefficient of the SF, defined as the percentage of the standard deviation, which can be calculated by the following expression:

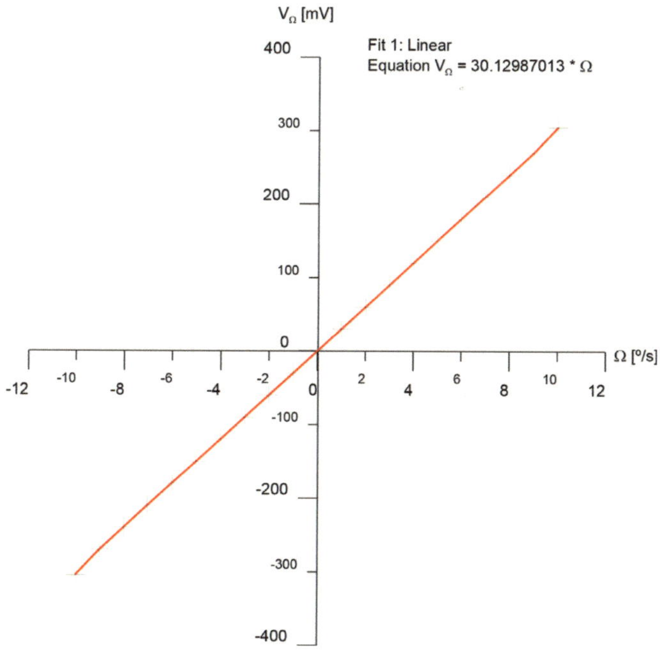

Figure 22. Output response curve V_Ω [mV] versus Ω [°/s] (in red color) and linear fitting (in green color) for 0 to ±10°/s (restricted dynamic range).

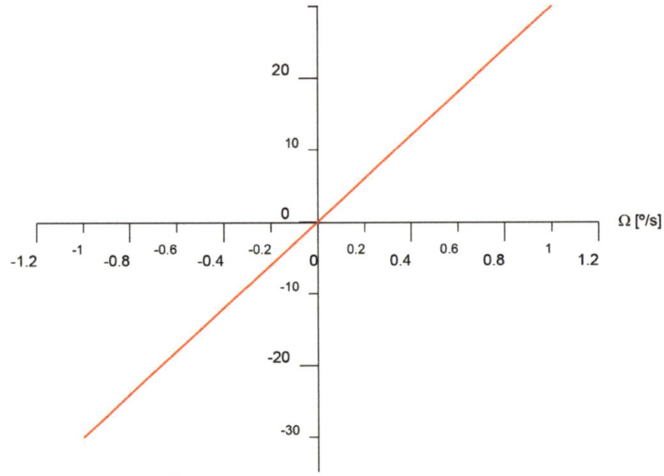

Figure 23. Output response curve V_{Ω} [mV] versus Ω [°/s] and linear fitting (in green) for 0 to ±1°/s).

$$SF - NonLinearity(\%) = \sqrt{\frac{1}{N}\sum_{n=1}^{N}\left|\frac{(\Delta V_{\Omega})_i}{(V_{\Omega\ lin})_i} \times 100\right|^2} \qquad (12)$$

so that in our case, for full dynamic range with N = 17 and taking the values obtained from **Table 2**, this expression yields a value of 4.386%. For **Table 3** Eq. (12) reaches the value of 0.838% and for **Table 4** its value is 0.0%. This values agree with these obtained for commercial IFOG units of similar characteristics.

5. Discussion of simulation results

The results obtained for the performance parameters of the gyroscope model designed in this article (threshold sensibility = 0.052°/h, dynamic range = ±78.19°/s, scale factor nonlinearity = 4.836%) are sufficient for intermediate grade gyroscopic applications, such as stabilization, pointing, and positioning of mobile platforms or inertial-navigation systems for terrestrial robots and automotive vehicles [31–36].

The effects of the different types of optical noise [37–50] which take place are not critical in the specific design of this sensor, since its operation works in a medium level of optical-power and the signal-to-noise-ratio (SNR) is relatively high at photodetector's optical-input (SNR > 100 dB). The most important type of optical-noise for this sensor is photon-shot-noise on photodetector, with a 3.31 pA noise-equivalent-current value, this value being much less than 100 μA that is the average photocurrent value for photodetector electrical output signal, in zero rotation-rate conditions. This type of noise is not susceptible of correction, since it owes to intrinsic quantum-mechanical mechanism of photoconductivity (electron–hole production by photonic shoot).

The relative intensity noise (RIN) is an important issue in this design, since it works at a medium-level of average optical-power coupled to photodetector optical-input (145.61 μW average optical-power value). This type of noise stems from two causes: (1) the two interfering optical waves do not come to photodetector with the same optical power level, due to polarization crosstalk between the two orthogonal polarizations states along the entire length of sensing fiber-coil (due to fiber-birefringence phenomenon), and (2) the light source is low-coherence (broadband source), thereby producing several beat wavelengths, which add at photodetector optical-input, causing a variation in relative intensity on every point of photodetector's response-curve. This noise can be minimized by reducing the optical power emitted by light source. But a very large reduction in optical power also lowers the SNR at the photodetector, so that to maintain it at a high level, the optical power emitted by light source cannot be reduced greatly.

The noise associated with the fiber nonlinear Kerr effect is based on the electro-optical phenomenon which consists in changes experienced by refractive index of the optical fiber caused when it is excited by an optical wave that varies in amplitude. This occurs by the fluctuation of the optical power level of light source. In the case of the gyroscopic system, this optical power variation coupled to the fiber coil causes changes on its refractive index, which results in a phase change in the optical wave propagated along the length of the optical fiber coil. This change can be evaluated as a phase-equivalent noise and could be diminished efficiently using a low coherence light source (broadband source). Another important aspect is providing the light source with a thermal stabilization system to achieve a constant level of optical power emission.

The Shupe thermal effect is due to local temperature gradients along the fiber coil length. These temperature gradients induce phase changes in the optical waves traveling through the fiber. This effect can be minimized performing an appropriate winding of fiber-coil, so that a uniform temperature distribution is achieved throughout its entire length. The quadrupolar winding (number of turns in each layer of coil equal to an integer multiple of four) fulfills this condition. Other minor optical noise sources with less effect on the optical signal detected by photodetector are due to backscattering and reflections phenomena along the length of the sensing fiber-coil. A serious disadvantage for this model design is that the results of optical simulation do not allow realizing the evaluation of the main sources of optical noise.

Regarding the electrical noise generated by the electronic circuits, the most important is white noise (thermal-noise or Johnson noise), which spreads equally over all the frequencies. An appropriate way for overcoming this noise source is performing a selective filtering at the frequency of the desired signal and fitting later the gain of the amplification stages to increase the electrical SNR at the output. In the case of the designed IFOG circuits, a strict design of LPF (Low-Pass-Filter) and BPF (Band-Pass-Filter) is necessary after photodetector-amplifier. It is crucial for obtaining a good scale-factor linearity of designed IFOG-model. It is due to the fact that it depends on linearity of the obtained V_{Ω} versus Ω graphical representation derived from signal demodulation process.

6. Conclusions

An IFOG prototype was theoretically designed by means of simulating tools. The conventional IFOG design with sinusoidal phase modulation is based on open-loop configuration. The main innovation of IFOG-design presented here is the use of a simple closed-loop electro-optical configuration, realized by means of optical and electronic cost competitive components. Furthermore, the proposed design also allows to reach substantial progress in stability and linearity of the Scale Factor (SF), dynamic range and threshold sensitivity of the gyroscope, compared to previous models proposed with the same fiber-optic coil length (L = 300 m). The cost advantage in optical subsystem is obtained by means of a design with depolarization of optical waves, by using two Lyot depolarizers, both realized in optical fiber. This allows using a sensing coil made in optical standard fiber, instead of a special polarization maintaining fiber, which is much more expensive. On the other hand, the electronic circuit subsystems (detection, demodulation, and feedback signal processing) is based on a conventional analog design, using classic electronic components which are high precision and cost competitive, so that it also contributes to achieving a reasonable cost and at the same time optimizing quality/price ratio of final device. An interesting observation regarding to cost is that if the entire volume occupied by the device does not suppose a major restriction (this condition is fulfilled in certain applications), it is possible to get an additional saving cost by means of a particular design of optical subsystem. This design can be based on a suitable selection of bulk optical components: the Integrated Optical Circuit (IOC) can be replaced with two 2 × 2 fiber optical couplers (SMF fiber), a fiber polarizer, and a fiber-based electro-optic phase modulator (PZT), since until the day the IOC is not standard-manufacture. The same way, for light source it is possible using a low-power and limited-consumption laser, as an Erbium-Doped-Fiber-Amplifier (EDFA), and for depolarization of the optical wave a new solution based on bulk optics can be adopted, such as crystal Lyot depolarizers.

Author details

Ramón José Pérez Menéndez

Address all correspondence to: ramonjose.perez@lugo.uned.es

UNED-Spain, Lugo, Spain

References

[1] Lefèvre HC. The Fiber-Optic Gyroscope. Second ed. Boston-London: Artech House; 2014

[2] Ezekiel S, Arditty HJ. Fiber-Optic Rotation Sensors. Berlin: Springer-Verlag; 1982

[3] Lawrence A. Modern Inertial Technology: Navigation, Guidance, and Control. 2nd ed. New-York: Springer-Verlag; 1992

[4] Sagnac G. L'ether luminex demontre par l'effect du vent relatif d'ether dans un interferometre en rotation uniforme. Comptes Rendus de l'Académie des Sciences. 1913;**95**:708-710

[5] Sagnac G. Sur la preuve de la realité de l'ether luminex par l'experience de l'interferographe tournant. Comptes Rendus de l'Académie des Sciences. 1913;**95**:1410-1413

[6] Hariharan P. Sagnac or Michelson-Sagnac interferometer. Applied Optics. 1975;**14**(10): 2319-2321

[7] Bergh RA, Lefevre HC, Shaw HJ. Overview of fiber-optic gyroscopes. IEEE Journal of Lightwave Technology. 1994;**LT-2**:91-107

[8] Bergh RA, Lefevre HC, Shaw HJ. All-single-mode fiber-optic gyroscope. Optics Letters. April 1981;**6**(4):198-200

[9] Moeller RP, Burns WK, Frigo NJ. Open-loop output and scale factor stability in a fiber-optic gyroscope. Journal of Lightwave Technology. Feb. 1989;**7**(2):262-269

[10] Bohm K, Petermann K. Signal processing schemes for the fiber-optic gyroscope. Proceedings of SPIE, Fiber Optic Gyros: 10th Anniversary Conf. 1986:36

[11] Burns WK et al. Fiber-optic gyroscopes with depolarized light. Journal of Lightwave Technology. 1992;**10**(7):992-998

[12] Szafraniec B et al. Theory of polarization evolution in Interferometric fiber-optic depolarized gyros. Journal of Lightwave Technology. 1999;**17**(4):579-590

[13] Kintner EC. Polarization control in optical-fiber gyroscopes. Optics Letters. 1981;**6**(3):154-156

[14] Sanders GA, Szafraniec B. Progress in fiber-optic gyroscope applications II with emphasis on the theory of depolarized gyros. AGARD/NATO Conf. Report on Optical Gyros and Their Applications, AGARDougraph. 1999;**339**:11/1-1142

[15] Pavlath GA. Productionization of fiber gyros at Litton guidance and control systems. SPIE Proceedings. 1991;**1585**:2-5

[16] Sanders GA, Szafraniec B, Liu R-Y, Laskoskie C, Strandjord L. Fiber optic gyros for space, marine and aviation applications. SPIE Proceedings. 1996;**2837**:61-71

[17] Patterson RA, Goldner EL, Rozelle DM, Dahlen NJ, Caylor TL. IFOG technology for embedded GPS/INS applications. SPIE Proceedings. 1996;**2837**:113-123

[18] Lefèvre HC et al. Integrated Optics: a Practical Solution For The Fiber-Optic Gyroscope. Proc. SPIE 0719, Fiber Optic Gyros: 10th Anniversary Conf. 101, March 11, 1987. 101-112. DOI: 10.1117/12.937545

[19] Kim BY et al. Response Of Fiber Gyros To Signals Introduced At The Second Harmonic Of The Bias Modulation Frequency. Proc. SPIE 0425, Single Mode Optical Fibers, 86, November 08 1983. DOI: 10.1117/12.936218

[20] Kim BY et al. Gated phase-modulation feedback approach to fiber gyroscopes. Optics Letters. 1984;**9**(6):263-265

[21] Kim BY et al. Gated phase-modulation feedback approach to fiber gyroscopes with linearized scale factor. Optics Letters. 1984;**9**:375-377

[22] Kim BY et al. Phase reading all-fiber-optic gyroscope. Optics Letters. 1984;**9**:378-380

[23] Böhm K et al. Signal processing schemes for the fiber-optic gyroscope. Proc. SPIE 0719, Fiber Optic Gyros: 10th Anniversary Conf. 101, March 11 1987. DOI: 10.1117/12.937536

[24] Moeller RP et al. Open-loop output and scale-factor stability in a fiber-optic-gyroscope. Journal of Lightwave Technology. 1989;**7**(2):262-269

[25] Ebberg A et al. Closed-loop fiber-optic gyroscope with a Sawtooth phase-modulated feedback. Optics Letters. 1985;**10**(6):300-302

[26] Kay CJ et al. Serrodyne modulator in a fibre-optic gyroscope. Optoelectronics, IEEE Proceedings Journal. 1985;**132**(5):259-264

[27] Yahalom R et al. Low-Cost, Compact Fiber-Optic Gyroscope For Super-Stable Line-Of-Sight Stabilization. Proceedings of the IEEE/ION Position Location and Navigation Symposium (PLANS), Indian Wells, CA(USA). May 2010. DOI: 10.1109/PLANS.2010.5507131

[28] Çelikel O et al. Establishment of all digital closed-loop Interferometric fiber-optic-gyroscope and scale factor comparison for open-loop and all digital closed-loop configurations. IEEE Sensors Journal. 2009;**9**(2):176-186

[29] Sandoval-Romero GE et al. Límite de Detección de Un Giroscopio de Fibra Óptica Usando Una Fuente de Radiación Superluminiscente. Rev. Mexicana de Física. 2002; **49**(2):155-165

[30] Medjadba H et al. Low-Cost Technique For Improving Open-Loop Fiber Optic Gyroscope Scale Factor Linearity. Proceedings of the International Conference on Information and Communication Technologies, 2006 (ICTTA'06), 2, Damascus (Syria). 1684718, 2057–2060, April 24–28, 2006. DOI: 10.1109/ICTTA.2006

[31] Bennett S et al. Fiber optic gyros for robotics. Service Robot: An International Journal. 1996;**2**(4)

[32] Emge S et al. Reduced Minimum Configuration Fiber Optic Gyro for Land Navigation Applications. Proc. SPIE 2837, Fiber Optic Gyros: 20th Anniversary Conf., Denver CO (USA), August 04, 1996

[33] Bennett SM et al. Fiber optic gyroscopes for vehicular use. Proceedings of the IEEE Conf. on Intelligent Transportation System (ITSC'97), Boston MA (USA), Nov. 9-12, 1997, 1053-1057. DOI: 10.1109/ITSC.1997.660619

[34] Cordova A, Patterson R, Rahu J, Lam L, Rozelle D. Progress in navigation grade FOG performance. SPIE Proceedings. 1996;**2837**:2017-2217

[35] Pavlath G. The LN200 fiber gyro based tactical grade IMU. Proc. Guidance, Navigation and Control, AIAA, pp. 898-904, 1993

[36] Kamagai T et al. Fiber optic gyroscopes for vehicle navigation systems. Fiber Optic Laser Sensors XI, Proceedings of SPIE. Sept. 1993;**2070**:181-191

[37] Hotate K, Tabe K. Drift of an optical fiber gyroscope caused by the faraday effect: Influence of the earth's magnetic field. Applied Optics. 1986;**25**:1086-1092

[38] Bohm K, Petermann K, Weidel E. Sensitivity of a fiber optic gyroscope to environmental magnetic fields. Optics Letters. 1982;**6**:180-182

[39] Blake JN. Magnetic field sensitivity of depolarized fiber optic gyros. SPIE Proc., 1367, Fiber Optic and Laser Sensors VIII. pp. 81-86, 1990

[40] Bergh RA, Lefevre HC, Shaw HJ. Compensation of the optical Kerr effect in fiber-optic gyroscopes. Optics Letters. June 1982;**7**:282-284

[41] Takiguchi K, Hotate K. Method to reduce the optical Kerr-effect-induced bias in an optical passive ring-resonator gyro. IEEE Photonics Technology Letters. Feb. 1992;**4**:2

[42] Bergh RA, Culshaw B, Cutler CC, Lefevre HC, Shaw HJ. Source statistics and the Kerr effect in fiber-optic gyroscopes. Optics Letters. 1982;**7**:563-565

[43] Shupe DM. Thermally induced nonreciprocity in the fiber-optic interferometer. Applied Optics. 1980;**19**(5):654-665

[44] Ruffin PB, Lofts CM, Sung CC, Page JL. Reduction of nonreciprocity in wound fiber optic interferometers. Optical Engineering. 1994;**33**(8):2675-2679

[45] Lofts CM, Ruffin PB, Parker M, Sung CC. Investigation of effects of temporal thermal gradients in fiber optic gyroscope sensing coils. Optical Engineering. Oct. 1995;**34**(10): 2853-2863

[46] Sawyer J, Ruffin PB, Sung CC. Investigation of effects of temporal thermal gradients in fiber optic gyroscope sensing coils, part II. Optical Engineering. Jan. 1997;**36**(1):29-34

[47] Ruffin PB, Sung CC, Morgan R. Analysis of temperature and stress effects in fiber optic gyroscopes. Fiber Optic Gyros, Proceedings of SPIE. Sept. 1991;**1585**:283-299

[48] Frigo NJ. Compensation of linear sources of non-reciprocity in Sagnac interferometers. Fiber Optic Laser Sensors I, SPIE Proceedings. 1983;**412**:268-271

[49] Ruffin PB, Baeder JS, Sung CC. Study of ultraminiature sensing coils and the performance of a depolarized interferometric fiber optic gyroscope. Optical Engineering. Apr. 2001; **40**(4):605-611

[50] Lefevre HC, Bergh RA, Shaw HJ. All-fiber gyroscope with inertial navigation short-term sensitivity. Optics Letters. 1982;**7**:454-456